New Insights into Virus Bioinformatics

New Insights into Virus Bioinformatics

Editor: Jake Mason

www.callistoreference.com

Callisto Reference,
118-35 Queens Blvd., Suite 400,
Forest Hills, NY 11375, USA

Visit us on the World Wide Web at:
www.callistoreference.com

ISBN: 978-1-64116-764-2 (Hardback)

Cataloging-in-Publication Data

New insights into virus bioinformatics / edited by Jake Mason.
p. cm.
Includes bibliographical references and index.
ISBN 978-1-64116-764-2
1. Virology--Data processing. 2. Bioinformatics. 3. Virology--Research. I. Mason, Jake.
QR370 .N49 2023
576.64--dc23

Table of Contents

Preface

Viruses are microscopic and infectious microorganisms that are made up of genetic material enclosed in a protein coating. They cause a variety of serious infectious diseases including influenza, hepatitis, common cold, rabies, AIDS, smallpox, dengue fever and measles. Viruses are classified into four types based on the host cell that they infect, and these include plant viruses, bacteriophages, animal viruses and fungal viruses. Virology is the scientific study of viruses and virus-like agents. The study of viral infections and outbreaks using methods of bioinformatics has become essential to virology research. These methods have been used effectively to diagnose, treat and prevent infections in both animals and humans. Bioinformatics investigations that can be performed on viruses include motif and epitope recognition, homology searching, identification of open reading frames, sequence alignment, and gene prediction. This book provides new insights into virus bioinformatics. It presents researches and studies performed by experts across the globe. The extensive content of this book provides the readers with a thorough understanding of the subject.

This book is a result of research of several months to collate the most relevant data in the field.

When I was approached with the idea of this book and the proposal to edit it, I was overwhelmed. It gave me an opportunity to reach out to all those who share a common interest with me in this field. I had 3 main parameters for editing this text:

1. Accuracy – The data and information provided in this book should be up-to-date and valuable to the readers.

2. Structure – The data must be presented in a structured format for easy understanding and better grasping of the readers.

3. Universal Approach – This book not only targets students but also experts and innovators in the field, thus my aim was to present topics which are of use to all.

Thus, it took me a couple of months to finish the editing of this book.

I would like to make a special mention of my publisher who considered me worthy of this opportunity and also supported me throughout the editing process. I would also like to thank the editing team at the back-end who extended their help whenever required.

Editor

1

Exploring the Papillomaviral Proteome to Identify Potential Candidates for a Chimeric Vaccine against Cervix Papilloma Using Immunomics and Computational Structural Vaccinology

Satyavani Kaliamurthi [1,2], Gurudeeban Selvaraj [1,2], Sathishkumar Chinnasamy [3], Qiankun Wang [3], Asma Sindhoo Nangraj [3], William CS Cho [4], Keren Gu [1,2] and Dong-Qing Wei [1,3,*]

1 Center of Interdisciplinary Science-Computational Life Sciences, College of Food Science and Engineering, Henan University of Technology, Zhengzhou 450001, China; satyavani.mkk@haut.edu.cn (S.K.); gurudeeb99@haut.edu.cn (G.S.); gkr@haut.edu.cn (K.G.)

2 College of Chemistry, Chemical Engineering and Environment, Henan University of Technology, Zhengzhou 450001, China

3 The State Key Laboratory of Microbial Metabolism, College of Life Sciences and Biotechnology, Shanghai Jiao Tong University, Shanghai 200240, China; sathishimb@gmail.com or sathishimb@sjtu.edu.cn (S.C.); wangqiankun@sjtu.edu.cn (Q.W.); sindhoo_sind@yahoo.com (A.S.N.)

4 Department of Clinical Oncology, Queen Elizabeth Hospital, Kowloon, Hong Kong; williamcscho@gmail.com or chocs@ha.org.hk

* Correspondence: dqwei@sjtu.edu.cn

Abstract: The human papillomavirus (HPV) 58 is considered to be the second most predominant genotype in cervical cancer incidents in China. HPV type-restriction, non-targeted delivery, and the highcost of existing vaccines necessitate continuing research on the HPV vaccine. We aimed to explore the papillomaviral proteome in order to identify potential candidates for a chimeric vaccine against cervix papilloma using computational immunology and structural vaccinology approaches. Two overlapped epitope segments (23–36) and (29–42) from the N-terminal region of the HPV58 minor capsid protein L2 are selected as capable of inducing both cellular and humoral immunity. In total, 318 amino acid lengths of the vaccine construct SGD58 contain adjuvants (Flagellin and RS09), two Th epitopes, and linkers. SGD58 is a stable protein that is soluble, antigenic, and non-allergenic. Homology modeling and the structural refinement of the best models of SGD58 and TLR5 found 96.8% and 93.9% favored regions in Rampage, respectively. The docking results demonstrated a HADDOCK score of -62.5 ± 7.6, the binding energy (-30 kcal/mol) and 44 interacting amino acid residues between SGD58-TLR5 complex. The docked complex are stable in 100 ns of simulation. The coding sequences of SGD58 also show elevated gene expression in *Escherichia coli* with 1.0 codon adaptation index and 59.92% glycine-cysteine content. We conclude that SGD58 may prompt the creation a vaccine against cervix papilloma.

Keywords: cellular immunity; codon frequency distribution; HPV58; minor capsid protein; TLR agonist; prophylaxis

1. Introduction

Currently, viral infection contributes to about 20% of the global burden of human cancer. Among other virus types, the human papillomavirus (HPV) is reported in about 5% of all human cancers, specifically infection associated with the cervix with 250,000 mortalities every year [1]. HPV with its double-stranded DNA contains a non-enveloped small virus, which infects a region of the cutaneous epithelial membrane (skin or integumentary system), or the mucous membrane (i.e., coated as an internal line in hollow spaced organs like the mouth, reproductive organ, urinary tract, or rectum) in the host system [2]. The nomenclature of HPV is distinguished by the International Committee on Taxonomy of Viruses (ICTV), and is based on the suggestion obtained from the study group of papillomavirus [3]. ICTV follows the practice of naming species after a specific virus, such as HPV16, while the related types, namely, the "type species," are designated as strains within the species [3,4]. For example, the frequently used term "HPV species alpha-9" is a synonym for the ICTV term "HPV16 species"; it contains the HPV types 16, 31, 33, -35, 52, 58, and 67 strains, respectively. According to ICTV, the species HPV16 belongs to the family of Papillomaviridae and the genus of Alphapapillomavirus [3]. The genomic relationship between different cancer types has demonstrated that more than 99% of cervical cancer patients are infected with 15 different types of α-clade HPV, defined as "high-risk" or "oncogenic" genital HPVs. The α-clade HPV (6 and 11) causes genital warts while the remaining strains of HPV are related to the risk of cervical cancer. HPV infection is attributed to more than 50% of oropharyngeal and anogenital cancers [5]. Generally, the human immune system can clear the pathogenic infection caused by HPV within 2 years, but this also depends on the efficiency of an individual's immune system and the invading type of HPV. However, in the case of a very weak immune system, it fails to remove the invading high-risk HPVs (hrHPV) that may lead to the development of cervical cancer [6,7]. hrHPV infections are responsible for causing more than 99% of precancerous cervical intraepithelial neoplasias (CIN) and invasive cervical cancers (ICC) [8–10]. In China, HPV-mediated cervical cancer is a substantial public health issue, with 1 million new cervical cancer incidences and 30,000 moralities registered every year [11,12]. In 2018, clinical, epidemiological, and clinicopathological studies reported HPV58 to be the second or third most predominant genotype in precancerous CIN I, II, III, and ICC lesions, a higher grade of squamous intraepithelial or cell carcinoma, and adenocarcinoma of HPV positive patients in different geographical regions of China [11]. Seven provinces of China have reported hrHPV-mediated cervical cancer incidences, namely, Guangdong, Liaocheng, Shanghai, Wenzhou, Wuhan, Southwestern China, and Western China [13–19]. Zhang et al. [20] reported that the HPV16 (6.4%) and HPV58 (5.3%) genotypes were predominantly found in males who had recently been involved in sex, in Shanghai.

Cervarix®, Gardasil®, and Gardasil 9® are the three non-infectious prophylactic Food and Drug Administration (FDA)-approved HPV licensed subunit vaccines in active usage. These vaccines were developed from the major capsid L1 virus-like particles (VLPs) using recombinant DNA technology. Cervarix is a bivalent vaccine based on Baculovirus fermentation and it provides ~70% protection against HPV (16 and 18)-mediated cervical cancer but not against genital warts [21]. Gardasil is a quadrivalent HPV (6, 11, 16, and 18) vaccine based on yeast fermentation technology. It is efficiently used for the prevention of genital warts and gives ~70% protection for cervical cancer [22]. In 2009, the FDA approved a nine-valent Gardasil 9® vaccine that provides protection to HPV types 6, 11, 16, 18, 31, 33, 45, 52, and 58. It has been used for both males and females in the age groups of 9–15 and 9–26 [23]. The new nine-valent vaccine exhibited a positive outcome in high-grade lesions in the absence of HPV (18 and 16) infections [24]. In October 2018, the FDA extended the use of Gardasil 9 to the age group of 27–45 among both the sexes. In addition, the L1 VLP (absence of viral genomic materials)-mediated vaccine production in the eukaryotic (ex. Baculovirus) host system is a complex and tedious process [25,26]. The main limitations of currently available prophylactic vaccines are as follows: they are strain specific, not therapeutic for patients already infected with HPV, they require multiple dosages, and are expensive [27,28]. In addition, the effective straightforward delivery of HPV vaccines can enhance the immunogenic potential against HPVs.

The implementation of the L2 minor capsid protein is a potential alternative in the HPV prophylactic vaccine production. Since the N-terminal region of the L2 protein is highly conserved in low-risk HPV (6 and 11) and 13 different hrHPVs, it is contrasted with the type-specific protection of

L1 prophylactic VLPs [29]. The single copy of the L2 protein ~473 amino acids (AA) is present in each L1 capsomere, resulting in 72 copies per virion [30]. Incidentally, the L2 protein plays a vital role in L1 assembly into the VLPs and enhances the encapsidation of the double-stranded ~8kb circular viral genome [31]. Moreover, the full-length or polypeptides (1–8 or 11–200 AA in length) of the L2 protein enhance the production of neutralizing antibodies in vaccinated experimental models including mice, cattle, and rabbit [32–34]. To date, no L2 VLP-derived prophylactic vaccines have been approved in clinical trials [26] due to their limitation of weak immunogenicity, which imitates the incapability of multimerizing into VLPs.

With this information, in this study, we aimed to design the novel chimeric vaccine from the N-terminal region of the L2 sequence of the HPV58 targets to hrHPVs. The immunomic tools namely, the immune epitope database (IEDB) and NetMHCv4.0, Tepitool, CTLPred, PAComplex, IFNepitope, ABCPred, AllerTOP, AllergenFPv 1.0, ANTIGENpro, program of protein information resource (PIR), and epitopes conservancy were implemented to discover the overlapped epitope segment that is required to induce B-cell and T-cell immunity. Then, the chimeric vaccine (SGD58) was constructed using the overlapped epitope segments, Toll-like receptor (TLR) adjuvants, Th epitopes, and amino acid linkers. The physiochemical and immunological properties of the chimeric vaccine were validated using Protparam, SolPro, VaxiJen, and ANTIGENpro tools. In addition, homology modeling using iterative threading ASSEmbly refinement (I-TASSER), the Robetta beta full chain protein structure prediction server, structural refinement (GalaxyRefine and 3DRefine), and structural validation (protein structure analysis [ProSA], Ramachandran plot, and ERRAT) were performed to obtain the best three-dimensional (3D) model of the chimeric vaccine and the target TLR5 receptor. Then, the interaction of the chimeric vaccine with TLR5 and stability of this complex were determined through protein–protein (PP) docking and molecular dynamic (MD) simulation. Moreover, the virtual cloning and gene expression of the chimeric vaccine in *Escherichia coli (E. coli)* were analyzed to obtain a low-cost HPV vaccine.

2. Materials and Methods

2.1. Protein Sequences

The L2 protein of HPV58 (Accession No.: P26538), the Flagellin protein of *Salmonella enterica* serovar *Dublin* (Accession No.: Q06971), and human TLR5 (Accession No.: O60602) sequences were obtained from the Swiss-Prot reviewed universal protein knowledgebase (UniProt) database [35]. The designed chimeric vaccine was named SGD58, using the name of the first and principal authors along with the strain number.

2.2. Immunomics Analysis

2.2.1. MHC-I Binding Epitope Segments Prediction

Two servers, IEDB and NetMHCv4.0, have been exploited for the identification of major histocompatibility complex class I (MHC-I) binding epitopes from the N-terminal region of the L2 sequence. Specific human MHC-I alleles such as the human leukocyte antigen (HLA)-A* (01:01, 02:01, 02:07, 11:01, 24:01), HLA-B* (46:01, 58:01) and HLA-C* (07:02, 12:03) were abundantly diagnosed in different regions of China, including Guizhou, Henan, Taihu River Basin, Tibetan, Yunnan, Wenzhou, and Wuhan. These alleles were selected for epitope prediction [36–42]. IEDB [43] is a freely available analysis resource with specified algorithms for the identification and determination of immunogenic epitopes. A consensus method was implemented to predict the MHC-I binding epitopes and its production pathway [44]. In this consensus method, three algorithms namely, the neural network (artificial), matrix method (stabilized), and peptide libraries (combinatorial) were combined to predict the promising CTL epitope segments. The epitopes involve proteasomal cleavage (pCle), a transporter associated with antigen processing (TAP), and the MHC-I binding pathway. The lowest percentile rank

(<10%) indicated the good binding efficiency of epitopes with the restricted alleles. NetMHCV4.0 [45] is another potential tool implemented to find MHC-I binding peptides with the best Pearson's correlation coefficient (PCC) of 0.895, based on the combined neural network. The strong and weak binding peptides were predicted based on the thresholds of <0.5 and <2, respectively.

2.2.2. CTL Epitope and TCR -Peptide/Peptide -MHC Interfaces Prediction

The CTLPred tool is a direct method for the prediction of CTL epitope segments instead of MHC binders. The prominent combined approaches were implemented to find the epitopes, based on both the artificial neural networks (ANN) trained by Stuttgart neural network simulator (SNNS) and support vector machine (SVM) methods. The combined methods demonstrate a higher level of accuracy (75.8 %) compared with other individual methods of prediction such as ANN (72.2%) and SVM (75.4%). The default cutoff scores of 0.51 of ANN and 0.36 of SVM were used to find the epitopes or non-epitopes at which the sensitivity and specificity of the predictions are almost similar [46]. A web server PAComplex provides access to examine and visualize the TCR-peptide and peptide–MHC interface (pMHC), respectively. For a given viral protein query sequence, the joint Z-value is obtained with a threshold 2.5. Moreover, it allows the selection of only limited allotypes of MHC class I such as HLA-A*(02:01), HLA-B*(08:01, 35:01, 35:08, 44:05), and HLA-E, respectively. The Z-value was calculated using the following formula:

$$Jz = Z_{MHC} \times Z_{TCR} \quad (1)$$

where Z_{MHC} and Z_{TCR} are the score of a TCR-pMHC complex, calculated by (E-μ)/σ. E denotes the interaction score, μ denotes the mean, and σ denotes the standard deviation from 10,000 random interfaces [47].

2.2.3. MHC Class-II Binding Epitopes Prediction

MHC-II alleles include DQB1*(03:01, 03:03, 06:01), DRB1*(07, 09, 14:01:01, 15:01, 15:07:01), and DPB1*(05:01,05:02:01), specific to the Henan, Taihu River Basin, Tibetan, Yunnan, Wenzhou, and Wuhan provinces of China, which have been selected for epitope prediction [36–42]. The IEDB consensus approaches were used to predict MHC-II binding epitope segments using the neural network-based alignment, stabilized matrix methods-based alignment, and combinatorial library-based algorithms [48]. The peptides with the lowest percentile rank were considered to be of a higher binding affinity. Tepitool [49] is a tool from IEDB analysis resources, which provides accession to the prediction of both class I and II binders. The peptides which show the lowest percentile rank (IC_{50}< or = 500nM) are potentially considered as higher affinity binding peptides.

2.2.4. Interferon-Gamma (INF-γ) Inducing Epitope Prediction

IFNepitope (http://webs.iiitd.edu.in/raghava/ifnepitope/) that is a potential server useful for the prediction and design of INF-γ inducing epitopes. INF-γ inducing epitopes were identified based on motif-based SVM or hybrid algorithms. The hybrid method using residue or dipeptide composition shows 81.39% accuracy [50].

2.2.5. Linear B-Cell Epitope Prediction

ABCPred is used to predict linear B-cell epitopes. It provides 65.93% of accuracy with the involvement of the recurrent neural network (RNN) algorithm. It consists of 700 B-cell and non-B-cell epitope segment datasets each with a length of 20 amino acids [51].

2.2.6. Allergenicity Prediction

AllerTOP is the first proper alignment-free allergenicity server. In this, five machine learning methods such as partial least squares, logistic regression, decision tree, naive Bayes, or *k* nearest neighbors (kNN = 1) were implemented to find the allergen. It shows 88.7%, 90.7%, and 86.7% accuracy, specificity, and sensitivity, respectively [52]. AllergenFPv 1.0 is another essential tool for allergenicity prediction based on novel descriptor fingerprint approaches. Twenty naturally existing amino acids in the protein sequences were classified into five descriptors (E) such as E1 (hydrophobicity), E2 (size), E3 (helix-forming propensity), E4 (relative abundance of amino acids), and E5 (β-strand forming propensity). Based on this, the strings were transformed into normal vectors by auto cross covariance (ACC) transformation to find the allergen protein. It exhibits accuracy (87%), specificity (89%), and sensitivity (86%) [53].

2.2.7. Antigenicity

ANTIGENpro is the potential alignment-free and sequence-based antigenicity prediction server with 79% accuracy and an area under curve (AUC) of 0.89. It shows results based on amino acid composition and the random-forest algorithm. The datasets were trained using 5-fold cross-validation. It consists of both protective antigen (193) and non-antigen (193) sequences. It predicts whether the given query epitope segments are antigenic or non-antigenic with their respective probability [54].

2.2.8. Cross-Reactivity Analysis with Human Proteomes

The presence or absence of similarity in predicted epitope segments with the human proteome was analyzed using the peptide-matching program of PIR [55].

2.2.9. Epitopes Conservancy Analysis

Epitopes conservancy (EC) and molecular evolutionary genetic analysis (MEGA) v7.0 tools were used to perform conservancy analysis. The EC tool [56] was employed to find the degree of conservancy of the epitope segments within the set of given hrHPV L2 protein sequences. The selected epitope segments of HPV58 with 15 hrHPV (16, 18, 31, 33, 35, 39, 45, 51, 52, 56, 59, 68, 69, 73, and 82) strains performed EC analysis. ClustalW-based multiple sequence alignment [57] was used to determine the sequence conservation of the overlapped epitope segments with 15 other hrHPV (16, 18, 31, 33, 35, 39, 45, 51, 52, 56, 59, 68, 69, 73, and 82) strains.

2.3. Chimeric Vaccine Design (SGD58) and Validation

2.3.1. Assessment of the Physicochemical Properties of SGD58

The complete chimeric vaccine was designed by joining the optimized epitope segments (02), TLR adjuvants (02), and Th epitopes (02) with suitable amino acid linkers. Moreover, it is required to find the solubility of the designed chimeric vaccine on overexpression in *E. coli*. SOLpro is a useful tool to find the solubility of protein based on the two-stage SVM algorithm. It achieves an overall accuracy of 74%, which develops on standard evaluation metrics with 10-fold cross-validation. It predicts the query protein to be soluble or insoluble at $P \geq 0.5$ [58]. A range of physiochemical characteristics of the designed chimeric vaccine was also determined through ProtParam [59].

2.3.2. Determination of Antigenicity

VaxiJen is the primary server used for the prediction of antigenicity of the input sequence against different targets such as virus, bacteria, fungi, parasites, and tumors. Antigenicity was calculated based on the physicochemical properties of the protein sequences. Every target organism dataset contained 100 antigens and non-antigens. Moreover, the model organisms were validated using leave-one-out cross-validation (LOO-CV); it provides 89% accuracy and an AUC of 0.964 respectively, at the threshold of 0.4 [60].

2.3.3. Analysis of the Tertiary Structure

2.3.3.1. Homology Modeling

For homology modeling, I-TASSER and Robetta were the servers used to design the 3D structure of SGD58 and TLR5. I-TASSER is a potential server that depends on the secondary-structure-mediated program of "Profile-Profile threading alignment (PPA) and iterative implementation of the TASSER." It has predicted a number of protein structures on request basis from 35 countries worldwide. For the query inputs, the user obtains the confidence score, TM score (topology similarity assessments of the two various protein structures), root-mean-square deviation (RMSD), and cluster density values. Nevertheless, the higher C-score (ranging from −5 to +2) determines the best model with a higher confidence level. Moreover, the 3D structure of the modeled protein was visualized using UCSF Chimera [61]. The Robetta beta server was used to predict the full chain protein structure. This server (http://robetta.bakerlab.org) gives automated tools for the analysis and prediction of the protein structure. Robetta provides both the *ab initio* and comparative models of protein domains. The comparative models are built from structures detected and aligned by HHSEARCH, SPARKS, and Raptor. The loop regions are assembled from fragments and optimized to fit the aligned template structures. The *de novo* models are built using the Robetta *de novo* protocol. For structure prediction, the submitted query sequences were analyzed minutely into putative domains. For domain prediction, a hierarchical screening method called "Ginzu" was used [62]. Besides, due to the unavailability of the crystal structure of TLR5, we have chosen TLR5 (PDB ID: 3J0A) as a template model to perform the homology modeling.

2.3.3.2. 3D Modeled Structure Refinement

The high C-score model of the designed vaccine from the I-TASSER and model 3 from Robetta beta was further refined using the GalaxyRefine and 3DRefine tools. The GalaxyRefine is a tool that is accessible in the GalaxyWeb server: it is useful to refine the structure of a protein from the given query sequences based on template-based modeling, and undergoes loop and terminus portion refinement through the *ab initio* modeling method. The ninth critical assessment of techniques for protein structure prediction (CASP9) optimizes refinement and produces consistent core structures [63]. Another tool is 3Drefine, which prompts iteration analysis for ~300 amino acid residues efficiently in less than 5 min. It performs post-refinement model analysis with both or single MolProbity and random walk (RW) plus methods. The results are visualized using JSmol [64]. The top five models of each tool were used for further validation.

2.3.3.3. 3D Refined Structures Validation

The refined 3D models from the above steps were validated using three interactive services namely, ProSA, Ramachandran plot analysis, and ERRAT. ProSA-web is a potential tool for the refinement, validation, prediction, and modeling of protein structures. It indicates the difference in the protein structures through the respective score and energy spot. It also facilitates the validation process of the protein structure that is acquired from X-ray scanning, nuclear magnetic resonance (NMR) spectroscopy analysis, and theoretical calculations. As an output, the Z-score corresponds to the overall feature of the validated model [65]. RAMPAGE tool is used to validate the percentage (%) of favored, allowed, and outlier region in the given query chimeric vaccine [66]. The statistics of noncovalent interactions between carbon, nitrogen, and oxygen atoms in the input sequence with best-resolution crystallographic structures were compared using the ERRAT tool. It implements an empirical atom-based approach for verification of the protein structure and is more sensitive to errors (1.5A) [67]. Similar steps were followed to validate the TLR5 model.

2.4. *Conformational B-Cell Epitopes Prediction*

DiscoTope 2.0 is a potential tool used to analyze the conformational (discontinuous) B-cell epitopes from the input sequence. It showed a highly significant prediction performance with an AUC of 0.824. The default −3.7 threshold limit provides significant specificity (0.75) and sensitivity (0.47). It was selected for the present analysis, and the final score was evaluated by the combined calculation of the propensity score (PS) and contact numbers [68].

2.5. *Investigation of the Interaction between SGD58 and TLR5*

2.5.1. Protein–Protein (PP) Interaction of the SGD58 with TLR5

The PP interactions are the midpoint for all the biochemical pathways that are involved in the biological functions. HADDOCK v2.2 is a server used for the docking of PP complexes [69]. The scoring function was executed based on the weighted sum of the various energy terms (Van der Waals energy, electrostatic energy, desolvation energy, restraints energy and buried surface area). In addition, intermolecular contacts such as hydrogen bonds (HB) and those that are non-bonded were determined by using the program for automatically plotting protein–protein interactions (LIGPLOT-DIMPLOT) v 4.5.3 [70]. The validated best model of the vaccine construct (Robetta model 3) and TLR5 (Robetta model 5) was chosen as a ligand and receptor, respectively.

2.5.2. MD Simulation

MD simulation determines the strength of the docked complex and the designed vaccine (SGD58). The GROMACS 5.1.2 package, with the CHARMM force field was used to perform MD simulation. The transferable intermolecular potential with three points (TIP3P) and simple point charge (SPC) in the cubic cell of the water model was resolved with protein, and with the addition of appropriate counter ions to satisfy electroneutrality. For the MD simulation, the system was implemented with volume-based canonical (NVT) and pressure-based isothermal-isobaric (NPT) ensembles [71]. The energy minimization of the system was performed with the steepest descent method, which facilitates 50,000 minimization steps and 1000 kJ mol^{-1} of tolerance. The Ewald method with a cutoff for short-range neighbor distance (1.0 nm), and Coulomb (1.0 nm) was used to calculate van der Waals (vdW) and electrostatic interactions [72]. SPC resolved the system and the final minimizations were calculated for a realistic structure concerning the geometry: and solvent orientations were used in the production of the MD simulation. SETTLE and LINCS algorithms were used to assist the geometry of the water molecules and bond angles [72,73]. The pressure of the system (300 K, 1bar) was embraced using the Parrinello–Rahman method and the temperature was regulated using the V-rescale method. Temperature and pressure equilibrated systems were employed for production run (100 ns) and time step (2 fs). The resulting structural coordinates were saved at every 2ps of an interval.

2.6. *Analysis of Virtual Gene Expression and Cloning*

EMBOSS Backtranseq v1.0 [74] is a suitable tool to uptake the query protein sequences, reverse translate, and return the optimizing coding sequences. Furthermore, the properties of the obtained coding sequences were analyzed to obtain the increased level of gene expression in the respective host. It is well known that the codon plays a crucial role in the expression of the recombinant proteins in various organisms (e.g., *E. coli*, *Homo sapiens*, *Saccharomyces cerevisiae*, *Mus musculus*, and *Rattus norvegicus*). The GenScript rare codon analysis [75] is a prominent tool for codon usage and its distribution analysis (codon adaptation index-CAI, glycine-cystine content-GC, and codon frequency distribution-CFD) in the individual expression host organism is based on the optimum Gene^{TM} algorithm. The endonuclease NdeI (N-terminal) and BamHI (C-terminal) restriction enzyme sites were added to the respective cloning sequences in the host (*E. coli*).

3. Results

3.1. Analysis of Selected Sequences

The retrieved L2 of HPV58, the Flagellin protein of *Salmonella enterica* serovar Dublin, and humanTLR5 contain 472, 505, and 858 amino acids residues, respectively. However, for the potential epitope prediction, an N-terminal region of 12-114 AA residues (highly conserved and virus surface exposed region) were selected from L2 protein sequences. In the case of the Flagellin protein, the N terminal (5–143 amino acids) and C-terminal (419–504 amino acids) regions were selected for the vaccine design.

3.2. Immunomics Analysis

3.2.1. MHC-I Binding Epitopes

The region-specific MHC-I alleles in the Chinese population, namely, the HLA-A* (01:01, 02:01, 02:07, 11:01, 24:01), HLA-B* (46:01, 58:01), and HLA-C* (07:02, 12:03) restricted epitopes, were obtained using the IEDB consensus and NetMHC v.4.0 tools. The epitope segments from the N-terminal region of HPV 58 were overlapped using two different tools as shown in Table S1.

3.2.2. CTL Epitopes and TCR-Peptide/Peptide-MHC Interfaces

The overlapped epitope segments obtained from the above MHC-I prediction were compared with the results of both CTLPred [46] and PAComplex [47] servers. Furthermore, the shared epitope segments obtained from the CTLPred and PAComplex were used for epitope selection, and vaccine design as shown in Table S1.

3.2.3. MHC-II Binding and IFN-γ Producing Epitopes

The lowest percentile rank with strong binding affinity epitope segments with human MHC-II alleles, namely, the DQB1*(03:01, 03:03, 06:01), DRB1*(07, 09, 14:01:01, 15:01, 15:07:01), and DPB1*(05:01, 05:02:01) restricted epitopes were obtained using IEDB consensus and Tepitool servers. The overlapping promiscuous epitope segments from the above prediction (Table S2) were selected and evaluated for their INF-γ production ability. The overlapped INF-γ producing CD4+ (MHC-II) epitope segments are as given in Table S2. Therefore, the shared MHC-II epitope segments could produce IFN-γ against viral infection. Interestingly, the above-obtained overlapped $CD4^+$ epitopes shared the $CD8^+$ epitope segments.

3.2.4. Continuous B-Cell Epitopes

ABCPred predicted the potential antigenic linear B-cell epitope segments. The overlapped B-cell epitopes and their respective position are given in Table S2.

3.2.5. Selection of the Overlapped Epitope Segments

According to the above results (MHC-1, CTL, MHC-11, INF-γ), only four potential antigenic epitope segments were selected from the HPV58 minor capsid protein. The epitope segments, namely,23–36, 30–43, 10–23, and 29–42 from the N-terminal region of the HPV58 L2 protein have been chosen for further studies (Table 1).

Table 1. Selection of overlapped epitope segments from the N-terminal region of HPV58.

Start	End	Overlapped Epitope Segments	IFN-γ Producing Epitopes (Potentiality/ Value)	AllergenFP	AllerTOP	AntigenPro	Cross-Reactivity with Human Proteomes
23	36	**KVEGTTIADQILRY**	+0.528	Non-allergen	Non-allergen	Antigen	Similarity level zero
30	43	ADQILRYGSLGVFF	+1.000	Non-allergen	-	Antigen	Similarity level zero
10	23	CKASGTCPPDVIPK	+1.000	-	-	Antigen	Similarity level zero
29	42	**IADQILRYGSLGVF**	+1.000	Non-allergen	Non-allergen	Antigen	Similarity level zero

The results of the analysis showed four overlapped epitope segments, among these the two epitope segments (23–36 and 29–42) indicated in bold were selected for the vaccine construction. The IFN-γ production ability of the overlapped epitope segments is presented in positive values.

3.2.6. Antigenicity, Allergenicity, and Cross-Reactivity of Selected Epitope Segments

The start and end positions, epitope segments, pro-inflammatory cytokines (INF-γ) productivity, allergenicity, antigenicity, and cross-reactivity with human proteomes of the selected overlapped HPV58 are given in Table 1. Among the four, only AllergenFP and AllerTOP declared segments (23–36 and 29–42) as non-allergen, respectively. AntigenPro shows the selected epitope segments having antigenic potential. In addition, the cross-reactivity result of the epitope segments with the human proteomes has a zero similarity level. It confirmed that there was no distinctive match of overlapped epitope segments as found in *Homo sapiens*. It indicated that these epitope segments would not cause or induce any autoimmune disease or disorders. Based on the above overall analysis, the 23–36 and 29–42 epitope segments were selected to design the vaccine construct.

3.2.7. Epitopes Conservancy

In epitopes-based vaccine design, the conserved epitope segments would be essential in order to give wider cross-protection against various hrHPV strains. Supplementary Table S3 gives the comprehensive analysis of overlapped EP ($\geq$30%), positions, subsequences identity, and hrHPV. The conservation of selected epitopes has cross-protection to the 15 hrHPV as shown in Supplementary Figure S1a. The overlapped epitope segments $KVEGTTIADQILRY_{23\text{-}36}$ and $IADQILRYGSLGVF_{29\text{-}42}$ with 15-hrHPV strains are illustrated in supplementary Figure S1b,c.

3.3. Vaccine Engineering

3.3.1. Designing of Chimeric Vaccine SGD58

The complete vaccine construct consists of (1) two selected epitope sequences (23–36 and 29–42); and (2) two different TLR adjuvants, Flagellin and RS09. Flagellin is recognized as the TLR5 agonist that involves the activation of innate immunity. The head and tail of the vaccine construct contain the N-terminal (5-143 amino acids) and C-terminal regions (419-504 amino acids) of the Flagellin. In addition, RS09, a synthetic short peptide of the TLR4 ligand, is also used; (3) two different T helper (Th) epitopes, namelyPADRE and TpD, are used in the construct. The pan HLA-DR-binding epitope (PADRE) is frequently employed for synthetic or recombinant vaccine development and another universal epitope, TpD, which has 31 amino acids, is also used; and (4) the seven different parts in the vaccine construct were associated with the linkers GPGPG, AAY, and EAAAK. Finally, the designed chimeric vaccine (SGD58) with 318 amino acid sequences was drawn using illustrator for biological sequences (IBS) v1.0 as illustrated in Figure 1.

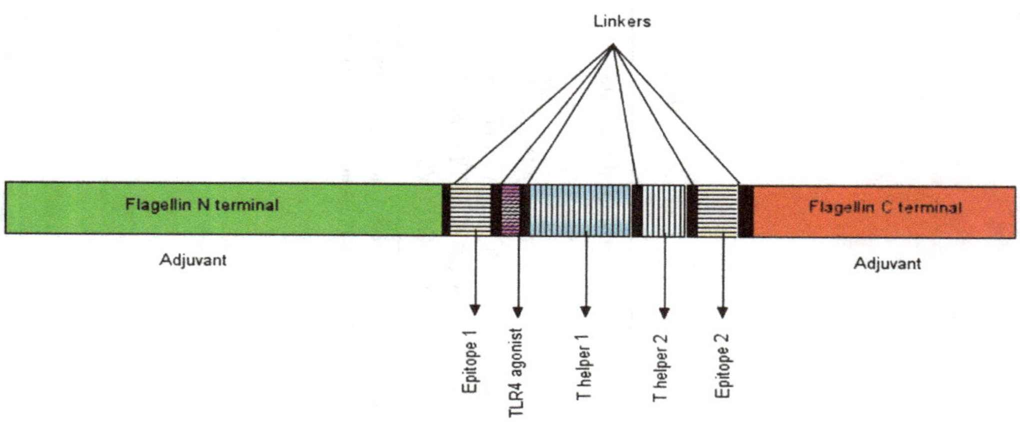

Figure 1. The designed vaccine construct 58 (SGD58). The SGD58 contains seven different segments, namelytwo different adjuvants (N and C-terminal region of Flagellin-TLR5 agonist); two epitope segments (23–36 and 29–42); adjuvant RS09 (TLR4 agonist) and two Th epitopes (PADRE and TpD). All the segments were joined together by using the following linkers (GPGPG, AAY or EAAAK).

3.3.2. Physiochemical and Immunological Properties of SGD58

The various physiochemical properties of the chimeric vaccine (molecular weight, theoretical pI, the total number of negatively and positively charged residues, extinction coefficient, estimated half-life, instability and aliphatic index, GRAVY, and solubility) are demonstrated in Table 2. The designed chimeric vaccine from the L2 protein of HPV58 was highly stable with the computed instability index of 35.88 (<40) and a molecular weight of 33.15 kDa. The highest amino acid composition in the chimeric vaccine is alanine (12.1%), serine (12.1%), leucine (9.3%), asparagine (8.4%), isoleucine (7.6%), glycine (7.0%), glutamine (6.5%), threonine (5.9%), aspartic acid (4.8%), valine (4.5%), glutamic acid (3.7%), proline (3.7%), arginine (3.4%), lysine (3.1%), phenylalanine (2.2%), tyrosine (2.2%), histidine (1.1%), methionine (1.4%), and cysteine (0.3%). The results of the SOLpro analysis indicate that the chimeric vaccine was soluble (0.62) on overexpression in *E. coli*. The antigenic score of the chimeric vaccine was demonstrated as 0.4301 (> threshold of 0.4) and 0.9438 using VaxiJen and ANTIGENpro (Table 2).

Table 2. Evaluation of the various physiochemical and immunological properties of the chimeric vaccine SGD58.

Properties	Results/Values
Number of amino acids	318
Molecular weight	33,394.15
Theoretical pI	8.00
Total number of negatively charged residues (Asp + Glu)	24
Total number of positively charged residues (Arg + Lys)	25
Extinction coefficient M-1 cm-1	12,950
Half-life	20 h (Mammalian reticulocytes, *in vitro*), 30 min (yeast, *in vivo*) and >10h (*E.coli*, *in vivo*)
Instability index	35.88 (Indicates protein as stable)
Aliphatic index	94.62
Grand average of hydropathicity (GRAVY)	−0.190
Solubility (SolPro)	Soluble with probability 0.621085
Antigenicity (VaxiJen)	0.4301(Probable antigen)
Antigenicity (AntigenPro)	0.943820 (Probable antigen)

3.4. Structural Analysis

3.4.1. Homology Modeling

I-TASSER homology modeling indicated that model 3 of the chimeric vaccine (SGD58) has the highest C-score of −0.39. In addition, the TM and RMSD scores of 0.66 and 7.2Å of model 3 represent the standard similarity and accuracy of the modeled structures. Notably, there is no 3D structure available for the TLR5 on the protein data bank. I-TASSER was used to model the 3D conformational structure of TLR5. The C-score, TM, and RMSD scores of the best model 1 of TLR5 are −0.35, 0.82 and 6.8Å, respectively. Therefore, model 1 is suggested as the best model with a higher confidence level. In addition, modeling with Robetta, we have obtained five different models for TLR5 and the chimeric vaccine candidate. Models 1 and 4 are *de novo* models while models 2, 3, and 5 are *ab initio* models. The Ginzu domain prediction confidence score for TLR5 is 0.9375, and for SGD58 is 0.6502. All the models were selected for structural refinement analysis.

3.4.2. Structural Refinement

The acquired best-modeled 3D structures of SGD58 and TLR5 underwent structural refinement using servers GalaxyWEB and 3D refine. The GalaxyRefine program gives five best-refined models for the whole SGD58 and TLR5. In addition, the lowest 3Drefine score represents the good quality of the refined model, based on the 3D refine force field. All the refined models (1–5) of both GalaxyWEB and 3D refine were used for further structural validation.

3.4.3. Structural Validation

The refined 3D structure obtained in the above section underwent quality improvement using three potential tools: ProSA-web, RAMPAGE, and the ERRAT. The Z-score (ProSA), overall quality factor (ERRAT), and the favored, allowed, and outlier region (RAMPAGE) of the validated 3D structure of SGD58 are given in Table S4 and TLR5 in Table S5. Figure 2 illustrates the Z-score, overall quality factor and the favored, allowed, and outlier region of the selected best SGD58 model. From overall comparison of the obtained results, the Robetta model 3 of SGD58 (Figure S2a) and the Robetta model 5 of TLR5 (Figure S2b) using UCSF Chimera were selected for further analysis.

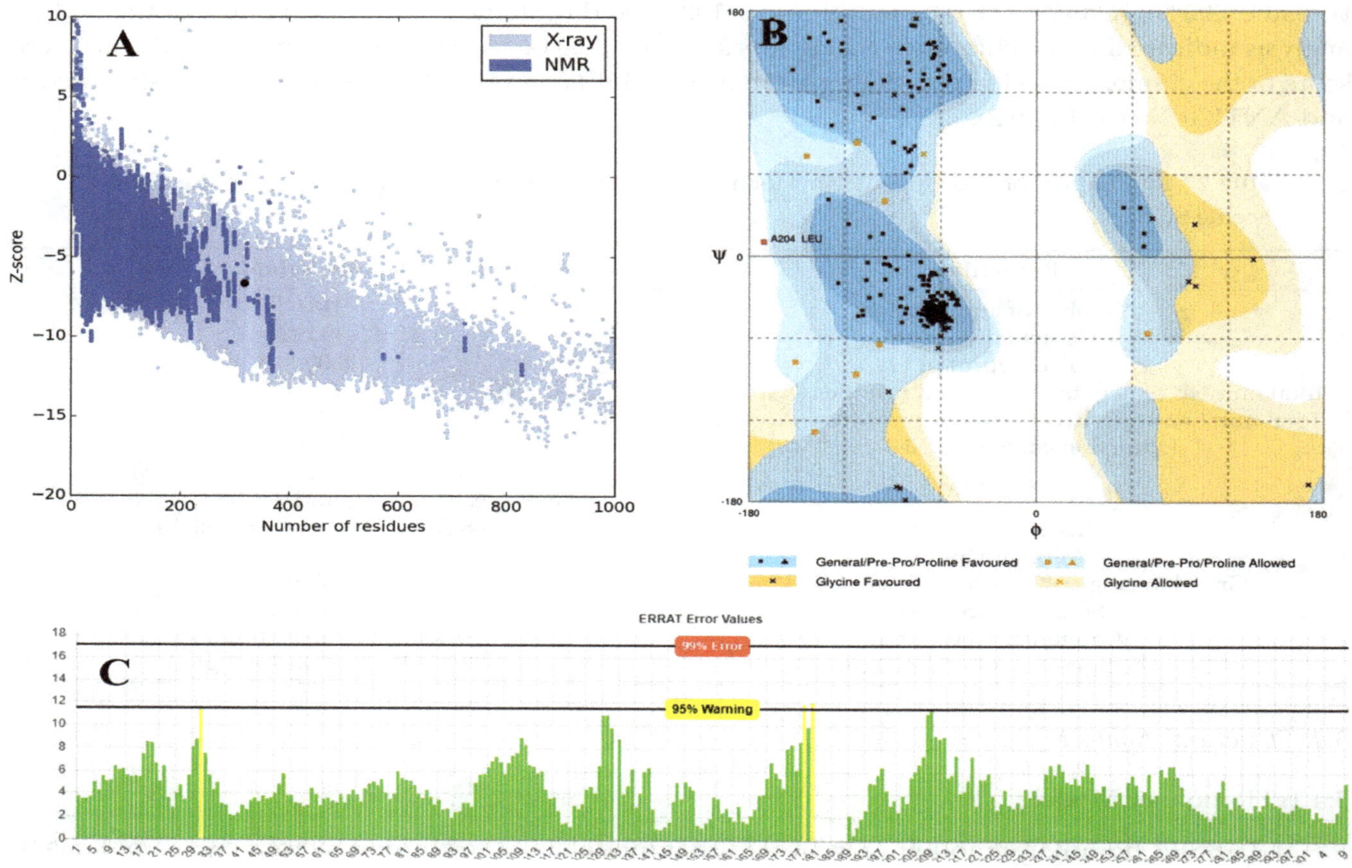

Figure 2. Validation outcome of the refined 3D structure of SGD58. (**A**)The ProSA Z-score of the refined model is -6.65, which is shownby the dark black color spot. The values are presented in the range of native protein conformation. The dark blue and light blue color region represents the Nuclear magnetic resonance and X-ray spectroscopy determination of the experimental protein chains in the protein database (PDB). The X- and Y-axis represent the number of amino acid residues and Z-scores respectively; (**B**) In the Ramachandran plot of the refined model, we illustrated the favored in green circle (96.8%), the allowed in triangle (2.8%) and the outlier in yellow shaded circle (0.3%) regions. (**C**) The ERRAT Plot shows that the overall high quality factor of the refined SGD58 is 99.0033. * On the error axis, two lines are drawn to indicate the confidence with which it is possible to reject regions that exceed that error value. ** Expressed as the percentage of the protein for which the calculated error value falls below the 95% rejection limit. Good high resolution structures generally produce values around 95% or higher. For lower resolutions (2.5 to 3A), the average overall quality factor is around 91%.

3.5. *Conformational B-Cell Epitopes*

The results of the Discotope 2.0 analysis demonstrated that 18 potential B-cell epitopes were obtained from 318 total residues. Table S6 explains the respective amino acid, residue with contact number, propensity, and Discotope score of the predicted B-cell epitopes.

3.6. *Investigation of the Interaction between SGD58 and TLR5*

3.6.1. PP Docking Interaction

From the above findings, the best-refined model of SGD58 (Robetta model 3) and TLR5 (Robetta model 5) was used to perform molecular docking using the HADDOCK server. The binding cavity of TLR5 and Flagellin was obtained from a previous report [76,77]. The input TLR5 receptor contains 858 amino acids and SGD58 contains 2923 amino acids. The human TLR5 sequence contains 21 different leucine-rich repeats (LRR) segments with 443 amino acids. Flagellin contains two D1/D0 TLR-binding domains in the N and C terminals of the sequence.

The HADDOCK method directly permits the integration of biophysical information about the protein–protein complex in order to constrain docking. In this study, we docked the target receptor (TLR5) chimeric vaccine candidate (SGD58) to observe the interaction between the complexes. The HADDOCK method clustered 116 structures into 12 clusters, which represents 58.0% of the water-refined models that HADDOCK generated. The TLR5-SGD58 complex shows the highest HADDOCK score of -62.5 ± 7.6, representing the good affinity level between the target and vaccine. The buried surface area (BSA) of cluster 4 of the TLR5-SGD58 complex is 1914.4 ± 124.4, which indicates close proximity and a less water-exposed protein surface.

The desolvation energy (43.1 ± 7.9), restraints violation energy (1192.9 ± 96.93), and BSA have a high-quality association with the docking score of the complex. The HADDOCK score, interaction energy, Van der Waals energy, electrostatic energy, desolvation energy, restraints violation energy, and BSA of the top tenclusters are given in Table S7. In all, there are 44 hydrogen bonds between cluster 4 of the TLR5-SGD58 complex. The following amino acids, namely, Lys 148, Asp258, Ser 145, Arg 62, Asn 65, Gln 96, Arg 122, Asn 150, Asn 266, Tyr 120, Lys 177, Gly 76, Glu 80, Gln 72, Thr 73, Ser69, Asn 65, Gln 96, Arg 122, Thr 58, Asp 118, Gly 119, Asp 93, Asn 65, Ser 69, Arg 122, Asn 123, Asp 258, Thr 73, Gln 72, Glu 80, Asn 83, Lys 125, Gln 72, Ser 69, Asn 65, Arg 122, Ser 145, Gln 254, Gly 119, Glu 80, Asn 83, Thr 73, Arg 247, Glu 171, Asp 118, and His 143 act as interacting residues present in the best four clusters of the TLR5-SGD58 complex(Figure 3). Thus, the TLR5-SGD58 complex docking analysis and the intermolecular hydrogen bonding patterns confirm that the interaction of the chimeric vaccine candidate with the target TLR5 can induce both cellular and humoral immunity and inhibit HPV progression.

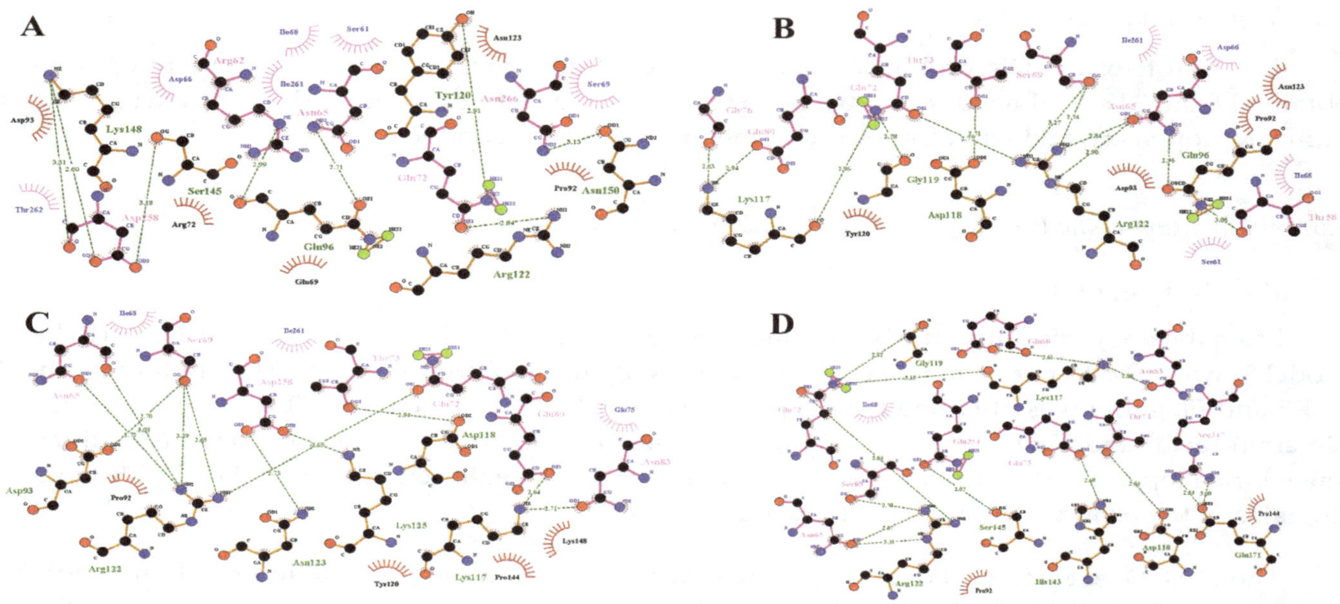

Figure 3. LIGPLOT prepared interacting residues in the TLR5_SGD58 complex. (**A**–**D**) represent the best structures of the TLR5_SGD58 cluster. The color-coding represents the TLR5 in brown color and the SGD58 in pink color. The dashed lines in green color denote hydrogen-bonding interactions.

3.6.2. MD Simulation

MD simulation demonstrated the stability of the TLR5_SGD58 docked complex in the active site of TLR5. RMSD is the known parameter by which to determine the obtained structure from the MD trajectory. This parameter was evaluated as a preliminary analysis of the backbone atoms of TLR5. The structure and dynamic proprieties of TLR5 were determined using the backbone RMSDs during the simulation period (Figure 4). The RMSD of TLR5 was gradually increased until 20 ns and a nearly 8–100 ns period with 3–4 nm of deviation, and after 30–80 ns, it was stable. The RMSD curve of SGD58 indicates an insignificant variation from 0–20 ns at 5.0 nm, and after 20 ns, it was stable. SGD58 has more a stable RMSD value compared to TLR5 (Figure 4A). The flexibility of each residue in the TLR5-SGD58 complex is calculated based on root mean square fluctuations (RMSF). Figure 4B shows that TLR5 has an insignificant variation of residues, which indicates that this molecule was stable during the MD simulation time of 100 ns. These residues have well-known interactions with the vaccine candidate. The ND1b domain (100–200 residues) of the Flagellin fragments of SGD58 shows a low flexibility, which can be attributed to their interaction with the TLR5 protein (Figure 4C). In addition, CD1 and CD0 domains have more fluctuations (200–250 residues). Figure 4D illustrates hydrogen bond interactions throughout the simulation period, to understand the binding efficiency of TLR5 with SGD58. The average number of hydrogen bond interactions was observed in 2.0 nm. Figure S3 illustrates that the potential energy (PE), temperature, total energy (TE), and pressure of SGD58 was stable during the simulation period. The average TE of SGD58 is −7207307343.324 with a standard deviation of 4279.598082. In addition, the average PE of SGD58 is −8985342.697 with a standard deviation of 3370.264894. PE and TE attained equilibrium at a temperature of 300K. The result of the radius of gyration (Rg) analysis is shown in Figure S4. The simultaneous changes in the Rg plots of the complex with TLR5 (Figure S4a) and SGD58 (Figure S4b) indicate that the substantial nature of the complex frequently increases. The Rg plots compression of SGD58 with TLR5 is similar to the RMSD parameter, which indicates the effort of SGD58 to reach internal configuration in TLR5.

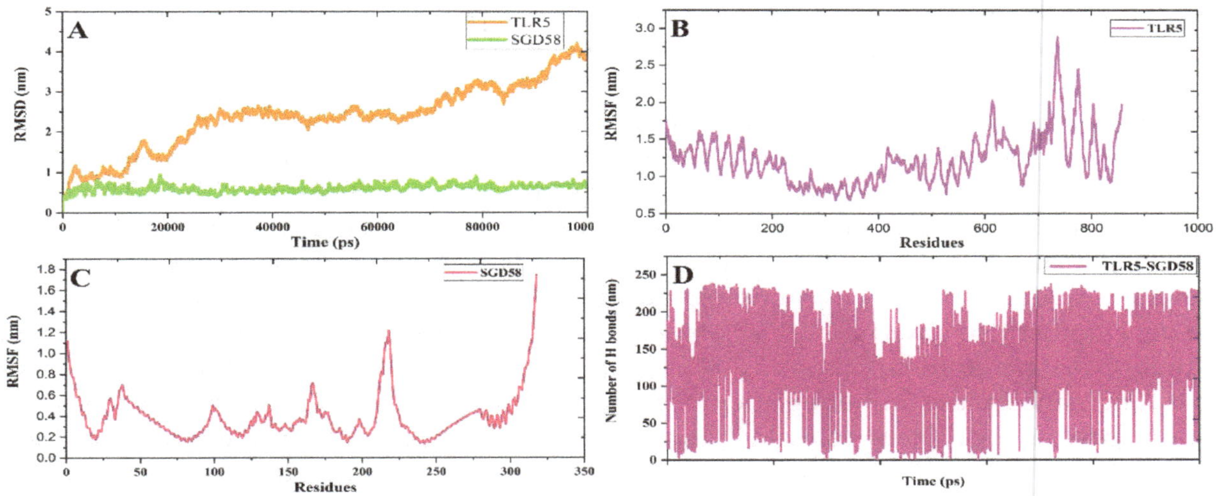

Figure 4. (**A**) The backbone root mean square deviation profile of TLR5 and SGD58 is depicted for the entire 100 ns. (**B**) The root mean square fluctuation of TLR5. (**C**) The root mean square fluctuation of SGD58. (**D**) The total number of intermolecular H bond interactions between SGD58 in complex with TLR5.

3.6.3. Virtual Gene Expression and Cloning

For the given SGD58, the optimized reverse-translated coding sequences were obtained using the EMBOSS Backtranseq tool. The maximal protein expression of this optimized coding sequence in the host (*E. coli*) was analyzed by the GenScript's OptimumGene^{TM} codon optimization tool. Figure S5 illustrates the CAI, GC, and CFD of the gene transcript. The gene (reverse translated coding sequence of the vaccine construct) having an ideal CAI value of 1.00 (>0.8) is more suitable for the above expression (*E. coli*) in the host organism. Moreover, 59.92% of ideal GC content is presented in the gene (between 30% and 70%). However, values outside of these peak ranges would severely inhibit the transcriptional and translational efficiency of the gene products. The CFD value of the gene is 100%, representing their highest codon frequency distribution in the preferred expression organism.

4. Discussion

Immunomics is an integrative area of computer science and experimental immunology and plays a vital role in vaccine development. Immunomics tools and databases are used to forecast the target epitope segments to enhance CTL or B-cell immunity in a cost-effective manner and less experimental time [78–82]. The computational vaccine design involves the engineering of potential non-pathogenic epitope segments with adjuvants to enhance the function of the human immune system against dreadful diseases, including cervical cancer. Unlike conventional vaccines, peptide or epitope vaccines have several advantages; they are synthetic (pathogen-free), have less unwanted side effects, minimize accidental allergenic reactions, and design and predict peptides with self or non-self antigen to elicit and balance the immune responses [83,84]. HPV58 is considered as the most predominant genotype causing cervical cancer incidences in China. HPV has type-restriction, non-targeted delivery, and a high-cost of existing vaccines, which makescontinuing research on HPV vaccine development necessary. Therefore, this study aimed at the design of a chimeric vaccine *via* targeting HPV through immunomics, PP docking, and MD simulation approaches.

Earlier reports suggest that the L2 protein is majorly buried or hidden under the surface of native and matured virions [32,85,86]. The initial interactions between L1/L2 are hydrophobic with coverage of small stretches of amino acid sequences. It exhibits potential effects during *in vitro* assembly [87]. However, the structural relation of L2 minor capsid protein to L1 in the virion particles is not clearly known. In another study, Henio et al. [88] reported that the L1/L2 proteins of HPV have various antigenic epitope segments such as 32–81, 212–231, 272–291, and 347–381 amino acids, and these could be accessible on the surface of L1/L2 virus-like particles. In particular, the N-terminal region of the L2 protein is highly conserved and has diverse functions: it mainly participates in the attachment of the virus particle and its genome assembly in the host system. The N-terminal region 1–12 amino acids are in the DNA binding domain, the 9–12 amino acids are in the furin-cleavage site, and the 22–45 amino acids are in the cyclophilin-B and β-actin-binding domain [89–91]. The L2 protein can act as the prospective target to design the next-generation prophylactic vaccines for HPV [92]. This strategy is prominently supported by the early evidence, such as the production of cross-neutralizing antibodies RG1 [93], K4L2, and K18L2 [94] against the target sites (17–36 AA and 20–38 AA) of L2 in various experimental models. The weak immunogenicity of L2 is a significant obstacles in epitope-based vaccine development; to date, no L2-VLPs-based prophylactic vaccine has been approved for clinical application [30]. Therefore, in this study, we selected the L2 protein to predict the potential epitope candidate and enhance the immunogenicity using adjuvants to design a chimeric vaccine.

Adjuvants have different roles including up-regulation of the immune response, increased action of neutralizing antibodies, processing of cytosolic MHC class-I restricted peptides and target presentation to a specific receptor, acting as an immunogen, and application in the preparation of a single dose [95]. To increase the immunogenicity of L2-derived peptides, the selection of a suitable adjuvant plays a vital role in the vaccine design. Instead of mixing the appropriate adjuvant directly, designing the peptide vaccine candidate using suitable linkers and adjuvants could be highly effective. Alhydrogel® adjuvant 2%, known as "alum," is frequently used as an adjuvant in diphtheria-tetanus-pertussis (DTP), HPV, and hepatitis vaccination [96]. Although the alum adjuvants induced the Th2-mediated immune response, they are ineffective to the pathogen, which is indeed of the Th1 immunity. Later, the emulsion-based incomplete Freund's adjuvant (IFA) was developed, which induces potential Th2, and little Th1-mediated immune responses [97]. However, the application of the emulsion-derived adjuvants is not supported well in the vaccination program due to the induction of autoimmune disorders and an unclear mode of action [98]. To overcome these issues TLR-based ligands were developed and achieved success in the generation of both Th1 and Th2 immune responses in the experimental models. Alphs et al. [99] achieved potential immunogenicity of the synthetic lipo-peptide (HPV16 L2) vaccine through fusion with the Th epitope and TLR ligand. Bacterial Flagellin is a potential TLR5 ligand, which can induce the production of both Th1 and Th2 immune responses.

It is frequently used as an adjuvant in the recombinant vaccine production when fusing with antigenic particles [100,101]. Flagellin, a TLR5 ligand, binds to the particular domain (Toll/interleukin-1 receptor) of the TLR5 receptor in humans. Notably, another newly developed and licensed adjuvant used for human vaccine development is the adjuvant system 04 (AS04). Moreover, AS04 was developed by a combination of 3-O-desacyl-4′-monophosphoryl lipid A (MPL), which is a prominent TLR4 agonist and aluminum salt. In the presence of Cervarix, the AS04 adjuvant induced the function of NF-kappa and cytokine synthesis in cancer cells and animal model. It leads to the appearance of increased antigen-loaded dendritic cells and monocytes followed by $CD4^+$ and B-cells in the injection site [102]. Moreover, as a result of two-dose schedule trial in young girls (9–14 years), the HPV16/18 plus AS04 adjuvant vaccine (Cervarix) was highly immunogenic and has been approved clinically for the prevention of HPV infection, precancerous CIN (I/II/II), and cervical cancer [103]. In 18–25-year-old Chinese women, AS04 adjuvant vaccines were reported as having immunogenicity and an acceptable harmless profile from the randomized-controlled trial [104]. RS09 (short synthetic peptide segments "APPHALS"), is a ligand to TLR4. RS09 does not contain any toxicological effects, and is devoid of skin irritation, serious eye damage, and carcinogenic properties, etc. It successfully enhances the nuclear factor of kappa-light-chain-enhancer of activated B cell (NF-κB) translocation pathways and enhances the pro-inflammatory cytokine and antibody serum concentration in macrophage cells and animal models [105]. TLR adjuvants play an advanced role in commercial vaccines [106]. Two universal Th epitopes (PADRE and TpD) were added to the chimeric vaccine to resolve the deficiency of Th responses. The pan HLA DR-binding epitope is known as "PADRE". It has a binding affinity with more than 15 MHC class-II allotypes and induces proliferative $CD4^+$ in peripheral blood mononuclear cells (PBMC) from humans [107]. In this manner, it explains the issue raised by HLA polymorphism in the human population [108]. It is extensively studied for synthetic peptide-based vaccine development in C57BL/6 cervical cancer models and Phase I/II clinical trials [109–111]. TpD is another universal memory T-cell helper peptide. The immunization of TpD produces a promising antibody and enhances long-term $CD4^+$ immune response in mice, *Rhesus macaques*, cynomolgus monkeys, and PBMC in humans [112]. Therefore, we have chosen two different Th epitopes (PADRE and TpD) and two TLR adjuvants (Flagellin and RS04) to enhance the immunogenicity in the chimeric vaccine.

A small flexible linker sequence was employed to join various segments of epitopes, the TLR agonist, and the Th epitopes in the vaccine construct. In this study, GGS linker was used to join the various segments in the vaccine design. The GGS linker facilitates the natural rotation or movement of the epitope segments and adjuvants and ameliorates their free identification by the surface receptor molecules [113,114]. GGS linkers contain nonpolar glycine (Gly) and polar serine (Ser) amino acid residues, which prohibit unnecessary complex formations between the linked partners and retain the function of the chimeric vaccine. A GGS spacer was presented in both natural and artificial linkers, to either increase the stability of the binding domain partners or stabilize the PP complex [115].

The targeted delivery of a vaccine can improve the efficiency and achieve a better outcome. In this study, weselected TLR5 as the target for the chimeric vaccine. Innate immunity-inducing TLR5 are mainly expressed on antigen presenting monocytes and dendritic cells while they encounter the entry of pathogenic microbes [116]. Moreover, TLRs are a well-categorized pattern recognition receptor (PPR) family, which involves the sensing of invading virulent pathogens entry into the host [117]. Horseshoe-shaped LRR are present in each TLRs conserved fold for binding to their respective legends [118]. Numerous studies report that after the binding of TLR5 to the specific ligand, it induces the myeloid differentiation gene 88 (MYD88), which triggers activation of the tumor downstream signaling pathways including NF-κB, mitogen-associated protein kinase (MAPK), and interferon regulatory factors (IRFs) [119]. Once the TLRs recognize their ligand, they become active and induces the production of pro-inflammatory cytokines such as tumor necrosis factor (TNF), interleukins (IL), and INFs [117,118]. In this manner, the host cell increases the capacity to eliminate the invading pathogens. Kim et al. [120] reported that TLR5 is a potential biomarker for the malignant transformation of cervical squamous cells. Therefore, in this study, we selected TLR5 as a target for the chimeric vaccine and its efficiency was determined using PP docking and MD simulation.

TLR5 is an excellent receptor for Flagellin, which is the major component of the bacterial Flagella [116]. Flagellin adjuvant has been extensively used in experimental HPV vaccination. Interestingly, when the host cell responds to Flagellin, TLR5 induces B-cell differentiation into the plasma producing B-cells. Earlier reports demonstrate the significance of Flagellin fused L2-multimer vaccines in experimental rabbits and mice [121,122]. Flagellin is madeup of four important domains: D0, D1, D2, and D3. The D0 and D1 domains are composed of highly conserved N-terminal (1 to 200 amino acids) and C-terminal (405–494 amino acids) regions, which is important for TLR5 agonist action. In addition, the centered hypervariable D2 and D3 regions show the vast differences, by their size and composition, among the various bacterial microorganisms [123–125]. Owing to the higher antigenicity and toxicity caused by the central D2/D3 domain, this antigenic part is removed or replaced by the optimized epitope segments or different adjuvants in the vaccine design. D2/D3 antigen replacement in Flagellin enhances mucosa-immunoglobulin productions in the experimental animal models through intranasal immunization [126]. Therefore, we selected the N and C-terminal regions of Flagellin in the design of the chimeric vaccine. Forstneric et al. [125] reported the appropriate identification of Flagellin by the homology modeled hTLR5 and mTLR5. The crystallographic complex structure of zebra fish TLR5 with the domains (D1, D2 or D3) of *Salmonella* sp. were also studied [77]. However, there was a lack of availability of the crystal structure of TLR5 until now. Therefore, in this study, homology modeling, structural refinement, and validation were performed and found using the Robetta model 5 of TLR5 obtained using the Robetta, 3DGalaxyRefine, ProSA-web, RAMPAGE, and ERRAT. Moreover,

TLR5 is greatly presented in vertebrates [77,125,127]. It facilitates the PP interaction analysis of TLR5 with Flagellin using HADDOCK, as shown in Figure 4b. It shows the interacting residues of this complex observed between the LRR region of TLR5 and the D0/D1 domain of Flagellin. This finding is significantly supported by earlier reports [125,127] concerning TLR5 recognition of the D0 of Flagellin by the inflammasome receptor, for preventing the immune escape of invading pathogenic strains. The MD simulation results depict a constant and stable interaction between SGD58 and TLR5. During the MD simulation study, TLR5 was stable after 15 ns, whereas SGD58 exhibited insignificant variations and was then stable after 10 ns. The structural changes were observed to have gained the optimal sustainability of SGD58 and TLR5. In addition, very slight changes were noticed in the D0/D1 domain regions of SGD58. Finally, the results of the codon optimization and virtual cloning confirms the translated chimeric vaccine sequence in *E. coli* to be capable of regulating the higher level of gene expression. The successful expression of the designed virus-like particles of HPV in *E. coli* is reported in earlier studies [128,129], which can enhance the production of the vaccine at a cheaper cost.

From this report, the new chimeric vaccine candidate was engineered using various immunomics tools, PP docking, and MD simulation, which can reduce the experimental cost and time. The designed chimeric vaccine SGD58, has appropriate structurally refined, physiochemical, and immunological properties that can produce humoral and cellular immune responses against HPV. The designed chimeric vaccine has cross-production with the 15 different hrHPV strains. Further experimental investigation is planned to determine the efficiency of the chimeric vaccine, especially allele-specific for the Chinese population.

Supplementary Materials:
Figure S1: Conservation of two overlapped epitope segments in fifteen hrHPV strains. Figure S2: Refined 3D structure of the SGD58 and TLR5 by using UCSF Chimera. (a) The 3D structure of the SGD58 was obtained through homology modeling by using Robetta. (b) The 3D structure of the TLR5 was obtained through homology modeling by using Robetta. Figure S3: (a) The total energy, (b) potential energy, (c) temperature and (d) pressure plots of MD simulation for the TLR5-SGD58 complex in simulations of 100 ns. Figure S4: (a) Rg plot of the TLR5 and (b) The Rg plot of SGD58 complex. Figure S5: Codon optimization and *in silico* cloning of the gene. Table S1: The overlapped epitope segments of MHC-I, CTL and TCR from N-terminal region of HPV58 were predicted by using different servers. Table S2: The overlapped epitope segments of MHC-II, INF-gamma producing and B-cell epitopes N-terminal region of HPV58 by using different servers. Table S3: Conservation across-hrHPV strains by the overlapped HPV58 epitope segments. Table S4: Validation of 3D structures of the designed SGD58 obtained by the I-TASSER (I-T) and Robetta and its refinement by the Galaxy Refine (Gal) and 3Drefine (3DR). Table S5: Validation of 3D structures of the TLR5 obtained by the I-TASSER (I-T) and Robetta and its refinement by the GalaxyRefine (Gal) and 3Drefine (3DR). Table S6: Dis-continuous B-cell epitopes identified in the refined 3D structure of designed vaccine constructs of HPV58 by using Discotope 2.0. Table S7: Statistical analysis of the TLR5-SGD58 docking result obtained by HADDOCK.

Author Contributions: G.S., S.K., and D.-Q.W. conceived and designed the experiment. S.K and G.S. performed the immunoinformatics, vaccine design, and molecular docking studies. S.C., Q.W., and A.S.N. performed the molecular dynamics simulation. W.C.C., S.K., G.S., S.C., and D.-Q.W. wrote the main manuscript text. S.K. and A.S.N. formatted the manuscript and figures according to the instructions. G.S., W.C.C., G.K., and D.-Q.W. critically reviewed the manuscript. All the authors approved the final manuscript.

Acknowledgments: The authors duly acknowledge the financial support of the Ministry of Science and Technology of China (Grant No.: 2016YFA0501703), Henan Natural Science (Grant No.: 162300410060), grants from the State Key Lab on Microbial Metabolism, and joint research funds for Medical and Engineering and Scientific Research at Shanghai Jiao Tong University (YG2017ZD14) to DQ.W.; Henan University of Technology (Grant No.: 21450004) to S.K.; Henan University of Technology (Grant No.: 21450003) and China Postdoctoral Science Foundation (Grant No.: 2018M632766) to G.S.; The simulations in this work were supported by the Center for High Performance Computing, Shanghai Jiao Tong University, China.

References

1. Frazer, I.H. Development and implementation of papillomavirus prophylactic vaccines. *J. Immunol.* **2014**, *192*, 4007–4011. [CrossRef] [PubMed]
2. McLaughlin-Drubin, M.E.; Munger, K. Oncogenic activities of human papillomaviruses. *Virus Res.* **2009**, *143*, 195–208. [CrossRef] [PubMed]

3. Bernard, H.U.; Burk, R.D.; Chen, Z.; Van Doorslaer, K.; ZurHausen, H.; de Villiers, E.M. Classification of papillomaviruses (PVs) based on 189 PV types and proposal of taxonomic amendments. *Virology* **2010**, *401*, 70–79. [CrossRef] [PubMed]
4. Fauquet, C.M.; Mayo, M.A.; Maniloff, J.; Desselberger, U.; Ball, LA. Family Papillomaviridae. In *Virus Taxonomy. The Eighth Report of the International Committee on Taxonomy of Viruses*; Elsevier: Amsterdam, The Netherlands, 2005; pp. 239–255.
5. Haedicke, J.; Iftner, T. Human papillomaviruses and cancer. *Radiother. Oncol.* **2013**, *108*, 397–402. [CrossRef]
6. Tao, G.; Yaling, G.; Zhan, G.; Pu, L.; Miao, H. Human papillomavirus genotype distribution among HPV-positive women in Sichuan province, Southwest China. *Arch. Virol.* **2018**, *163*, 65–72. [CrossRef]
7. Dunne, E.F.; Unger, E.R.; Sternberg, M.; McQuillan, G.; Swan, D.C.; Patel, S.S.; Markowitz, L.E. Prevalence of HPV infection among females in the United States. *JAMA* **2007**, *297*, 813–819. [CrossRef]
8. Kenter, G.G.; Welters, M.J.; Valentijn, A.R.; Lowik, M.J.; Berends-van der Meer, D.M.; Vloon, A.P.; Drijfhout, J.W.; Wafelman, A.R.; Oostendorp, J.; Fleuren, G.J.; et al. Phase I immunotherapeutic trial with long peptides spanning the E6 and E7 sequences of high-risk human papillomavirus 16 in end-stage cervical cancer patients shows low toxicity and robust immunogenicity. *Clin. Cancer Res.* **2008**, *14*, 169–177. [CrossRef]
9. de Vos van Steenwijk, P.J.; Ramwadhdoebe, T.H.; Lowik, M.J.; van der Minne, C.E.; Berends-van der Meer, D.M.; Fathers, L.M.; Valentijn, A.R.; Oostendorp, J.; Fleuren, G.J.; Hellebrekers, B.W.; et al. A placebo-controlled randomized HPV16 synthetic long-peptide vaccination study in women with high-grade cervical squamous intraepithelial lesions. *Cancer Immunol. Immunother.* **2012**, *61*, 1485–1492. [CrossRef]
10. Rerucha, C.M.; Caro, R.J.; Wheeler, V.L. Cervical Cancer Screening. *Am. Fam. Physician* **2018**, *97*, 441–448.
11. Hong, Y.; Zhang, C.; Li, X.; Lin, D.; Liu, Y. HPV and cervical cancer related knowledge, awareness and testing behaviors in a community sample of female sex workers in China. *BMC Public Health* **2013**, *13*, 696. [CrossRef]
12. Chen, W.; Zheng, R.; Baade, P.D.; Zhang, S.; Zeng, H.; Bray, F.; Jemal, A.; Yu, X.Q.; He, J. Cancer statistics in China, 2015. *CA Cancer J. Clin.* **2016**, *66*, 115–132. [CrossRef] [PubMed]
13. Zhou, H.L.; Zhang, W.; Zhang, C.J.; Wang, S.M.; Duan, Y.C.; Wang, J.X.; Yang, H.; Wang, X.Y. Prevalence and distribution of human papillomavirus genotypes in Chinese women between 1991 and 2016: A systematic review. *J. Infect.* **2018**, *76*, 522–528. [CrossRef] [PubMed]
14. You, W.; Li, S.; Du, R.; Zheng, J.; Shen, A. Epidemiological study of high-risk human papillomavirus infection in subjects with abnormal cytological findings in cervical cancer screening. *Exp. Ther. Med.* **2018**, *15*, 412–418. [CrossRef] [PubMed]
15. Long, W.; Yang, Z.; Li, X.; Chen, M.; Liu, J.; Zhang, Y.; Sun, X. HPV-16, HPV-58, and HPV-33 are the most carcinogenic HPV genotypes in Southwestern China and their viral loads are associated with severity of premalignant lesions in the cervix. *Virol. J.* **2018**, *15*, 94. [CrossRef] [PubMed]
16. Zhang, C.; Huang, C.; Zheng, X.; Pan, D. Prevalence of human papillomavirus among Wenzhou women diagnosed with cervical intraepithelial neoplasia and cervical cancer. *Infect. Agent. Cancer* **2018**, *13*. [CrossRef] [PubMed]
17. Dai, X.; Chen, L.; Li, J.; Wu, Y.; Hu, Y.; Xiang, F.; Guan, Q. Distribution characteristics of different human papillomavirus genotypes in women in Wuhan, China. *Cancer Med.* **2018**, *32*, e22581. [CrossRef]
18. Zhao, P.; Liu, S.; Zhong, Z.; Hou, J.; Lin, L.; Weng, R.; Su, L.; Lei, N.; Hou, T.; Yang, H. Prevalence and genotype distribution of human papillomavirus infection among women in northeastern Guangdong Province of China. *J. Clin. Lab. Anal.* **2018**, *18*, 204. [CrossRef] [PubMed]
19. Liu, S.; Zhong, Z.; Hou, J.; Lin, L.; Weng, R.; Su, L.; Lei, N.; Hou, T.; Yang, H.; Li, K.; et al. Analysis of HPV distribution in patients with cervical precancerous lesions in Western China. *BMC Infect. Dis.* **2017**, *96*, e7304. [CrossRef]
20. Zhang, C.; Zhang, C.; Huang, J.; Shi, W. The Genotype of Human Papillomavirus and Associated Factors Among High Risk Males in Shanghai, China: A Molecular Epidemiology Study. *Med. Sci. Monit.* **2018**, *24*, 912–918. [CrossRef]
21. FDA licensure of bivalent human papillomavirus vaccine (HPV2, Cervarix) for use in females and updated HPV vaccination recommendations from the Advisory Committee on Immunization Practices (ACIP). *MMWR Morb. Mortal. Wkly. Rep.* **2010**, *59*, 626–629.

22. Recommendations on the use of quadrivalent human papillomavirus vaccine in males–Advisory Committee on Immunization Practices (ACIP), 2011. *MMWR Morb. Mortal. Wkly. Rep.* **2011**, *60*, 1705–1708.
23. Petrosky, E.; Bocchini, J.A., Jr.; Hariri, S.; Chesson, H.; Curtis, C.R.; Saraiya, M.; Unger, E.R.; Markowitz, L.E. Use of 9-valent human papillomavirus (HPV) vaccine: Updated HPV vaccination recommendations of the advisory committee on immunization practices. *MMWR Morb. Mortal. Wkly. Rep.* **2015**, *64*, 300–304. [PubMed]
24. Paz-Zulueta, M.; Alvarez-Paredes, L.; Rodriguez Diaz, J.C.; Paras-Bravo, P.; Andrada Becerra, M.E.; Rodriguez Ingelmo, J.M.; Ruiz Garcia, M.M.; Portilla, J.; Santibanez, M. Prevalence of high-risk HPV genotypes, categorised by their quadrivalent and nine-valent HPV vaccination coverage, and the genotype association with high-grade lesions. *BMC Cancer* **2018**, *18*, 112. [CrossRef] [PubMed]
25. Jiang, R.T.; Schellenbacher, C.; Chackerian, B.; Roden, R.B. Progress and prospects for L2-based human papillomavirus vaccines. *Expert Rev. Vaccines* **2016**, *15*, 853–862. [CrossRef] [PubMed]
26. Chroboczek, J.; Szurgot, I.; Szolajska, E. Virus-like particles as vaccine. *ActaBiochim. Pol.* **2014**, *61*, 531–539.
27. Pandhi, D.; Sonthalia, S. Human papilloma virus vaccines: Current scenario. *Indian J. Sex Transm. Dis. AIDS* **2011**, *32*, 75–85. [CrossRef] [PubMed]
28. Monie, A.; Hung, C.F.; Roden, R.; Wu, T.C. Cervarix: A vaccine for the prevention of HPV 16, 18-associated cervical cancer. *Biologics* **2008**, *2*, 97–105.
29. Angioli, R.; Lopez, S.; Aloisi, A.; Terranova, C.; De Cicco, C.; Scaletta, G.; Capriglione, S.; Miranda, A.; Luvero, D.; Ricciardi, R.; et al. Ten years of HPV vaccines: State of art and controversies. *Crit. Rev. Oncol. Hematol.* **2016**, *102*, 65–72. [CrossRef]
30. Karanam, B.; Jagu, S.; Huh, W.K.; Roden, R.B. Developing vaccines against minor capsid antigen L2 to prevent papillomavirus infection. *Immunol. Cell Biol.* **2009**, *87*, 287–299. [CrossRef]
31. Schiller, J.T.; Muller, M. Next generation prophylactic human papillomavirus vaccines. *Lancet Oncol.* **2015**, *16*, e217–e225. [CrossRef]
32. Wang, J.W.; Roden, R.B. L2, the minor capsid protein of papillomavirus. *Virology* **2013**, *445*, 175–186. [CrossRef] [PubMed]
33. Chandrachud, L.M.; Grindlay, G.J.; McGarvie, G.M.; O'Neil, B.W.; Wagner, E.R.; Jarrett, W.F.; Campo, M.S. Vaccination of cattle with the N-terminus of L2 is necessary and sufficient for preventing infection by bovine papillomavirus-4. *Virology* **1995**, *211*, 204–208. [CrossRef] [PubMed]
34. Gaukroger, J.M.; Chandrachud, L.M.; O'Neil, B.W.; Grindlay, G.J.; Knowles, G.; Campo, M.S. Vaccination of cattle with bovine papillomavirus type 4 L2 elicits the production of virus-neutralizing antibodies. *J. Gen. Virol.* **1996**, *77 (Pt 7)*, 1577–1583. [CrossRef]
35. UniProt: The universal protein knowledgebase. *Nucleic Acids Res.* **2017**, *45*, D158–D169. [CrossRef] [PubMed]
36. Chen, S.; Hong, W.; Shao, H.; Fu, Y.; Liu, X.; Chen, D.; Xu, A. Allelic distribution of HLA class I genes in the Tibetan ethnic population of China. *Int. J. Immunogenet.* **2006**, *33*, 439–445. [CrossRef] [PubMed]
37. Chen, S.; Hu, Q.; Xie, Y.; Zhou, L.; Xiao, C.; Wu, Y.; Xu, A. Origin of Tibeto-Burman speakers: Evidence from HLA allele distribution in Lisu and Nu inhabiting Yunnan of China. *Hum. Immunol.* **2007**, *68*, 550–559. [CrossRef]
38. Chen, S.; Ren, X.; Liu, Y.; Hu, Q.; Hong, W.; Xu, A. Human leukocyte antigen class I polymorphism in Miao, Bouyei, and Shui ethnic minorities of Guizhou, China. *Hum. Immunol.* **2007**, *68*, 928–933. [CrossRef]
39. Wang, X.C.; Sun, L.Q.; Ma, L.; Li, H.X.; Wang, X.L.; Wang, X.; Yun, T.; Meng, N.L.; Lv, D.L. Prevalence and genotype distribution of human papillomavirus among women from Henan, China. *Asian Pac. J. Cancer Prev.* **2014**, *15*, 7333–7336. [CrossRef] [PubMed]
40. Zhang, H.Y.; Fei, M.D.; Jiang, Y.; Fei, Q.Y.; Qian, H.; Xu, L.; Jin, Y.N.; Jiang, C.Q.; Li, H.X.; Tiggelaar, S.M.; et al. The diversity of human papillomavirus infection among human immunodeficiency virus-infected women in Yunnan, China. *Virol. J.* **2014**, *11*, 202. [CrossRef] [PubMed]
41. Lu, J.F.; Shen, G.R.; Li, Q.; Chen, X.; Ma, C.F.; Zhu, T.H. Genotype distribution characteristics of multiple human papillomavirus in women from the Taihu River Basin, on the coast of eastern China. *BMC Infect. Dis.* **2017**, *17*, 226. [CrossRef]
42. Wang, Y.; Xue, J. Distribution and role of high-risk human papillomavirus genotypes in women with cervical intraepithelial neoplasia: A retrospective analysis from Wenzhou, southeast China. *Cancer Med.* **2018**. [CrossRef] [PubMed]

43. Dai, X.; Chen, L.; Li, J.; Wu, Y.; Hu, Y.; Vita, R.; Overton, J.A.; Greenbaum, J.A.; Ponomarenko, J.; Clark, J.D.; et al. The immune epitope database (IEDB) 3.0. *Cancer Med.* **2015**, *43*, D405–D412.
44. Moutaftsi, M.; Peters, B.; Pasquetto, V.; Tscharke, D.C.; Sidney, J.; Bui, H.H.; Grey, H.; Sette, A. A consensus epitope prediction approach identifies the breadth of murine T(CD8+)-cell responses to vaccinia virus. *Nat. Biotechnol.* **2006**, *24*, 817–819. [CrossRef] [PubMed]
45. Nielsen, M.; Lundegaard, C.; Worning, P.; Lauemoller, S.L.; Lamberth, K.; Buus, S.; Brunak, S.; Lund, O. Reliable prediction of T-cell epitopes using neural networks with novel sequence representations. *Protein Sci.* **2003**, *12*, 1007–1017. [CrossRef] [PubMed]
46. Bhasin, M.; Raghava, G.P. Prediction of CTL epitopes using QM, SVM and ANN techniques. *Vaccine* **2004**, *22*, 3195–3204. [CrossRef] [PubMed]
47. Liu, I.H.; Lo, Y.S.; Yang, J.M. PAComplex: A web server to infer peptide antigen families and binding models from TCR-pMHC complexes. *Nucleic Acids Res.* **2011**, *39*, W254–W260. [CrossRef]
48. Wang, P.; Sidney, J.; Dow, C.; Mothe, B.; Sette, A.; Peters, B. A systematic assessment of MHC class II peptide binding predictions and evaluation of a consensus approach. *PLoS Comput. Biol.* **2008**, *4*, e1000048. [CrossRef]
49. Paul, S.; Sidney, J.; Sette, A.; Peters, B. TepiTool: A Pipeline for Computational Prediction of T Cell Epitope Candidates. *Curr. Protoc. Immunol.* **2016**, *114*, 18–19. [CrossRef]
50. Dhanda, S.K.; Vir, P.; Raghava, G.P. Designing of interferon-gamma inducing MHC class-II binders. *Biol. Direct* **2013**, *8*, 30. [CrossRef]
51. Saha, S.; Raghava, G.P. Prediction of continuous B-cell epitopes in an antigen using recurrent neural network. *Proteins* **2006**, *65*, 40–48. [CrossRef]
52. Dimitrov, I.; Bangov, I.; Flower, D.R.; Doytchinova, I. AllerTOP v.2—A server for in silico prediction of allergens. *J. Mol. Model.* **2014**, *20*, 2278. [CrossRef] [PubMed]
53. Dimitrov, I.; Naneva, L.; Doytchinova, I.; Bangov, I. Allergen FP: Allergenicity prediction by descriptor fingerprints. *Bioinformatics* **2014**, *30*, 846–851. [CrossRef] [PubMed]
54. El-Manzalawy, Y.; Dobbs, D.; Honavar, V. Predicting protective bacterial antigens using random forest classifiers. In Proceedings of the ACM Conference on Bioinformatics, Computational Biology and Biomedicine, Orlando, FL, USA, 7–10 October 2012; pp. 426–433.
55. Chen, C.; Li, Z.; Huang, H.; Suzek, B.E.; Wu, C.H. A fast Peptide Match service for UniProt Knowledgebase. *Bioinformatics* **2013**, *29*, 2808–2809. [CrossRef] [PubMed]
56. Bui, H.H.; Sidney, J.; Li, W.; Fusseder, N.; Sette, A. Development of an epitope conservancy analysis tool to facilitate the design of epitope-based diagnostics and vaccines. *BMC Bioinform.* **2007**, *8*, 361. [CrossRef] [PubMed]
57. Kumar, S.; Stecher, G.; Tamura, K. MEGA7: Molecular Evolutionary Genetics Analysis Version 7.0 for Bigger Datasets. *Mol. Biol. Evol.* **2016**, *33*, 1870–1874. [CrossRef] [PubMed]
58. Magnan, C.N.; Randall, A.; Baldi, P. SOLpro: Accurate sequence-based prediction of protein solubility. *Bioinformatics* **2009**, *25*, 2200–2207. [CrossRef]
59. Gasteiger, E.; Hoogland, C.; Gattiker, A.; Duvaud, S.E.; Wilkins, M.R.; Appel, R.D.; Bairoch, A. Protein Identification and Analysis Tools on the ExPASy Server. In *The Proteomics Protocols Handbook*; Walker, J.M., Ed.; Humana Press: Totowa, NJ, USA, 2005; pp. 571–607.
60. Doytchinova, I.A.; Flower, D.R. VaxiJen: A server for prediction of protective antigens, tumour antigens and subunit vaccines. *BMC Bioinform.* **2007**, *8*, 4. [CrossRef]
61. Zhang, Y. I-TASSER server for protein 3D structure prediction. *BMC Bioinform.* **2008**, *9*, 40. [CrossRef] [PubMed]
62. Kim, D.E.; Chivian, D.; Baker, D. Protein structure prediction and analysis using the Robetta server. *Nucleic Acids Res.* **2004**, *32*, W526–W531. [CrossRef]
63. Ko, J.; Park, H.; Heo, L.; Seok, C. GalaxyWEB server for protein structure prediction and refinement. *Nucleic Acids Res.* **2012**, *40*, W294–W297. [CrossRef]
64. Bhattacharya, D.; Cheng, J. i3Drefine software for protein 3D structure refinement and its assessment in CASP10. *PLoS ONE* **2013**, *8*, e69648. [CrossRef] [PubMed]
65. Wiederstein, M.; Sippl, M.J. ProSA-web: Interactive web service for the recognition of errors in three-dimensional structures of proteins. *Nucleic Acids Res.* **2007**, *35*, W407–W410. [CrossRef] [PubMed]

66. Lovell, S.C.; Davis, I.W.; Arendall, W.B., 3rd; de Bakker, P.I.; Word, J.M.; Prisant, M.G.; Richardson, J.S.; Richardson, D.C. Structure validation by Calpha geometry: phi, psi and Cbeta deviation. *Proteins* **2003**, *50*, 437–450. [CrossRef] [PubMed]
67. Colovos, C.; Yeates, T.O. Verification of protein structures: Patterns of nonbonded atomic interactions. *Protein Sci.* **1993**, *2*, 1511–1519. [CrossRef] [PubMed]
68. Kringelum, J.V.; Lundegaard, C.; Lund, O.; Nielsen, M. Reliable B cell epitope predictions: Impacts of method development and improved benchmarking. *PLoS Comput. Biol.* **2012**, *8*, e1002829. [CrossRef] [PubMed]
69. van Zundert, G.C.P.; Rodrigues, J.P.G.L.M.; Trellet, M.; Schmitz, C.; Kastritis, P.L.; Karaca, E.; Melquiond, A.S.J.; van Dijk, M.; de Vries, S.J.; Bonvin, A.M.J.J. The HADDOCK2.2 Web Server: User-Friendly Integrative Modeling of Biomolecular Complexes. *J. Mol. Biol.* **2016**, *428*, 720–725. [CrossRef] [PubMed]
70. Wallace, A.C.; Laskowski, R.A.; Thornton, J.M. LIGPLOT: A program to generate schematic diagrams of protein-ligand interactions. *Protein Eng.* **1996**, *8*, 127–134. [CrossRef]
71. Lei, D.; Zhang, X.; Jiang, S.; Cai, Z.; Rames, M.J.; Zhang, L.; Ren, G.; Zhang, S. Structural features of cholesteryl ester transfer protein: A molecular dynamics simulation study. *Proteins* **2013**, *81*, 415–425. [CrossRef]
72. Schuttelkopf, A.W.; van Aalten, D.M. PRODRG: A tool for high-throughput crystallography of protein-ligand complexes. *ActaCrystallogr. D Biol. Crystallogr.* **2004**, *60*, 1355–1363. [CrossRef]
73. Kirubakaran, P.; Kothandan, G.; Cho, S.J.; Muthusamy, K. Molecular insights on TNKS1/TNKS2 and inhibitor-IWR1 interactions. *Mol. Biosyst.* **2014**, *10*, 281–293. [CrossRef]
74. Rice, P.; Longden, I.; Bleasby, A. EMBOSS: The European Molecular Biology Open Software Suite. *Trends Genet.* **2000**, *16*, 276–277. [CrossRef]
75. The GenScript Rare Codon Analysis. Available online: https://www.genscript.com/tools/rare-codon-analysis (accessed on 1 October 2018).
76. Song, W.S.; Jeon, Y.J.; Namgung, B.; Hong, M.; Yoon, S.I. A conserved TLR5 binding and activation hot spot on flagellin. *Sci. Rep.* **2017**, *7*, 40878. [CrossRef] [PubMed]
77. Yoon, S.I.; Kurnasov, O.; Natarajan, V.; Hong, M.; Gudkov, A.V.; Osterman, A.L.; Wilson, I.A. Structural basis of TLR5-flagellin recognition and signaling. *Science* **2012**, *335*, 859–864. [CrossRef] [PubMed]
78. Negahdaripour, M.; Eslami, M.; Nezafat, N.; Hajighahramani, N.; Ghoshoon, M.B.; Shoolian, E.; Dehshahri, A.; Erfani, N.; Morowvat, M.H.; Ghasemi, Y. A novel HPV prophylactic peptide vaccine, designed by immunoinformatics and structural vaccinology approaches. *Infect. Genet. Evol.* **2017**, *54*, 402–416. [CrossRef] [PubMed]
79. Kaliamurthi, S.; Selvaraj, G.; Kaushik, A.C.; Gu, K.R.; Wei, D.Q. Designing of CD8(+) and CD8(+)-overlapped CD4(+) epitope vaccine by targeting late and early proteins of human papillomavirus. *Biologics* **2018**, *12*, 107–125.
80. Khan, A.; Junaid, M.; Kaushik, A.C.; Ali, A.; Ali, S.S.; Mehmood, A.; Wei, D.Q. Computational identification, characterization and validation of potential antigenic peptide vaccines from hrHPVs E6 proteins using immunoinformatics and computational systems biology approaches. *PLoS ONE* **2018**, *13*, e0196484. [CrossRef]
81. Zhao, M.; Wei, D.Q. Rare Diseases: Drug Discovery and Informatics Resource. *Interdiscip. Sci.* **2018**, *10*, 195–204. [CrossRef]
82. Sabah, S.N.; Gazi, M.A.; Sthity, R.A.; Husain, A.B.; Quyyum, S.A.; Rahman, M.; Islam, M.R. Designing of Epitope-Focused Vaccine by Targeting E6 and E7 Conserved Protein Sequences: An Immuno-Informatics Approach in Human Papillomavirus 58 Isolates. *Interdiscip. Sci.* **2018**, *10*, 251–260. [CrossRef]
83. Kaliamurthi, S.; Selvaraj, G.; Junaid, M.; Khan, A.; Gu, K.; Wei, D.Q. Cancer Immunoinformatics: A Promising Era in the Development of Peptide Vaccines for Human Papillomavirus induced Cervical cancer. *Curr. Pharm. Des.* **2018**. [CrossRef]
84. Purcell, A.W.; McCluskey, J.; Rossjohn, J. More than one reason to rethink the use of peptides in vaccine design. *Nat. Rev. Drug Discov.* **2007**, *6*, 404–414. [CrossRef]
85. Bywaters, S.M.; Brendle, S.A.; Tossi, K.P.; Biryukov, J.; Meyers, C.; Christensen, N.D. Antibody Competition Reveals Surface Location of HPV L2 Minor Capsid Protein Residues 17-36. *Viruses* **2017**, *9*, 336. [CrossRef] [PubMed]
86. Buck, C.B.; Cheng, N.; Thompson, C.D.; Lowy, D.R.; Steven, A.C.; Schiller, J.T.; Trus, B.L. Arrangement of L2 within the papillomavirus capsid. *J. Virol.* **2008**, *82*, 5190–5197. [CrossRef] [PubMed]

87. Finnen, R.L.; Erickson, K.D.; Chen, X.S.; Garcea, R.L. Interactions between papillomavirus L1 and L2 capsid proteins. *J. Virol.* **2003**, *77*, 4818–4826. [CrossRef] [PubMed]
88. Heino, P.; Skyldberg, B.; Lehtinen, M.; Rantala, I.; Hagmar, B.; Kreider, J.W.; Kirnbauer, R.; Dillner, J. Human papillomavirus type 16 capsids expose multiple type-restricted and type-common antigenic epitopes. *J. Gen. Virol.* **1995**, *76*, 1141–1153. [CrossRef]
89. Yang, R.; Yutzy, W.H.T.; Viscidi, R.P.; Roden, R.B. Interaction of L2 with beta-actin directs intracellular transport of papillomavirus and infection. *J. Biol. Chem.* **2003**, *278*, 12546–12553. [CrossRef]
90. Fay, A.; Yutzy, W.H.T.; Roden, R.B.; Moroianu, J. The positively charged termini of L2 minor capsid protein required for bovine papillomavirus infection function separately in nuclear import and DNA binding. *J. Virol.* **2004**, *78*, 13447–13454. [CrossRef]
91. Richards, R.M.; Lowy, D.R.; Schiller, J.T.; Day, P.M. Cleavage of the papillomavirus minor capsid protein, L2, at a furin consensus site is necessary for infection. *Proc. Natl. Acad. Sci. USA* **2006**, *103*, 1522–1527. [CrossRef]
92. Tyler, M.; Tumban, E.; Chackerian, B. Second-generation prophylactic HPV vaccines: Successes and challenges. *Expert Rev. Vaccines* **2014**, *13*, 247–255. [CrossRef]
93. Gambhira, R.; Jagu, S.; Karanam, B.; Gravitt, P.E.; Culp, T.D.; Christensen, N.D.; Roden, R.B. Protection of rabbits against challenge with rabbit papillomaviruses by immunization with the N terminus of human papillomavirus type 16 minor capsid antigen L2. *J. Virol.* **2007**, *81*, 11585–11592. [CrossRef]
94. Rubio, I.; Seitz, H.; Canali, E.; Sehr, P.; Bolchi, A.; Tommasino, M.; Ottonello, S.; Muller, M. The N-terminal region of the human papillomavirus L2 protein contains overlapping binding sites for neutralizing, cross-neutralizing and non-neutralizing antibodies. *Virology* **2011**, *409*, 348–359. [CrossRef]
95. Cox, J.C.; Coulter, A.R. Adjuvants—A classification and review of their modes of action. *Vaccine* **1997**, *15*, 248–256. [CrossRef]
96. Marrack, P.; McKee, A.S.; Munks, M.W. Towards an understanding of the adjuvant action of aluminium. *Nat. Rev. Immunol.* **2009**, *9*, 287–293. [CrossRef] [PubMed]
97. Chan, P.K.; Liu, S.J.; Cheung, T.H.; Yeo, W.; Ngai, S.M.; Cheung, J.L.; Chong, P.; Man, S. T-cell response to human papillomavirus type 58 L1, E6, And E7 peptides in women with cleared infection, cervical intraepithelial neoplasia, or invasive cancer. *Clin. Vaccine Immunol.* **2010**, *17*, 1315–1321. [CrossRef] [PubMed]
98. Ott, G.; Barchfeld, G.L.; Chernoff, D.; Radhakrishnan, R.; van Hoogevest, P.; Van Nest, G. MF59. Design and evaluation of a safe and potent adjuvant for human vaccines. *Pharm. Biotechnol.* **1995**, *6*, 277–296.
99. Alphs, H.H.; Gambhira, R.; Karanam, B.; Roberts, J.N.; Jagu, S.; Schiller, J.T.; Zeng, W.; Jackson, D.C.; Roden, R.B. Protection against heterologous human papillomavirus challenge by a synthetic lipopeptide vaccine containing a broadly cross-neutralizing epitope of L2. *Proc. Natl. Acad. Sci. USA* **2008**, *105*, 5850–5855. [CrossRef] [PubMed]
100. Huleatt, J.W.; Jacobs, A.R.; Tang, J.; Desai, P.; Kopp, E.B.; Huang, Y.; Song, L.; Nakaar, V.; Powell, T.J. Vaccination with recombinant fusion proteins incorporating Toll-like receptor ligands induces rapid cellular and humoral immunity. *Vaccine* **2007**, *25*, 763–775. [CrossRef] [PubMed]
101. Mizel, S.B.; Bates, J.T. Flagellin as an adjuvant: Cellular mechanisms and potential. *J. Immunol.* **2010**, *185*, 5677–5682. [CrossRef] [PubMed]
102. Didierlaurent, A.M.; Morel, S.; Lockman, L.; Giannini, S.L.; Bisteau, M.; Carlsen, H.; Kielland, A.; Vosters, O.; Vanderheyde, N.; Schiavetti, F.; et al. AS04, an aluminum salt- and TLR4 agonist-based adjuvant system, induces a transient localized innate immune response leading to enhanced adaptive immunity. *J. Immunol.* **2009**, *183*, 6186–6197. [CrossRef]
103. Pinto, L.A.; Dillner, J.; Beddows, S.; Unger, E.R. Immunogenicity of HPV prophylactic vaccines: Serology assays and their use in HPV vaccine evaluation and development. *Vaccine* **2018**, *36 Pt A*, 4792–4799. [CrossRef]
104. Zhu, F.C.; Hu, S.Y.; Hong, Y.; Hu, Y.M.; Zhang, X.; Zhang, Y.J.; Pan, Q.J.; Zhang, W.H.; Zhao, F.H.; Zhang, C.F.; et al. Efficacy, immunogenicity, and safety of the HPV-16/18 AS04-adjuvanted vaccine in Chinese women aged 18–25 years: Event-triggered analysis of a randomized controlled trial. *Cancer Med.* **2017**, *6*, 12–25. [CrossRef]
105. Shanmugam, A.; Rajoria, S.; George, A.L.; Mittelman, A.; Suriano, R.; Tiwari, R.K. Synthetic Toll like receptor-4 (TLR-4) agonist peptides as a novel class of adjuvants. *PLoS ONE* **2012**, *7*, e30839. [CrossRef] [PubMed]

106. Reed, S.G.; Hsu, F.C.; Carter, D.; Orr, M.T. The science of vaccine adjuvants: Advances in TLR4 ligand adjuvants. *Curr. Opin. Immunol.* **2016**, *41*, 85–90. [CrossRef] [PubMed]
107. Alexander, J.; Sidney, J.; Southwood, S.; Ruppert, J.; Oseroff, C.; Maewal, A.; Snoke, K.; Serra, H.M.; Kubo, R.T.; Sette, A.; et al. Development of high potency universal DR-restricted helper epitopes by modification of high affinity DR-blocking peptides. *Immunity* **1994**, *1*, 751–761. [CrossRef]
108. Rosa, D.S.; Tzelepis, F.; Cunha, M.G.; Soares, I.S.; Rodrigues, M.M. The pan HLA DR-binding epitope improves adjuvant-assisted immunization with a recombinant protein containing a malaria vaccine candidate. *Immunol. Lett.* **2004**, *92*, 259–268. [CrossRef] [PubMed]
109. Ressing, M.E.; van Driel, W.J.; Brandt, R.M.; Kenter, G.G.; de Jong, J.H.; Bauknecht, T.; Fleuren, G.J.; Hoogerhout, P.; Offringa, R.; Sette, A.; et al. Detection of T helper responses, but not of human papillomavirus-specific cytotoxic T lymphocyte responses, after peptide vaccination of patients with cervical carcinoma. *J. Immunother.* **2000**, *23*, 255–266. [CrossRef] [PubMed]
110. Wu, C.Y.; Monie, A.; Pang, X.; Hung, C.F.; Wu, T.C. Improving therapeutic HPV peptide-based vaccine potency by enhancing CD4+ T help and dendritic cell activation. *J. Biomed. Sci.* **2010**, *17*, 88. [CrossRef] [PubMed]
111. Daftarian, P.; Mansour, M.; Benoit, A.C.; Pohajdak, B.; Hoskin, D.W.; Brown, R.G.; Kast, W.M. Eradication of established HPV 16-expressing tumors by a single administration of a vaccine composed of a liposome-encapsulated CTL-T helper fusion peptide in a water-in-oil emulsion. *Vaccine* **2006**, *24*, 5235–5244. [CrossRef]
112. Fraser, C.C.; Altreuter, D.H.; Ilyinskii, P.; Pittet, L.; LaMothe, R.A.; Keegan, M.; Johnston, L.; Kishimoto, T.K. Generation of a universal CD4 memory T cell recall peptide effective in humans, mice and non-human primates. *Vaccine* **2014**, *32*, 2896–2903. [CrossRef]
113. Li, G.; Huang, Z.; Zhang, C.; Dong, B.J.; Guo, R.H.; Yue, H.W.; Yan, L.T.; Xing, X.H. Construction of a linker library with widely controllable flexibility for fusion protein design. *Appl. Microbiol. Biotechnol.* **2016**, *100*, 215–225. [CrossRef]
114. Yu, K.; Liu, C.; Kim, B.G.; Lee, D.Y. Synthetic fusion protein design and applications. *Biotechnol. Adv.* **2015**, *33*, 155–164. [CrossRef]
115. van Rosmalen, M.; Krom, M.; Merkx, M. Tuning the Flexibility of Glycine-Serine Linkers to Allow Rational Design of Multidomain Proteins. *Biochemistry* **2017**, *56*, 6565–6574. [CrossRef]
116. Hoang, T.X. Identification and Characterization of a Splicing Variant in the 5′ UTR of the Human TLR5 Gene. *BioMed Res. Int.* **2017**, *2017*, 8727434. [CrossRef] [PubMed]
117. Takeuchi, O.; Akira, S. Pattern recognition receptors and inflammation. *Cell* **2010**, *140*, 805–820. [CrossRef]
118. Bell, J.K.; Mullen, G.E.; Leifer, C.A.; Mazzoni, A.; Davies, D.R.; Segal, D.M. Leucine-rich repeats and pathogen recognition in Toll-like receptors. *Trends Immunol.* **2003**, *24*, 528–533. [CrossRef]
119. Kopp, E.; Medzhitov, R. Recognition of microbial infection by Toll-like receptors. *Curr. Opin. Immunol.* **2003**, *15*, 396–401. [CrossRef]
120. Kim, W.Y.; Lee, J.W.; Choi, J.J.; Choi, C.H.; Kim, T.J.; Kim, B.G.; Song, S.Y.; Bae, D.S. Increased expression of Toll-like receptor 5 during progression of cervical neoplasia. *Int. J. Gynecol. Cancer* **2008**, *18*, 300–305. [CrossRef] [PubMed]
121. Cao, L.; Zhang, T.; Zhu, J.; Li, A.; Zheng, K.; Zhang, N.; Su, B.; Xia, W.; Wu, H.; Li, N.; et al. Polymorphism of TLR5 rs5744174 is associated with disease progression in Chinese patients with chronic HBV infection. *Apmis* **2017**, *125*, 708–716. [CrossRef]
122. Kalnin, K.; Tibbitts, T.; Yan, Y.; Stegalkina, S.; Shen, L.; Costa, V.; Sabharwal, R.; Anderson, S.F.; Day, P.M.; Christensen, N.; et al. Low doses of flagellin-L2 multimer vaccines protect against challenge with diverse papillomavirus genotypes. *Vaccine* **2014**, *32*, 3540–3547. [CrossRef]
123. Eaves-Pyles, T.D.; Wong, H.R.; Odoms, K.; Pyles, R.B. Salmonella flagellin-dependent proinflammatory responses are localized to the conserved amino and carboxyl regions of the protein. *J. Immunol.* **2001**, *167*, 7009–7016. [CrossRef]
124. Murthy, K.G.; Deb, A.; Goonesekera, S.; Szabo, C.; Salzman, A.L. Identification of conserved domains in Salmonella muenchenflagellin that are essential for its ability to activate TLR5 and to induce an inflammatory response in vitro. *J. Biol. Chem.* **2004**, *279*, 5667–5675. [CrossRef]
125. Forstneric, V.; Ivicak-Kocjan, K. The role of the C-terminal D0 domain of flagellin in activation of Toll like receptor 5. *PLoS Pathog.* **2017**, *13*, e1006574. [CrossRef] [PubMed]

126. Yang, J.; Zhong, M.; Zhang, Y.; Zhang, E.; Sun, Y.; Cao, Y.; Li, Y.; Zhou, D.; He, B.; Chen, Y.; et al. Antigen replacement of domains D2 and D3 in flagellin promotes mucosal IgA production and attenuates flagellin-induced inflammatory response after intranasal immunization. *Hum. Vaccin. Immunother.* **2013**, *9*, 1084–1092. [CrossRef] [PubMed]
127. Velova, H.; Gutowska-Ding, M.W.; Burt, D.W.; Vinkler, M. Toll-like receptor evolution in birds: Gene duplication, pseudogenisation and diversifying selection. *Mol. Biol. Evol.* **2018**. [CrossRef] [PubMed]
128. Hu, Y.M.; Huang, S.J.; Chu, K.; Wu, T.; Wang, Z.Z.; Yang, C.L.; Cai, J.P.; Jiang, H.M.; Wang, Y.J.; Guo, M.; et al. Safety of an Escherichia coli-expressed bivalent human papillomavirus (types 16 and 18) L1 virus-like particle vaccine: An open-label phase I clinical trial. *Hum. Vaccin. Immunother.* **2014**, *10*, 469–475. [CrossRef] [PubMed]
129. Xie, M.; Li, S.; Shen, W.; Li, Z.; Zhuang, Y.; Mo, X.; Gu, Y.; Wu, T.; Zhang, J.; Xia, N. Expression, purification and immunogenicity analysis of HPV type 18 virus-like particles from *Escherichia coli*. *Sheng Wu Gong Cheng XueBao* **2009**, *25*, 1082–1087.

2

Assembly and Characterization of HBc Derived Virus-Like Particles with Magnetic Core

Jakub Dalibor Rybka [1,†,*], Adam Aron Mieloch [1,2,†], Alicja Plis [1], Marcin Pyrski [3], Tomasz Pniewski [3] and Michael Giersig [1,4]

1 Center for Advanced Technology, Adam Mickiewicz University in Poznań, Umultowska 89C, 61-614 Poznań, Poland; amieloch@amu.edu.pl (A.A.M.); alkaplis@o2.pl (A.P.); giersig@amu.edu.pl (M.G.)
2 Faculty of Chemistry, Adam Mickiewicz University in Poznań, Umultowska 89B, 61-614 Poznań, Poland
3 Institute of Plant Genetics, Polish Academy of Sciences, Strzeszyńska 34, 60-479 Poznań, Poland; mpyr@igr.poznan.pl (M.P.); tpni@igr.poznan.pl (T.P.)
4 Institute of Experimental Physics, Freie Universität Berlin, Arnimallee 14, 14195 Berlin, Germany
* Correspondence: jrybka@amu.edu.pl
† These authors contributed equally to this work.

Abstract: Core-virus like particles (VLPs) assembly is a kinetically complex cascade of interactions between viral proteins, nanoparticle's surface and an ionic environment. Despite many in silico simulations regarding this process, there is still a lack of experimental data. The main goal of this study was to investigate the capsid protein of hepatitis B virus (HBc) assembly into virus-like particles with superparamagnetic iron oxide nanoparticles (SPIONs) as a magnetic core in relation to their characteristics. The native form of HBc was obtained via agroinfection of *Nicotiana benthamiana* with pEAQ-HBc plasmid. SPIONs of diameter of 15 nm were synthesized and functionalized with two ligands, providing variety in ζ-potential and hydrodynamic diameter. The antigenic potential of the assembled core-VLPs was assessed with enzyme-linked immunosorbent assay (ELISA). Morphology of SPIONs and core-VLPs was evaluated via transmission electron microscopy (TEM). The most successful core-VLPs assembly was obtained for SPIONs functionalized with dihexadecyl phosphate (DHP) at SPIONs/HBc ratio of 0.2/0.05 mg/mL. ELISA results indicate significant decrease of antigenicity concomitant with core-VLPs assembly. In summary, this study provides an experimental assessment of the crucial parameters guiding SPION-HBc VLPs assembly and evaluates the antigenicity of the obtained structures.

Keywords: virus-like particles; VLPs; hepatitis B virus capsid protein; HBc; viral self-assembly; magnetic core; HBcAg

---

1. Introduction

Virus-like particles (VLPs) are non-infectious and non-replicating supramolecular assemblies composed of single or multiple viral proteins, which closely resemble native virions [1]. VLPs display a unique set of immunological characteristics that render them highly potent for vaccine development such as: nanometer range size, multivalent and highly repetitive surface geometry, the ability to elicit both innate and adaptive immune response [2]. Due to favorable surface morphology and a wide range of possible modifications, VLPs have been successfully used as a platform for multivalent vaccine creation [3–5]. Several VLP-based vaccines are currently commercially available (e.g., Cervarix®, Gardasil®, Sci-B-Vac™, Mosquirix™) with more undergoing clinical trials [6]. VLPs' applicability is not limited to their immunogenic properties. Some of the use cases include: highly selective and sensitive nanobiosensor for troponin I detection, light-harvesting VLPs for use in photovoltavic or photocatalytic devices, nanofiber-like VLPs for tissue regenerating materials, nanocontainers and nanoreactors [7–11].

Hepatitis B virus (HBV) is an enveloped, icosahedral, cDNA virus that belongs to the Hepadnaviridae family. The virion has a diameter of 42 nm and is composed of a lipid envelope with hepatitis B virus surface antigen (HBsAg) and inner nucleocapsid consisting of hepatitis B virus capsid protein—HBc (named also HB core antigen, HBcAg) [12]. HBc consists of 183–185 amino acids of which 149 N-terminal amino acids form an assembly domain and 34 amino acids form C-terminal arginine-rich domain (CTAD) required for the packaging of nucleic acid [13]. HBc is a homodimeric protein that has the ability to self-assemble into icosahedral and fenestrated T = 4 (120 dimers) and T = 3 (90 dimers) capsids with respective outer diameter of 34 and 30 nm [14]. T = 4 capsid is a dominant product of a wild type HBc in vitro self-assembly (~95%) [15]. HBc capsid is highly immunogenic and has been shown to induce both B- and T-cell response [16].

Superparamagnetic iron oxide nanoparticles (SPIONs) exhibit properties, such as high magnetic susceptibility, high saturation magnetization and low toxicity [17–19]. Due to the aforementioned properties, high-yield synthesis methods and a wide array of available surface modifications, SPIONs can be utilized in: magnetic bioseparation, magnetic hyperthermia, targeted drug delivery, in diagnostics as magnetic resonance imaging (MRI) contrast agents, etc. [20–23].

Introduction of SPIONs as the core of VLPs has been performed successfully with several viral proteins of different origin [24–26]. Magnetic core adds a multitude of advantageous properties. It allows for post-assembly magnetic bioseparation, which may be crucial for large scale production [27]. It also improves cellular uptake and magnetic relaxivities resulting in higher resolution MRI images, which combined with in vivo tracking may provide essential data regarding VLPs biodistribution [28]. Functionalized core can act as a substitute for native nucleic acid, and therefore, govern the process of protein recruitment and organization during self-assembly. Rational core design can be used to facilitate the assembly and enhance such parameters as, e.g., physicochemical stability, mechanical elasticity, capacity to withstand desiccation and long-term storage. On the other hand, HBc VLPs have been shown to be potent epitope carriers [3,29].

In the study by Shen et al., HBc was genetically engineered into a truncated version, deprived of 34 C-terminal amino acids responsible for nucleic acid packaging. The removed part was replaced by six consecutive histidine residues (His-tag). Fe_3O_4 nanoparticles functionalized with nickel-nitrilotriacetic acid (nickel-NTA) chelate were used as the core. The VLPs assembly was driven by the affinity of histidine tags to the nickel-NTA chelate [28]. This study prompted us to investigate whether native HBc VLPs assembly can be successful, without resorting to genetic engineering of HBc protein.

The main objective of this study was to investigate whether provided SPION surface modification is sufficient for SPION-HBc assembly. Even though HBc subunits exhibit an ability to assemble in the absence of genetic material, electrostatic interactions between positively charged CTAD of the capsid and negatively charged nucleic acid have a major influence on the assembly process [30]. Therefore, to mimic native electrostatic interactions, negatively charged ligands were chosen for SPIONs functionalization: dihexadecyl phosphate (DHP) and PL-PEG-COOH. Both compounds were successfully used in our previous study regarding the influence of ligand charge and length on the assembly of Brome mosaic virus derived virus-like particles with magnetic core [31].

2. Materials and Methods

2.1. Reagents

Oleic acid (technical grade 90%), Iron (III) acetylacetonate (97%), Sodium chloride (99%), Dihexadecyl phosphate (90%), 1-Octadecene (90%), 2-Butanol (95.5%), Trizma®hydrochloride (99%), Calcium chloride (97%), Magnesium sulfate (99.5%), Glycine (99%), Glycerol (99%), Urea (98%), 2-(N-Morpholino)ethanesulfonic acid (99%), Sucrose (99.5%), Sigma-Aldrich (Poznan, Poland). Toluene (99.5%), n-Hexane (99%), Chloroform (98.5%) and Hydrochloric acid (30–35%), Avantor (Gliwice,

Poland). 1,2-Distearoyl-sn-glycero-3-phosphoethanolamine-N-[carboxy-(polyethyleneglycol)-2000] (ammonium salt) (PL-PEG-COOH, 2000 Da PEG (99%), Avanti, Alabaster, AL, USA). Snakeskin®Dialysis Tubing, 10K MWCO, 22 mm, Thermo Fisher Scientific (Waltham, MA, USA). All chemicals were used as received. Water was purified with Hydrolab HLP5 instrument (0.09 μS/cm, Straszyn, Poland).

2.2. *Superparamagnetic Iron Oxide Nanoparticles (SPIONs) Synthesis*

Spherical iron oxide nanoparticles were synthesized via thermal decomposition of iron (III) acetylacetonate $Fe(acac)_3$ [32]. Briefly, 6 mmol of $Fe(acac)_3$ and 18 mmol of oleic acid were dissolved in 40 mL of 1-octadecene. The reaction was performed with continuous stirring and nitrogen flow. Temperature of the solution was increased to 220 °C and maintained for 1 h. Subsequently, the temperature was increased further to 320 °C and maintained for 1 h. After synthesis, the solution was left to cool down to ambient temperature and 200 mL of washing solution (3:1 v/v of 2-butanol and toluene) was added. The obtained mixture was placed on a neodymium magnet and left overnight to allow nanoparticles to precipitate. Supernatant was discarded and replaced with fresh washing solution. Sonicating bath was used to resuspend nanoparticles. The washing step was performed thrice. In the final step, nanoparticles were suspended in 20 mL of chloroform. Concentration of the nanoparticles was estimated by dried sample weighing.

2.3. *SPIONs Functionalization*

PL-PEG-COOH functionalization was performed as per a method published elsewhere [24], with minor modifications. Briefly, 3.0 mg of PL-PEG-COOH were added to 5 mL of 1.0 mg/mL SPIONs chloroform solution. The sample was briefly sonicated in a sonic bath and left open for chloroform evaporation. The obtained waxy solid was heated for 1 min in an 80 °C water bath. The following step was adding 5 mL of miliQ water and vortexing the sample to enhance micelles formation. Subsequently, the sample was washed thrice with chloroform to remove unbound PL-PEG-COOH. Finally, water phase containing functionalized SPIONs was collected and filtered through 0.22 μm pores. Concentration of the SPION-PEG nanoparticles was measured via thermogravimetric analysis described below.

DHP functionalization was performed as per a method published elsewhere [31], with minor modifications. Briefly, 10.0 mg of dihexadecyl phosphate were added to 20 mL of hexane and dissolved with heat-assisted magnetic stirring (75 °C, ca. 10 min). After DHP dissolution, a chloroform solution containing 10.0 mg of synthesized iron oxide nanoparticles coated with oleic acid was added. The mixture was shortly sonicated and 80 mL of water were added. Subsequently, the obtained two phase solution was briefly vortexed and sonicated until the water phase became turbid. In the next step, the solution was placed in a sonicating bath for 3–4 h with no temperature control exercised. After functionalization, the solution was left overnight to allow for phase separation. The Bobtom phase was collected and placed near neodymium magnet for 24 h to separate functionalized nanoparticles from the solution. The obtained precipitate was collected, suspended in 2 mL of miliQ water and filtered through 0.22 μm pores. Concentration of the SPION-DHP nanoparticles was measured via thermogravimetric analysis described below.

2.4. *Concentration Measurement via Thermogravimetric Analysis*

Thermogravimetric analysis was performed to measure concentrations of the functionalized SPIONs. The analysis was performed on TGA 4000 System (Perkin Elmer apparatus, Waltham, MA, USA). Briefly, a 20 μL sample was taken for measurement. Each sample was measured in triplicate. The sample was heated from 20 to 150 °C at 10 °C/min in nitrogen atmosphere. The lowest mass was taken as fully dried sample and used for further calculations. The obtained mass was normalized for 20 mg of the initial sample mass. The mean of three measurements was calculated. Density was derived from weighing 5 × 15 μL of the sample and dividing the mean mass by volume. Final concentrations were: SPION-DHP = 3.52 mg/mL and SPION-PEG = 3.81 mg/mL.

2.5. HBc Production and Preparation

HBc was produced in plants via a transient expression system based on agroinfiltration. HBc expression vector was constructed on the basis of pEAQ-HT plasmid, developed by Peyret and Lomonossoff [33]. The coding sequence of HBcAg of 552 bp in length derived from HBV subtype *adw4* (GenBank: Z35717), was cloned into the vector *Age* I and *Xho* I restriction sites using sites *Age* I and compatible ends of *Sal* I, respectively, introduced by PCR using the following primers:

Forward: AACCGGTATGGACATTGACCCTTATAAAGAATTTG

Reverse: TGTCGACTGCAGTTAACATTGAGATTCCCGAGATTGAG

Complete vector pEAQ-HBc was introduced into *Agrobacterium tumefaciens* EHA105 and LBA4404 strains via electroporation.

Agroinfection was performed with *Agrobacterium* strains grown overnight on selective liquid LB medium supplemented with kanamycin (50 mg/l) and used to infiltrate leaves of 5–7 week-old *Nicotiana benthamiana* plants, cultivated in growth chamber under 5–6 klx light intensity, 16/8 h photoperiod and at a 22/16 °C temperature regime. *Agrobacterium* cells were centrifuged at 2000 g for 3 min at 4 °C and resuspended in MES buffer (10 mM 2-(N-morpholino)ethanesulphonic acid, 10 mM $MgSO_4$, pH 5.7) to optical density at a 600 nm wavelength (OD_{600}) 0.6 or 0.1 for infiltration by syringe or exsiccator, respectively. *Agrobacterium* suspension, 0.5 mL per leaf, was injected with a syringe into the bottom side of the leaves. Alternatively, whole plants were inverted and immersed in 2 L of *Agrobacterium* suspension in exsiccator (Lab Companion VDP-25G, Seoul, Korea). Pump (AGA Labor PL2, Poznań, Poland) was then applied to reach underpressure (−0.08 MPa) for approximately 1 min. The vacuum was released and applied again to ensure infiltration of the whole leaves. After 10 days following the agroinfiltration concentration of HBc in plant tissue reached approximately 1 mg/g of fresh weight (data not shown). HBc was then extracted and partially purified using sucrose density gradient as described previously [33]. The concentration of HBc directly after purification was fixed to 0.1 mg/mL. Prior to SPION encapsulation, HBc was diluted twice in a disassembly buffer.

2.6. SPION-HBc Preparation

VLPs were prepared in line with a slightly modified procedure described elsewhere [28].

HBc dissociation: 300 μL of 0.1 mg/mL HBc were diluted with 300 μL of denaturant solution (5 M urea, 300 mM NaCl, 100 mM tris-HCl) and incubated at 25 °C for 3h.

SPION-HBc assembly: The solution of dissociated HBc was divided into 100 μL aliquots (HBc conc. 0.05 mg/mL). To each aliquot, functionalized SPIONs were added to a final concentration of 0.5, 1.0 and 2.0 mg/mL. Obtained solutions were dialyzed twice against 400 mL of assembly buffer (150 mM NaCl, 10 mM $CaCl_2$, 1% *w*/*v* glycine, 10% *v*/*v* glycerol, 50 mM tris-HCl, pH = 8) for 24 h at 4 °C.

2.7. VLPs Antigenicity

Antigenicity of HBc VLPs was assessed via enzyme-linked immunosorbent assay (ELISA). HBc assembled with functionalized SPION-PEG and SPION-DHP at different concentrations (mg/mL) in comparison to the standard protein (recombined in *E. coli*, Cat No. R8A120, Meridian Life Science Inc., Memphis, TN, USA). Antigenicity defined as absorbance at 405 nm of two-fold dilution series of VLPs (from 1:160 to 1:81,920) and standard protein (from 0.5 to 0.004 μg/mL).

2.8. Statistical Analysis

Results of SPION-HBc formation were analyzed using a two-way ANOVA followed by a Duncan test; differences were considered significant at $p \leq 0.05$. Statistical analysis was performed using the Statistica 8.0 statistical software package (StatSoft Inc., Tulsa, OK, USA).

2.9. Characterization Methods

Transmission electron microscopy (TEM) images were acquired with Hitachi TEM HT7700 microscope (Tokyo, Japan). Grids were made of copper coated with a carbon film, mesh 300. Samples were prepared by placing 15 μL drop on the grid and draining the excess solution with blotting paper and left for 15 min. to dry. Subsequently, samples were negatively stained with 10 μL of 2% uranyl acetate. Particle size analysis was performed with free ImageJ software version 1.51w (NIH, Bethesda, MD, USA).

Dynamic light scattering (DLS) and ζ-potential measurements were performed on Malvern Zetasizer Nano ZS90 (Worcestershire, UK) in a Folded Capillary Zeta Cell DTS1070. Prior to measurement, samples were briefly sonicated, diluted to optimal concentration and filtered with a 0.2 μm syringe filter (Merck Millipore, Burlington, MA, USA). Measurements were repeated in triplicate.

ELISA was performed on the assembled HBc VLPs, with or without SPION core. The assay was performed in line with a procedure described previously [34]. MaxiSorp (NUNC) 96-well microplate was coated overnight at 4 °C with of HBc-specific mAb (0.5 mg/mL) (Cat. No. C31190 M, Meridian Life Science Inc., Memphis, TN, USA) in carbonate buffer pH 9.6. Each step following the coating was preceded by three washes with PBST buffer (phosphate buffered saline with additional 0.05% *v/v* Tween20, Sigma, Saint Louis, MO, USA). The coated wells were blocked for 1 h with 5% (*w/v*) fat-free milk/PBS, followed by incubation with 100 μL of antibody solution for 1 h at 25 °C. The samples were added to the PBS-filled wells and two-fold serially diluted. HBc produced in *E. coli* (Cat. No. R8A120, Meridian Life Science) was used as the reference. Rabbit polyclonal PBST antibody specific to HBc (Cat. No. LS-C67451/18649, Life Span Biosciences, Seattle, VA, USA) 0.125 mg/mL and goat anti-rabbit whole-molecule polyclonal antibody AP-conjugated (Sigma) 1:10,000 dilutions were premixed and added as the primary and secondary antibody. Finally, the substrate for alkaline phosphatase (pNPP, Sigma) was added and the reaction was developed at 25 °C for at least 30 min. The absorbance was measured at 405 nm using a microplate reader (Model 680, Bio-Rad, Hercules, CA, USA).

3. Results

3.1. SPIONs Synthesis and Functionalization

Monodispersed superparamagnetic iron oxide nanoparticles (SPIONs) of 15 nm diameter were synthesized via thermal decomposition of iron (III) acetylacetonate $Fe(acac)_3$ (Figure S1). The approximate diameter of the HBc VLP internal cavity is 25 nm for the T = 4 particles and 21 nm for the T = 3 [14]. Therefore, both structures provide sufficient marginal space to accommodate the ligands. In order to obtain negative surface charge, SPIONs were functionalized with short and long chain ligands: dihexadecyl phosphate (DHP) and 1,2-distearoyl-sn-glycero-3-phosphoethanolamine-N-[carboxy-(polyethylene glycol)-2000] (PL-PEG-COOH) (Figure 1a,b). Functionalized SPIONs will be denoted as SPION-DHP and SPION-PEG, respectively.

DHP functionalization was performed according to our previously described method [31]. PL-PEG-COOH functionalization was achieved by a slightly modified protocol by Huan et al. [24]. Both SPION-DHP and SPION-PEG were analyzed via ζ-potential and dynamic light scattering (DLS) measurements (Table 1). In both cases, functionalization is driven by hydrophobic interactions between oleic acid residues present on the surface of as-obtained SPIONs and alkyl chains of the ligands. More detailed characterization of both functionalizations can be found in our previous work [31].

Despite rather small differences in surface charge, hydrodynamic radius differs substantially. Counterintuitively, long-chain PEG ligand provided smaller hydrodynamic radius than DHP, which may be partially caused by differences in ζ-potential. Another plausible explanation is the interplay of surface charge and multilayered micelle structure formation.

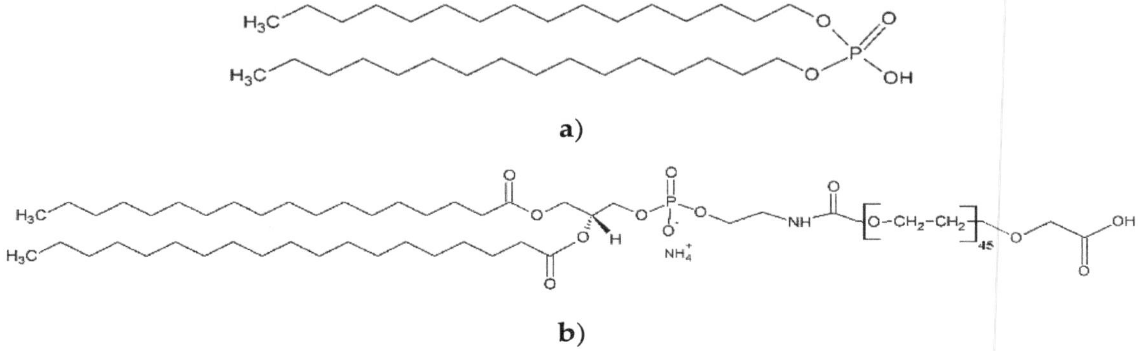

Figure 1. Molecular structure of ligands used for superparamagnetic iron oxide nanoparticles (SPIONs) functionalization. (**a**) dihexadecyl phosphate (DHP); (**b**) 1,2-distearoyl-sn-glycero-3-phosphoethanolamine-N-[carboxy-(polyethylene glycol)-2000] (PL-PEG-COOH).

Table 1. ζ-potential and hydrodynamic radius of the functionalized superparamagnetic iron oxide nanoparticles (SPIONs) obtained by dynamic light scattering (DLS) (Figures S2 and S3).

	SPION-DHP	SPION-PEG
ζ-potential	-44.0 ± 3.4 mV	-37.3 ± 2.9 mV
Hydrodynamic diameter	53.75 ± 1.93 nm	29.69 ± 1.57 nm

3.2. VLPs-SPION Assembly

The assembly rates and core encapsidation efficiency are strictly dependent on surface charge density, capsid protein concentration and core/capsid protein stoichiometric ratio [35]. Therefore, HBc concentration was fixed at 0.05 mg/mL while SPIONs concentrations were varied between: 0.05, 0.1 and 0.2 mg/mL. It is important to note that due to differences in ligands' molecular weight and probable differences in functionalization densities, equal w/v concentrations of SPION-PEG and SPION-DHP do not represent the same amount of particles in the solution. The most successful core-VLP assembly was obtained at following concentrations: 0.2 mg/mL SPION-DHP and 0.05 mg/mL SPION-PEG (Figure 2a,b).

TEM images were used to measure the diameter of the assembled VLPs. Mean diameter of SPION-DHP-HBc was 28.4 ± 1.2 nm while SPION-PEG-HBc mean diameter was 29.9 ± 1.5 nm (Table 2). The obtained measurements indicate that in both cases capsids assembled into T = 3 symmetry (native size of T = 3 capsid is 30 nm). Nonetheless, to assess the VLPs symmetry with certainty, crystallographic studies would be required. The obtained results are concordant with thermodynamic studies of nanospheres encapsulated in virus capsids reveling, in that core surface charge and its radius determine the size of the capsid formed around the nanoparticle [36]. In this case, despite T = 4 symmetry being a predominant form of in vitro HBc self-assembly (~95%), introduction of SPION-DHP and SPION-PEG facilitated assembly into a smaller, presumably T = 3 form. This may suggest that negative surface charge density was high enough to drive the assembly into less energetically-favorable capsid morphology. Studying TEM images, SPION-DHP assembly displayed higher efficiency than SPION-PEG and resulted in minority of empty capsids and unassembled cores. In comparison, SPION-PEG assembly produced a multitude of empty capsids along with unassembled cores.

Table 2. VLPs diameter measurements obtained from transmission electron microscopy (TEM) images. Measured with ImageJ software.

	SPION-DHP-HBc	SPION-PEG-HBc
Diameter	28.4 ± 1.2 nm	29.9 ± 1.5 nm
Number of measured VLPs	41	19

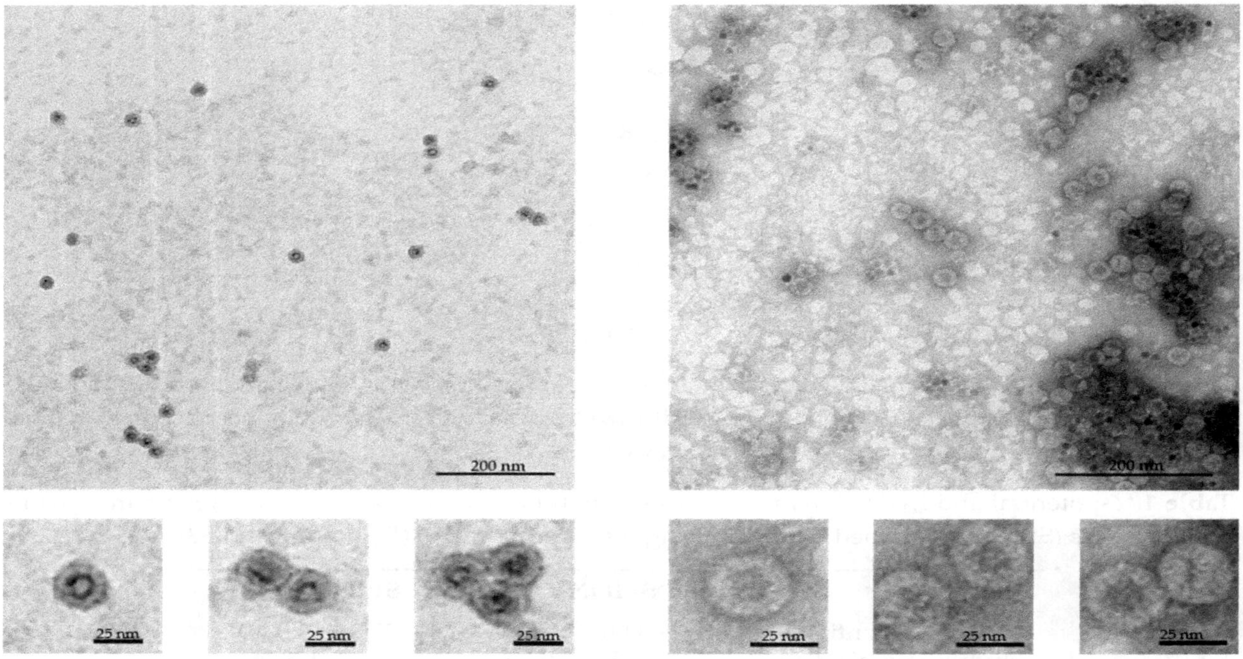

a) b)

Figure 2. Transmission electron microscopy (TEM) images of the assembled VLPs with magnetic cores, negatively stained with 2% uranyl acetate. (**a**) SPION-DHP-HBc VLPs obtained at 0.2 mg/mL of SPION-DHP and 0.05 mg/mL of HBc; (**b**) SPION-PEG-HBc VLPs obtained at 0.05 mg/mL of SPION-PEG and 0.05 mg/mL of HBcAg.

3.3. VLPs ELISA

Antigenicity of the obtained VLPs was assessed via ELISA (Figures 3 and S4). Despite unvaried HBc protein concentrations, for all VLPs variants decreased signals of HBc detection in comparison to control (initial HBc used) were observed, as well as some significant differences in signals among VLPs variants were found. All concentrations of SPION-PEG displayed a decrease in signal intensity; however, the differences between 0.05 and 0.1 mg/mL concentrations were not statistically significant. Additionally, in comparison to analogous variants of SPION-DHP. For SPION-PEG, the highest observed amount of core-VLPs was found at 0.05 mg/mL (Figure 2B), which also resulted in decreased signal intensity, although insignificantly different from other SPION-PEG concentrations. Finally, among all variants of SPION-HBc VLPs, the significantly lowest signal, 61.6% of HBc was recorded for 0.2 mg/mL SPION-DHP, the same concentration at which the highest number of core-VLPs was observed. The obtained results indicate that core introduction into HBc derived VLPs may decrease antigenicity. This phenomenon could be explained by the proclivity of HBcAg to assemble into smaller T = 3 capsids in the presence of the SPIONs, which in turn results in higher antigen density on the VLPs surface and competitive binding of antibodies. A study by Wu et al. has shown that the amount of antibodies bound to the capsid depends on its morphology and is significantly decreased for T = 3 capsids [37]. Moreover, steric hindrance has been proven to be a crucial factor for antibody binding to surface antigens [38,39]. These results provide a great starting point for further investigations of the relationship between core properties, capsid morphology, antigenicity and biological activity of the HBc derived VLPs.

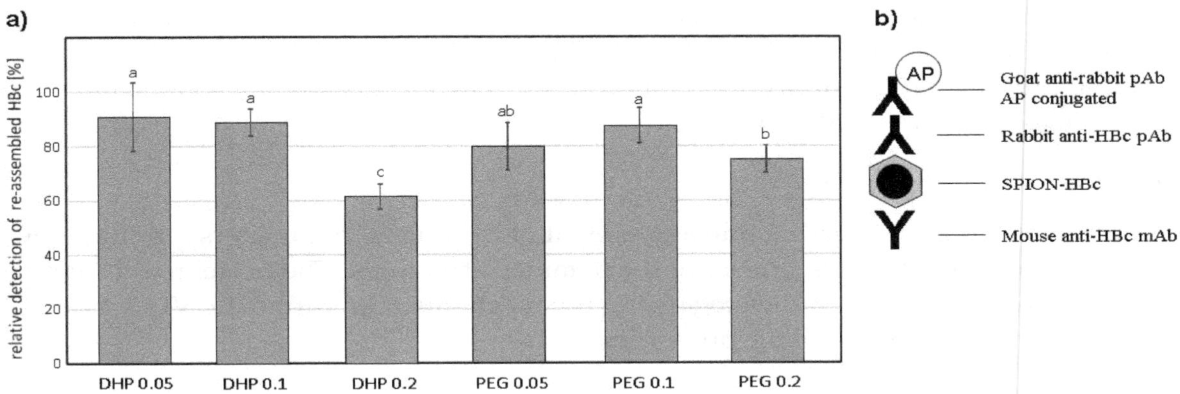

Figure 3. (**a**) HBc re-assembly on SPIONs functionalized with DHP or PEG in different concentrations (mg/mL) in comparison to the initial preparation of plant-derived antigen (100%). Statistically significant differences marked by a letter indexes; (**b**) Scheme of enzyme-linked immunosorbent assay (ELISA) test used for assay of VLP-assembled HBc and SPION-HBc VLPs. AP—alkaline phosphatase.

4. Discussion

Core-VLPs assembly, here SPION-HBc, is a kinetically complex cascade of interactions between viral proteins, nanoparticle surface and an ionic environment. In silico modeling predicts that core introduction provides a plethora of advantages such as increased assembly rates and efficiency over wider set of conditions, stimulation of the assembly below critical subunit concentration (CSC), possibility of templating VLPs morphology. Nonetheless, computational modeling results are not always confirmed in experimental studies. For example, the predicted increase of assembly efficiency driven by the increase of surface charge density has been overestimated in comparison to experimental data [35]. Moreover, simulations predominantly assume cores geometry perfectly commensurate with capsid interior. In our case, functionalized 15 nm SPIONs were not perfectly fitted into the capsid, which could elicit the existence of kinetic traps, not predicted by the computational studies [40]. The most successful core assembly was achieved at a 0.2/0.05 mg/mL core/protein ratio with SPION-DHP. Slightly lower ζ-potential concomitant with larger hydrodynamic diameter in comparison to SPION-PEG might indicate that surface charge density was higher in case of SPION-DHP. SPION-PEG assembly at 0.05/0.05 mg/mL core/protein ratio resulted in partially successful core assembly along with multitude of empty capsids and unassembled cores. This result indicates that one or several assembly parameters were suboptimal; however, due to the complexity of the process, we are unable to pinpoint the exact cause of lower efficiency. It is possible that interactions between less negatively charged SPION-PEG and positively charged domains of the HBc were insufficient to win competition over subunit-subunit attraction, resulting in the empty capsids. Additionally, it is important to note that the ligands used differed in length which could also affect assembly kinetics. In both cases, core introduction resulted in slightly smaller VLPs diameters even for T = 3 capsid morphology, which could be attributed to more compact HBc dimer-dimer spacing resulted from strong electrostatic core-HBc dimer interactions. This thesis stands in agreement with our experimental data showing smaller core-VLPs diameters for lower values of the core's ζ-potential (Table 2). VLPs morphology driven antigenicity is a crucial aspect determining its potential application, especially in the area of vaccinology. In that respect, our study demonstrated successful assembly together with substantially retained HBc antigenicity, although decreased in comparison to native HBc protein preparation containing mainly T = 4 capsids (Figure 3). This may indicate competitive binding of antibodies and/or steric clashes due to increased antigen surface density, stemming from the decreased VLPs diameter. Crucially, for many medical applications such as cell- or tissue-specific targeting, decreased immunogenicity of core antigen may be desirable. Nonetheless, this theory requires more in depth experimental investigation to be confirmed. Surface charge density is one of the most important parameters guiding core assembly. However, it is not easily measureable by standard lab equipment, which impedes rational design of physicochemical properties of the core. Therefore, we propose the use of ζ-potential and hydrodynamic diameter, as two parameters encompassing surface charge density. This approach would simplify and unify core's surface electrostatic characterization, providing more accessible tool for core-VLPs design.

5. Conclusions

This study provides an experimental assessment of the crucial parameters guiding SPION-HBc VLPs assembly and evaluates antigenicity of the obtained structures. The presented results highlight potential directions for further studies regarding the mechanism guiding HBc VLPs assembly with metallic cores as well as their antigenic properties.

Author Contributions: Conceptualization, J.D.R., A.A.M. and T.P.; Data curation, J.D.R., A.A.M., A.P. and M.P.; Formal analysis, J.D.R., A.A.M. and M.P.; Funding acquisition, J.D.R. and M.G.; Investigation, J.D.R., A.A.M., A.P. and M.P.; Methodology, J.D.R., A.A.M., M.P. and T.P.; Project administration, J.D.R.; Supervision, J.D.R., T.P. and M.G.; Validation, J.D.R. and A.A.M.; Visualization, A.A.M.; Writing—original draft, J.D.R. and A.A.M.; Writing—review & editing, J.D.R., A.A.M., M.P. and T.P.

Acknowledgments: We are grateful to: Monika Kręcisz, CAT, UAM, for her contribution; Aleksandra Gryciuk, IPG PAS, for construction of vector for plant transient expression. In addition, we are grateful to Professor George Lomonossoff, John Innes Centre, Norwich, UK, for kindly providing the pEAQ-HT vector for transient expression of HBcAg in plants and to Plant Bioscience Ltd for allowing the material to be used.

References

1. López-Sagaseta, J.; Malito, E.; Rappuoli, R.; Bottomley, M.J. Self-assembling protein nanoparticles in the design of vaccines. *Comput. Struct. Biotechnol. J.* **2016**, *14*, 58–68. [CrossRef] [PubMed]
2. Bachmann, M.F.; Jennings, G.T. Vaccine delivery: a matter of size, geometry, kinetics and molecular patterns. *Nat. Rev. Immunol.* **2010**, *10*, 787–796. [CrossRef] [PubMed]
3. Sominskaya, I.; Skrastina, D.; Dislers, A.; Vasiljev, D.; Mihailova, M.; Ose, V.; Dreilina, D.; Pumpens, P. Construction and immunological evaluation of multivalent hepatitis B virus (HBV) core virus-like particles carrying HBV and HCV epitopes. *Clin. Vaccine Immunol.* **2010**, *17*, 1027–1033. [CrossRef] [PubMed]
4. Galarza, J.M.; Latham, T.; Cupo, A. Virus-Like Particle (VLP) Vaccine Conferred Complete Protection against a Lethal Influenza Virus Challenge. *Viral Immunol.* **2005**, *18*, 244–251. [CrossRef] [PubMed]
5. Jackwood, D.J. Multivalent Virus-Like–Particle Vaccine Protects Against Classic and Variant Infectious Bursal Disease Viruses. *Avian Dis.* **2013**, *57*, 41–50. [CrossRef] [PubMed]
6. Mohsen, M.O.; Zha, L.; Cabral-Miranda, G.; Bachmann, M.F. Major findings and recent advances in virus–like particle (VLP)-based vaccines. *Semin. Immunol.* **2017**, *34*, 123–132. [CrossRef]
7. Park, J.-S.; Cho, M.K.; Lee, E.J.; Ahn, K.-Y.; Lee, K.E.; Jung, J.H.; Cho, Y.; Han, S.-S.; Kim, Y.K.; Lee, J. A highly sensitive and selective diagnostic assay based on virus nanoparticles. *Nat. Nanotechnol.* **2009**, *4*, 259–264. [CrossRef]
8. Stephanopoulos, N.; Carrico, Z.M.; Francis, M.B. Nanoscale Integration of Sensitizing Chromophores and Porphyrins with Bacteriophage MS_2. *Angew. Chem. Int. Ed.* **2009**, *48*, 9498–9502. [CrossRef]
9. Merzlyak, A.; Indrakanti, S.; Lee, S.-W. Genetically Engineered Nanofiber-Like Viruses For Tissue Regenerating Materials. *Nano Lett.* **2009**, *9*, 846–852. [CrossRef]
10. De la Escosura, A.; Nolte, R.J.M.; Cornelissen, J.J.L.M. Viruses and protein cages as nanocontainers and nanoreactors. *J. Mater. Chem.* **2009**, *19*, 2274. [CrossRef]
11. Uchida, M.; Klem, M.T.; Allen, M.; Suci, P.; Flenniken, M.; Gillitzer, E.; Varpness, Z.; Liepold, L.O.; Young, M.; Douglas, T. Biological Containers: Protein Cages as Multifunctional Nanoplatforms. *Adv. Mater.* **2007**, *19*, 1025–1042. [CrossRef]
12. Liang, T.J. Hepatitis B: The virus and disease. *Hepatology* **2009**, *49*, S13–S21. [CrossRef]
13. Nassal, M. The arginine-rich domain of the hepatitis B virus core protein is required for pregenome encapsidation and productive viral positive-strand DNA synthesis but not for virus assembly. *J. Virol.* **1992**, *66*, 4107–4116. [PubMed]
14. Crowther, R.A.; Kiselev, N.A.; Böttcher, B.; Berriman, J.A.; Borisova, G.P.; Ose, V.; Pumpens, P. Three-dimensional structure of hepatitis B virus core particles determined by electron cryomicroscopy. *Cell* **1994**, *77*, 943–950. [CrossRef]

15. Zlotnick, A.; Palmer, I.; Kaufman, J.D.; Stahl, S.J.; Steven, A.C.; Wingfield, P.T. Separation and crystallization of T = 3 and T = 4 icosahedral complexes of the hepatitis B virus core protein. *Acta Crystallogr. Sect. D Biol. Crystallogr.* **1999**, *55*, 717–720. [CrossRef]
16. Milich, D.R.; McLachlan, A.; Moriarty, A.; Thornton, G.B. Immune response to hepatitis B virus core antigen (HBcAg): localization of T cell recognition sites within HBcAg/HBeAg. *J. Immunol.* **1987**, *139*, 1223–1231. [PubMed]
17. Haracz, S.; Mróz, B.; Rybka, J.D.; Giersig, M.; Mrõz, B.; Rybka, J.D.; Giersig, M. Magnetic behaviour of non-interacting colloidal iron oxide nanoparticles in physiological solutions. *Cryst. Res. Technol.* **2015**, *50*, 791–796. [CrossRef]
18. Kręcisz, M.; Rybka, J.D.; Strugała, A.J.; Skalski, B.; Figlerowicz, M.; Kozak, M.; Giersig, M. Interactions between magnetic nanoparticles and model lipid bilayers—Fourier transformed infrared spectroscopy (FTIR) studies of the molecular basis of nanotoxicity. *J. Appl. Phys.* **2016**, *120*, 124701. [CrossRef]
19. Haracz, S.; Hilgendorff, M.; Rybka, J.D.; Giersig, M. Effect of surfactant for magnetic properties of iron oxide nanoparticles. *Nucl. Instruments Methods Phys. Res. Sect. B Beam Interact. Mater. Atoms* **2015**, *364*, 120–126. [CrossRef]
20. Mahmoudi, M.; Sant, S.; Wang, B.; Laurent, S.; Sen, T. Superparamagnetic iron oxide nanoparticles (SPIONs): Development, surface modification and applications in chemotherapy. *Adv. Drug Deliv. Rev.* **2011**, *63*, 24–46. [CrossRef]
21. Neuberger, T.; Schöpf, B.; Hofmann, H.; Hofmann, M.; von Rechenberg, B. Superparamagnetic nanoparticles for biomedical applications: Possibilities and limitations of a new drug delivery system. *J. Magn. Magn. Mater.* **2005**, *293*, 483–496. [CrossRef]
22. Teja, A.S.; Koh, P.-Y. Synthesis, properties, and applications of magnetic iron oxide nanoparticles. *Prog. Cryst. Growth Charact. Mater.* **2009**, *55*, 22–45. [CrossRef]
23. Malyutin, A.G.; Easterday, R.; Lozovyy, Y.; Spilotros, A.; Cheng, H.; Sanchez-Felix, O.R.; Stein, B.D.; Morgan, D.G.; Svergun, D.I.; Dragnea, B.; et al. Viruslike nanoparticles with maghemite cores allow for enhanced mri contrast agents. *Chem. Mater.* **2015**, *27*, 327–335. [CrossRef]
24. Huang, X.; Bronstein, L.M.; Retrum, J.; Dufort, C.; Tsvetkova, I.; Aniagyei, S.; Stein, B.; Stucky, G.; McKenna, B.; Remmes, N.; et al. Self-assembled virus-like particles with magnetic cores. *Nano Lett.* **2007**, *7*, 2407–2416. [CrossRef] [PubMed]
25. Okuda, M.; Eloi, J.-C.; Jones, S.E.W.; Verwegen, M.; Cornelissen, J.J.L.M.; Schwarzacher, W. Pt, Co–Pt and Fe–Pt alloy nanoclusters encapsulated in virus capsids. *Nanotechnology* **2016**, *27*, 095605. [CrossRef]
26. Gubin, S.P.; Koksharov, Y.A.; Khomutov, G.B.; Yurkov, G.Y. Magnetic nanoparticles: preparation, structure and properties. *Russ. Chem. Rev.* **2005**, *74*, 489–520. [CrossRef]
27. Fatima, H.; Kim, K.S. Magnetic nanoparticles for bioseparation. *Korean J. Chem. Eng.* **2017**, *34*, 589–599. [CrossRef]
28. Shen, L.; Zhou, J.; Wang, Y.; Kang, N.; Ke, X.; Bi, S.; Ren, L. Efficient Encapsulation of Fe_3O_4 Nanoparticles into Genetically Engineered Hepatitis B Core Virus-Like Particles Through a Specific Interaction for Potential Bioapplications. *Small* **2015**, *11*, 1190–1196. [CrossRef]
29. Roose, K.; De Baets, S.; Schepens, B.; Saelens, X. Hepatitis B core-based virus-like particles to present heterologous epitopes. *Expert Rev. Vaccines* **2013**, *12*, 183–198. [CrossRef]
30. Belyi, V.A.; Muthukumar, M. Electrostatic origin of the genome packing in viruses. *Proc. Natl. Acad. Sci. USA* **2006**, *103*, 17174–17178. [CrossRef]
31. Mieloch, A.A.; Kręcisz, M.; Rybka, J.D.; Strugała, A.; Krupiński, M.; Urbanowicz, A.; Kozak, M.; Skalski, B.; Figlerowicz, M.; Giersig, M. The influence of ligand charge and length on the assembly of *Brome mosaic virus* derived virus-like particles with magnetic core. *AIP Adv.* **2018**, *8*, 035005. [CrossRef]
32. Giersig, M.; Hilgendorff, M. Magnetic nanoparticle superstructures. *Eur. J. Inorg. Chem.* **2005**, *2005*, 3571–3583. [CrossRef]
33. Peyret, H.; Lomonossoff, G.P. The pEAQ vector series: the easy and quick way to produce recombinant proteins in plants. *Plant Mol. Biol.* **2013**, *83*, 51–58. [CrossRef] [PubMed]
34. Pyrski, M.; Rugowska, A.; Wierzbiński, K.R.; Kasprzyk, A.; Bogusiewicz, M.; Bociąg, P.; Samardakiewicz, S.; Czyż, M.; Kurpisz, M.; Pniewski, T. HBcAg produced in transgenic tobacco triggers Th1 and Th2 response when intramuscularly delivered. *Vaccine* **2017**, *35*, 5714–5721. [CrossRef] [PubMed]

35. Hagan, M.F. A theory for viral capsid assembly around electrostatic cores. *J. Chem. Phys.* **2009**, *130*, 114902. [CrossRef] [PubMed]
36. Šiber, A.; Zandi, R.; Podgornik, R. Thermodynamics of nanospheres encapsulated in virus capsids. *Phys. Rev. E Stat. Nonlinear Soft Matter Phys.* **2010**, *81*, 051919. [CrossRef]
37. Wu, W.; Chen, Z.; Cheng, N.; Watts, N.R.; Stahl, S.J.; Farci, P.; Purcell, R.H.; Wingfield, P.T.; Steven, A.C. Specificity of an anti-capsid antibody associated with Hepatitis B Virus-related acute liver failure. *J. Struct. Biol.* **2013**, *181*, 53–60. [CrossRef]
38. Kent, S.P.; Ryan, K.H.; Siegel, A.L. Steric hindrance as a factor in the reaction of labeled antibody with cell surface antigenic determinants. *J. Histochem. Cytochem.* **1978**, *26*, 618–621. [CrossRef]
39. De Michele, C.; De Los Rios, P.; Foffi, G.; Piazza, F. Simulation and Theory of Antibody Binding to Crowded Antigen—Covered Surfaces. *PLoS Comput. Biol.* **2016**, *12*, e1004752. [CrossRef]
40. Hagan, M.F. Controlling viral capsid assembly with templating. *Phys. Rev. E Stat. Nonlinear Soft Matter Phys.* **2008**, *77*, 051904. [CrossRef]

3

Gram-Positive Bacteria-Like DNA Binding Machineries Involved in Replication Initiation and Termination Mechanisms of Mimivirus

Motohiro Akashi * and Masaharu Takemura

Laboratory of Biology, Department of Liberal Arts, Faculty of Science, Tokyo University of Science, Shinjuku, Tokyo 162-8601, Japan; takemura@rs.kagu.tus.ac.jp
* Correspondence: motoa@rs.tus.ac.jp

Abstract: The detailed mechanisms of replication initiation, termination and segregation events were not yet known in *Acanthamoeba polyphaga mimivirus* (APMV). Here, we show detailed bioinformatics-based analyses of chromosomal replication in APMV from initiation to termination mediated by proteins bound to specific DNA sequences. Using GC/AT skew and coding sequence skew analysis, we estimated that the replication origin is located at 382 kb in the APMV genome. We performed homology-modeling analysis of the gamma domain of APMV-FtsK (DNA translocase coordinating chromosome segregation) related to FtsK-orienting polar sequences (KOPS) binding, suggesting that there was an insertion in the gamma domain which maintains the structure of the DNA binding motif. Furthermore, UvrD/Rep-like helicase in APMV was homologous to *Bacillus subtilis* AddA, while the chi-like quartet sequence 5′-CCGC-3′ was frequently found in the estimated *ori* region, suggesting that chromosomal replication of APMV is initiated via chi-like sequence recognition by UvrD/Rep-like helicase. Therefore, the replication initiation, termination and segregation of APMV are presumably mediated by DNA repair machineries derived from gram-positive bacteria. Moreover, the other frequently observed quartet sequence 5′-CGGC-3′ in the *ori* region was homologous to the mitochondrial signal sequence of replication initiation, while the comparison of quartet sequence composition in APMV/*Rickettsia*-genome showed significantly similar values, suggesting that APMV also conserves the mitochondrial replication system acquired from an ancestral genome of mitochondria during eukaryogenesis.

Keywords: DNA replication; *ori*; mitochondria; Rickettsia; gram-positive bacteria; APMV; Mimivirus; giant virus; eukaryogenesis

1. Introduction

Understanding the mechanism of genomic replication for all organisms, including the "giant viruses", is an important scientific endeavor. Mimivirus, the first giant virus to be discovered, has a 750-nm-long virion and a 1.2 Mb linear dsDNA genome [1,2]. The method of replication termination for *Acanthamoeba polyphaga mimivirus* (APMV) has been previously hypothesized [3]. The first model suggested that the replication of the lagging strand of APMV's linear genome is mediated by homologous recombination of approximately 617 bp located on both ends of the viral chromosome, similar to T4 phage replication, and is processed with Mimivirus R555 recombinase (Mre11/Rad50 fusion protein) [3,4]. Recently, the second model of replication termination and segregation of APMV was proposed [5]. In this model, the FtsK-like protein (also called packaging ATPase), binds FtsK orienting polar sequences (KOPS) and is localized to both ends of the nucleosome, resulting in chromosome segregation by the recombination of *dif* sequences [5]. The KOPS is the recognition sites of FtsK protein, and this protein controls the chromosome segregation in bacteria [6]. The second model reinforces the first model regarding homologous recombination between chromosomal ends. However, it should be noted that the bacteria do not perfectly conserve KOPS among species [6–8], suggesting that APMV could have KOPS of its own, which might be similar to bacterial KOPS.

The initiation of DNA replication of bacteria is mainly driven by DnaA-DnaA box interaction, which subsequently unwinds dsDNA via DnaB [9]. Rep protein, related DNA helicase such as DnaB, initiates plasmid replication [10]. The Rep protein relates not only to plasmids but also to chromosome replication in *Escherichia coli* [11]. The homologues of these helicases in gram-positive bacteria are PcrA and AddAB; both are involved in DNA repair, and the former is involved in rolling-circle replication [12–14]. Especially, AddAB mediates the homologous recombination with recognizing the five-nucleotide sequence called chi [12], so the recombination is sequence specific. The sequence chi is known as a site-specific recombination site that is catalyzed by the RecBCD pathway in *E. coli*, and RecB is a homologous helicase to AddA [15,16]. Chi sequence varies among bacterial species: *E. coli*: 5′-GCTGGTGG-3′; *Bacillus subtilis*: 5′-AGCGG-3′; *Lactococcus lactis*: 5′-GCGCGTG-3′ [12,15–17]. Based on the model of replication termination in APMV [5], the DNA replication initiation mechanism might also be homologous to that of bacteria. If so, the pair of bacteria-like DNA sequences and their recognition proteins which are related to the DNA replication initiation and segregation could be found in APMV. Plotting the nucleotide composition bias called genomic GC skew is a tool for visualizing the bias of the nucleotide composition on the genome, which is able to determine the origin of replication since the values of GC skew switch across the replication origin and its terminus [18], and which has been determined in bacteria with some improvements of the technique [19–21]. This nucleotide composition bias shaping the genomic polarity is thought to have results of the mutation and selection pressure against the different replication mechanism of leading/lagging strand [22,23]. Although the origin of replication remains putative in Mimivirus, the GC skew analysis against the Mimivirus genome with high resolution may facilitate the detection of the signal sequence of DNA replication initiation. Therefore, using bioinformatics, we analyzed the APMV genome to determine the DNA replication initiation/termination segregation mechanism in detail, starting with the GC skew analysis.

2. Materials and Methods

2.1. GC/AT, Coding Sequence (CDS) Skew Analysis

GC and AT skew of the APMV genome (AY653733.1) was analyzed using a method described previously [20,21]. Each index was calculated using the following formulas: GC skew = [G-C]/[G+C]; AT skew = [A-T]/[A+T] (window size: 10,001 bp; step size: 1000 bp), and the GC/AT skew and cumulative graph of these were plotted. To calculate coding sequence (CDS) skew, we indexed the CDS direction on the APMV genome (direction of the gene (D): positive: +1; negative: -1) and the CDS length (L). Subsequently, the CDS skew index was calculated against every CDS using the following formula: CDS skew = [D] × [L] (Figure 2b). The CDS skew and cumulative CDS skew were plotted with the CDS start positions.

2.2. Correlation Analysis of the CDS Length of Left/Right Side from the Estimated Ori Region

The CDS length of the left or right sides from the estimated origin (380,698 bp), and the CDS length of the positive or negative direction against the APMV genome (AY653733.1), were plotted using a box plot. The statistical differences between each of two groups were calculated by two-sample Kolmogorov–Smirnov (KS) test using "lawstat package" of R software (https://www.r-project.org) with default options.

2.3. Paralogous Gene Localization Analysis

We confirmed the three kinds of paralogous gene locations found on the two GC skew shift points (296,000 bp and 882,000 bp): *ankyrin repeat, serine/threonine protein kinase*, and *collagen triple helix repeat containing protein*. First, we made a gene list of the CDS information annotated as "ankyrin containing protein", "serine/threonine protein kinase", or "collagen triple helix repeat containing protein" in the APMV genome (AY653733.1), containing the data of locus tags, start/end physical positions, gene length and gene direction and GC content (Supplementary Data). Every pair of each of the three genes was listed and the distance between these pairs from the start position was calculated. This pair of directions on the physical position of the APMV genome was plotted using the Circos v0.69-6 software [24]. The pair of gene direction was also labeled (matched direction: forward-forward/reverse-reverse; mismatched direction: forward-reverse/reverse-forward). When drawing the graph with Circos, color coding was used to display the three kinds of genes, and states of matched/mismatched gene directions were displayed with shades of colors (ankyrin containing protein: blue; serine/threonine protein kinase: green; collagen triple helix repeat containing protein: red; matched: light; mismatched: shade). The columns of id, gene1_start_pos, gene2_start_pos, line_colour, and line_thickness in the Supplementary Data without header (Supplementary Data; the pane "Paralogous gene direction") can be used for loading the data to Circos. The distance between every two CDSs in each of the three genes was plotted by the CDS direction-matching patterns on the APMV genome ("match" or "mismatch"). The statistical differences between the two groups were calculated by two-sample KS test using the "lawstat package" in the R software with default options.

2.4. Sequence Alignment Analysis Neighbor-Net Network Analysis

The accessions of analyzed FtsK with positions of motor (ATPase) domain and gamma domain were as follows: APMV: AAV50705.1, 5-215 aa, 216-284 aa; *E. coli*: NP_415410.1, 868-1242 aa, 1268-1324 aa; *L. lactis*: NP_267812.1, 312-668 aa, 695-749 aa; *Pseudomonas aeruginosa*: Q9I0M3, 343-716 aa, 749-803 aa; *B. subtilis*: WP_003231869.1, 346-703 aa, 582-723 aa. The accession numbers of analyzed UvrD/Rep-like helicase were as follows: APMV: AKI80299.1; *E. coli*: YP_026251.1 (Rep), AAA67609.1 (UvrD), NP_417297.1 (RecB); *L. lactis*: WP_003132060.1 (PcrA), WP_010905024.1 (AddA); *B. subtilis*: WP_003233919.1 (PcrA), WP_003233100.1 (AddA). The dataset without the APMV sequence was used for alignment. Poor-quality sequences were masked using Prequal software [25], and sequence alignment analyses was performed using the MAFFT v7.222 software with "—auto" option [26]. Subsequently, the APMV sequence was added and realigned with MAFFT. To determine the alignment of the chi-binding site of AddA with APMV, two sequences of AddA from *L. lactis* and *B. subtilis* (WP_010905024.1 and WP_003233100.1) were recursively added with the APMV sequence. For the Neighbor-Net network analysis, alignment data containing APMV sequence were trimmed using trimAl 1.2rev59 software with "-strictplus" option [27], and the numbers of aligned residues used were: FtsK ATPase domain: 322 aa; FtsK gamma domain: 49 aa; UvrD/Rep-like helicase: 483 aa. The Neighbor-Net network tree was drawn by SplitTree4 (version 4.14.8) with 1000 bootstrap replicates [28,29].

2.5. KOPS Distribution in the Genome

The bacterial KOPS distributions on APMV and the bacterial genomes were plotted using the following genome data: APMV: AY653733.1; *L. lactis* Il1403: NC_002662.1; *E. coli* MG1655: NC_000913.3; *B. subtilis* 168: NC_000964.3. The information on the *ori* and *ter* positions of these bacteria was provided by Genome projector website (http://www.g-language.org/g3/) [30]. KOPS of each bacteria were listed as follows: *L. lactis*: 5′-GAGAAG-3′; *B. subtilis*: 5′-GAGAAGGG-3′; *E. coli*: 5′-GGGNAGGG-3′ [6–8].

2.6. Quartet Sequence Composition Analysis

Every quartet sequence compositions on APMV and the bacterial genomes were confirmed with the compseq program of the EMBOSS 6.6.0.0 software [31] with the "-word 4 " option, using the following genome data: APMV: AY653733.1; *L. lactis* Il1403: NC_002662.1; *E. coli* MG1655: NC_000913.3; *B. subtilis* 168: NC_000964.3; *Rickettsia prowazekii*: NC_000963.1; *Homo sapiens* mitochondria (MT): CM001971.1. To compare the compositions between the estimated *ori* region and the whole genome of APMV, sequence composition was also confirmed on the 375–385 kb region, and the ratio of *ori* region per whole genome (quartet nucleotide composition ratio) was calculated as follows: quartet nucleotide composition ratio = [*ori* observed/expected frequency]/[all genome observed/expected frequency]. The Grubbs test was performed on this data to detect the highly observed sequence on the *ori* region with the "outliers" package of the R software (options: type = 11; opposite = TRUE; two.sided = TRUE). The mutual observed frequency of quartet sequence among APMV, APMV *ori* region, and bacteria were plotted, while statistical differences were confirmed by two-sample KS test using the "lawstat package" of the R software with default options.

2.7. Homology Modeling Analysis

Homology modeling analyses of the gamma domain of packaging ATPase (FtsK-like protein, 216-284 aa region of AAV50705.1) were performed using the I-TASSER server [32]. Template structures of *P. aeruginosa* (2J5O and 2VE9) were used for modeling [33,34].

3. Results

3.1. GC/AT Skew and CDS Skew Analyses

The GC skew plot showed the two highest and lowest peaks at the symmetrical position point of the genome (296 kb and 882 kb), and the plot was able to separate the three regions by these positions (Figure 1a). Both end regions (<296 kb, >882 kb) were increasing; however, the former value was negative and the latter value was positive, indicating that the 5′ end contained a C nucleotide rather than G nucleotide, and the 3′ end contained a G nucleotide rather than a C nucleotide. The middle region of the graph (from 296 kb to 882 kb) was almost flat (Figure 1b), and the cumulative GC skew plot of this region was increasing (Figure 1b), indicating that the number of G and C nucleotides in this region was slightly skewed to the G nucleotide. The AT skew graph showed the shift point of the value, which corresponded to the peak of the valley of the cumulative AT skew graph at 382,000 $\pm$ 5000 bp region, suggesting that the origin of DNA replication is located in this region (Figure 1). However, the cumulative AC skew did not show any peaks (Figure S1).

CDS skew analysis showed that the shift point and cumulative CDS skew analysis exhibited a V-shaped graph, similar to the AT skew/cumulative AT skew (Figure 2a). Furthermore, the valley of this cumulative CDS skew graph was located at 382,698 bp, and the gene direction faced outward from this peak. Therefore, we concluded that the estimated *ori* region is located at the 382 kb position of APMV genome. The location of the estimated *ori* region is biased toward the 5′ end from the center of the genome; however, there were no significant differences in CDS length between the right and left sides of the estimated *ori* region and between the positive and negative strands (Figure S2). We found the same paralogous genes on the GC skew shift points (296,000 bp, 882,000 bp): ankyrin repeat, serine/threonine protein kinase, and collagen triple helix repeat containing protein. Paralogues of these genes were located on the line of symmetry position in the genome, especially *collagen triple helix repeat containing protein* gene, which exhibited the exact positions of the GC skew shift points with opposite gene direction, suggesting that the nucleotide compositions of these paralogues formed the shift points (Figure 3a). Additionally, we analyzed the gene-to-gene distances between each pair of paralogues for each of the three genes, indicating that the distance between the gene direction matched pair was shorter than that between the mismatched pair (fold change: 2.6, $p < 0.05$, Figure 3b). Thus, APMV forms the double-folding structure and is the cause of paralogue generation by homologous recombination (Figure S3).

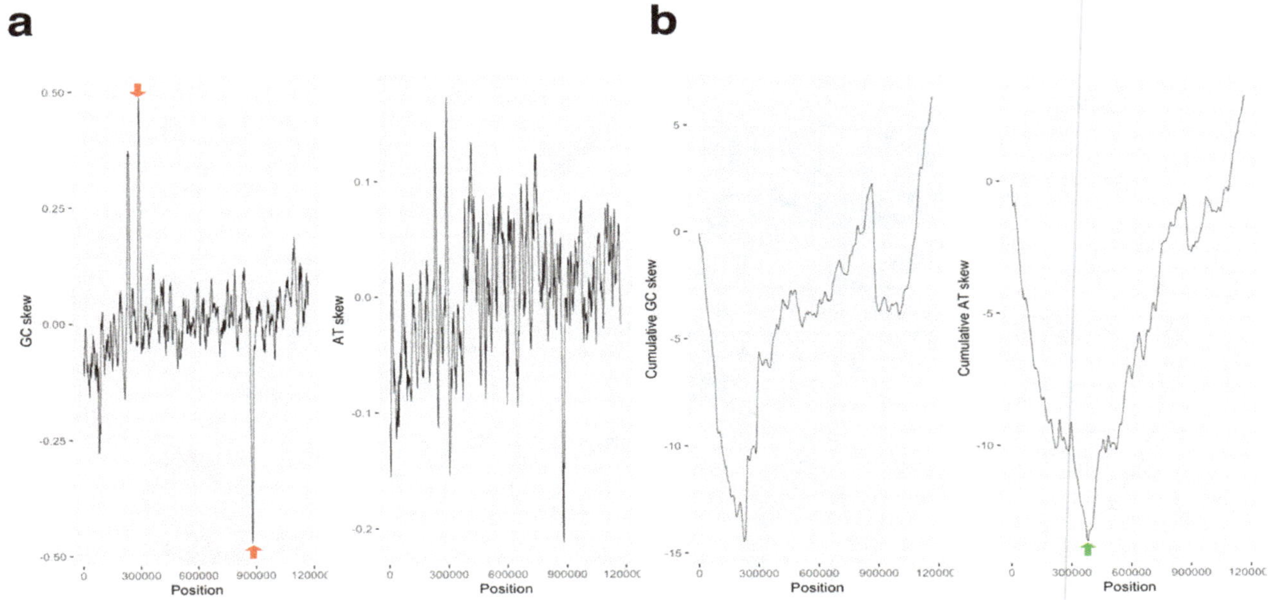

Figure 1. Analyses of the GC/AT skew in the *Acanthamoeba polyphaga mimivirus* (APMV) genome. (**a**) GC and AT skew plot of APMV genome (AY653733.1, step size: 1000 bp; window size: 10,001 bp). Red arrows on the GC skew plot indicate the highest/lowest peaks, which are located at the symmetrical position of the genome (GC skew: 0.488234 at 296,000 bp; -0.472989 at 882,000 bp). (**b**) Cumulative GC and AT skew plots corresponding to the graphs on the panel (a). The green arrow on the graph on the cumulative AT skew plot shows the lowest valley point on this graph, which was estimated in the genomic region as an origin of replication (382,000 ± 5000 bp).

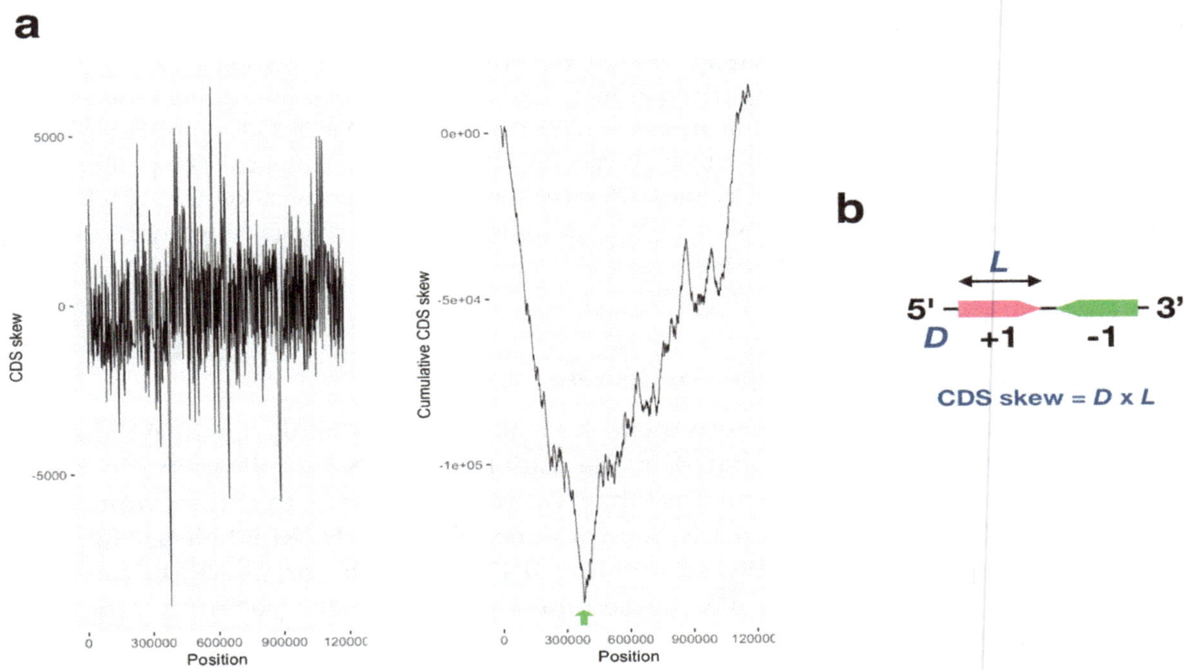

Figure 2. Analyses of the coding sequence (CDS) skew of the APMV genome. (**a**) CDS skew plot (left) and cumulative CDS skew plot (right) of the APMV genome (AY653733.1). The green arrow on the graph on the cumulative CDS skew plot shows the lowest valley point, which was estimated to be genomic region of the origin of 382,698 bp. (**b**) Calculation of the CDS skew index. D: Direction values of the CDS against the APMV genome (AY653733.1); L: CDS length (bp). Pink and green arrows indicate the positive and negative CDS direction on the APMV genome, respectively.

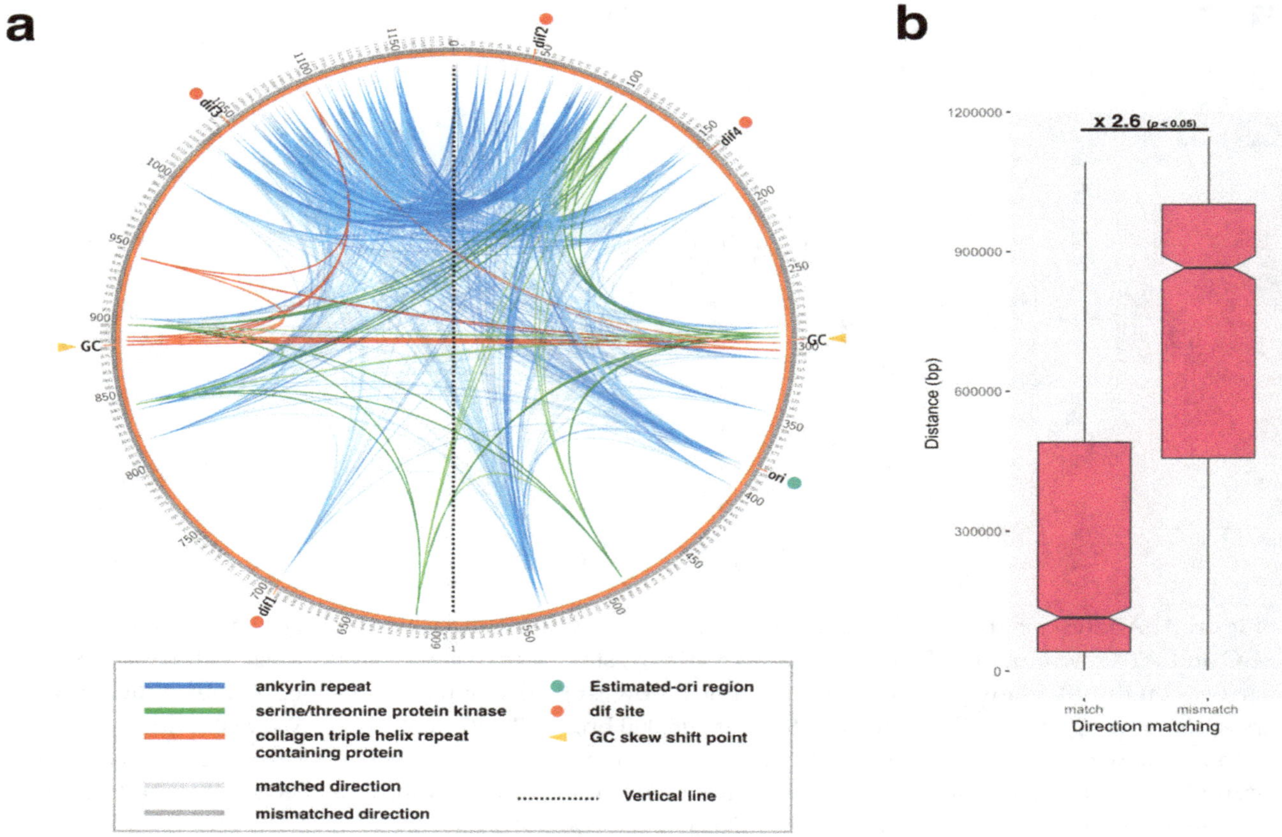

Figure 3. Paralogous gene localization found at the GC skew shift point. (**a**) The three genes found on the GC skew shift point were plotted (ankyrin repeat, serine/threonine kinase, and collagen triple helix repeat containing protein). Four *dif* sites (*dif*1-4) were described previously (see text). Each of the three genes were separately categorized as matching or mismatching the coding sequence (CDS) direction on the *Acanthamoeba polyphaga mimivirus* (APMV) genome (AY653733.1) with shades of colors (light: matched direction; dark: mismatched direction). (**b**) Distances between each pair of CDSs in each of the three genes were plotted by the CDS direction matching patterns on the APMV genome ("match" or "mismatch"). Labels above the box plots indicate the fold change between the two values with the resulting *p*-value from KS-tests.

3.2. Initiation of DNA Replication

3.2.1. Sequence Analysis of UvrD/Rep-Like Helicase

Since the replication initiation mechanism had not yet been analyzed in APMV, we sought to determine the protein that participated in the initiation of DNA replication. In doing so, we discovered a similar sequence to Rep helicase that was related to the both chromosome and plasmid replication [10,35]. This protein has already been annotated as "UvrD/REP helicase family protein", suggesting that this protein is possibly an initiator of DNA replication in APMV (accession ID of NCBI protein database: AKI80299.1). We aligned the UvrD/REP helicase family protein of APMV (AKI80299.1) with the bacterial homologues: Rep, UvrD, and RecB of *E. coli*; PcrA and AddA of *B. subtilis*; and *L. lactis* [10,12,13]. The alignment and phylogenetic analysis showed that the UvrD/Rep-like helicase of APMV is a close relative to the AddA rather than Rep (Figure 4a,c). Seven regions that were known to be conserved among AddA and other helicases [36] were also highly conserved in the UvrD/Rep-like helicase (Figure 4a). AddA forms a heterodimer with AddB, which recognizes the chi sequence of *B. subtilis* [37]. Therefore, we compared the homologous region of the chi binding site of AddA with UvrD/Rep-like helicase of APMV. The results showed that there were three regions containing chi binding sites in UvrD/Rep-like helicase of APMV, and two out of seven sites were conserved: Q1115 and I1157 of *Bacillus*-AddA corresponded to the Q922 and I924 of UvrD/Rep-like helicase of APMV; and Y1204 of *Bacillus*-AddA presumably corresponded to the Y973 of UvrD/Rep-like helicase of APMV, which was shifted one amino acid residue to the C-terminus from the homologous site against Y1204 in *B. subtilis* (Figure 4b). Although we have aligned the other region containing four DNA-binding residues between the 1012–1019 aa region of *Bacillus*-AddA with the 610–617 aa region of UvrD/Rep-like helicase, both of which are constructed with polar amino acids, we could not find any conserved sequences of DNA binding sites (K1013, S1015, V1016, and S1017 in *Bacillus* AddA, Figure 4b). These four residues in *Bacillus*-AddA bind the phosphate at the 3′ end of the chi sequence [37], so that the content of polar amino acids in this region, rather than the exact amino acid sequence, is important for *Bacillus*-AddA to bind to the chi sequence. Therefore, we estimated that the 610–617 aa region of UvrD/Rep-like helicase of APMV could also bind the 3′ end of the chi-like sequence in the APMV genome. Altogether, the UvrD/Rep-like helicase of APMV was similar to the AddA of gram-positive bacteria, and thus this protein would presumably recognize the chi sequence.

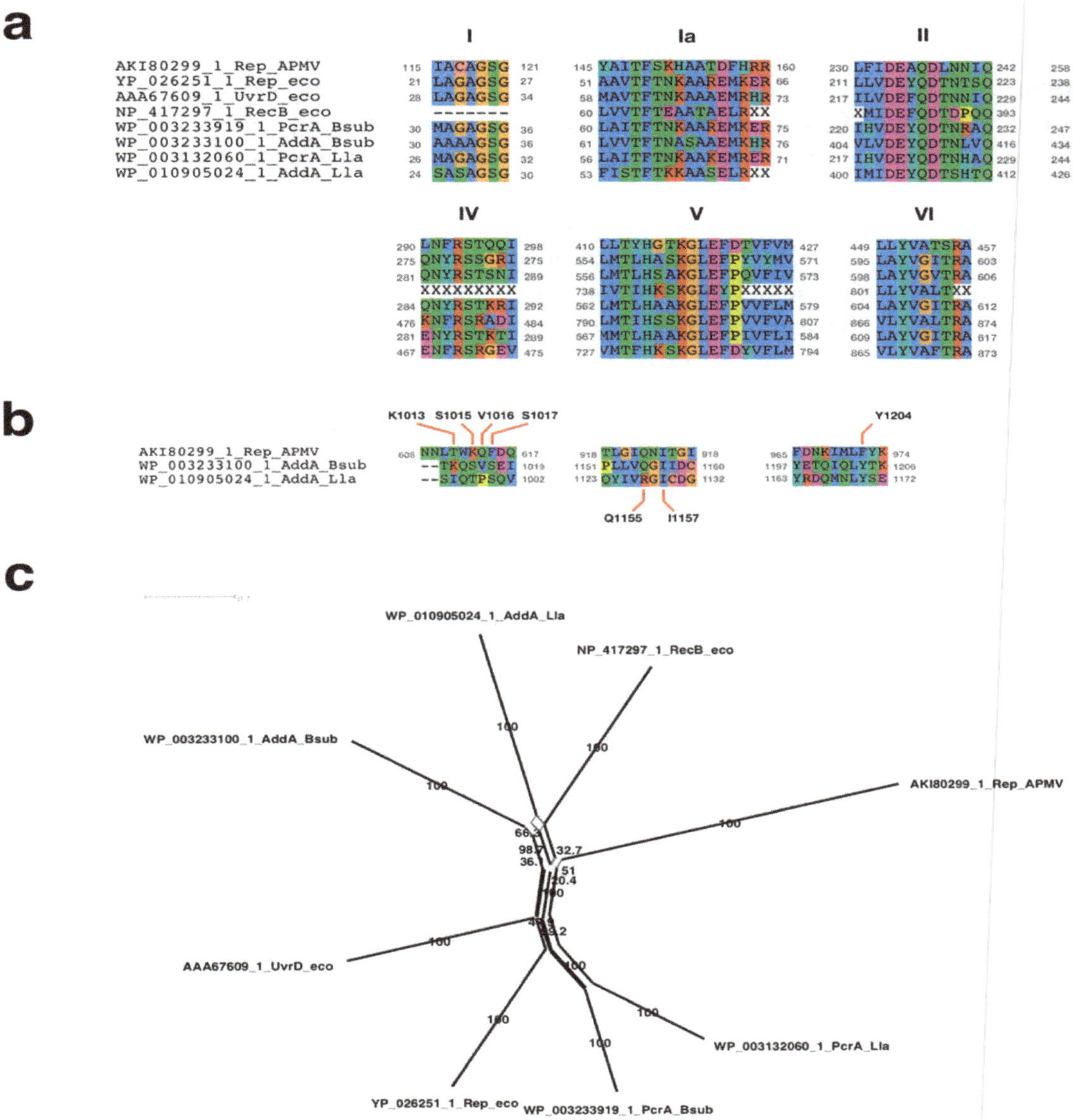

Figure 4. Sequence analyses of UvrD/Rep-like helicase of APMV. (**a**) Sequence alignment of seven conserved regions among APMV and three bacteria. Character "X" on the alignment indicates the

residues masked by the prequel program [25]. Accession number and species name are labeled on the left side of the alignments. Grey colored numbers, which are located on both sides of the alignments, indicate the physical positions of each sequence. (**b**) Sequence alignment of chi sequence binding sites. Accession number and species name are labeled on the left side of the alignment. Grey colored numbers, which are located on both sides of the alignments, indicate the physical positions of each sequence. Labels above/under the alignment indicate the amino acids related to the chi sequence binding sites determined using *B. subtilis*-AddA [37]. (**c**) Neighbor-Net network tree of the UvrD/Rep-like helicase. Accession number and species name corresponding to the sequence alignment are labeled at the end of the branch. Numbers on nearby branches indicate bootstrap test values with 1000 replicates. Scale bar: number of substitutions per site.

3.2.2. Signal Sequence of the Initiation of DNA Replication

Subsequently, we sought to find the sequence that was related to the origin of recognition on the probable *ori* region. To accomplish this, we first calculated the quartet nucleotide composition ratio of the *ori*-containing region per total genome (Figure 5). For details of the calculation of the ratio, see Materials and Methods. The four types of sequences that had a ratio greater than 2 contained 75–100% of GC nucleotides, and two of four sequences, CGGC and CCGC, exhibited significantly different ratios from the other sequences ($p < 0.05$, Figure 5a). Furthermore, two sequences (CCGC and GCGG) were complementary to each other (Figure 5a). This pair of sequences was identical to four out of five nucleotides on the chi sequence, which was recognized by *B. subtilis*-AddA (5′-AGCGG-3′) [12]. Additionally, the sequence CGGC is known to be a part of a complementary sequence that is known as a replication signal sequence of mitochondrial genome in humans, 3′-GGCCG-5′ [38]. Therefore, we analyzed the distribution of these quartet sequence in the APMV genome. The sequence densities of CCGC and GCGG in the APMV genome showed that both sequences had two peaks, and one of each peak was located on the *ori*-containing region, while the others were on the axial symmetric position (about 800,000 bp, Figure 5b).

Similar to this, those of CGGC and GCCG exhibited two peaks and the peak on the 5′ side in GCCG located on the *ori*-containing region, although that of the CGGC slightly skewed from the *ori*-containing region to the 5′ end of the genome (Figure 5c). Each type of quartet sequence composition frequencies was calculated in APMV, four kinds of bacteria (*B. subtilis*, *L. lactis* and *E. coli*, *and R. prowazekii*), and human mitochondrial genome (*H. sapiens* MT). *R. prowazekii* is considered as an ancestor of mitochondria [39], while the mitogenome sequence similarity against the Mimivirus genome has been reported recently [40]. These frequencies were then plotted in every pair of bacteria, mitochondria, APMV, and the *ori*-containing region of APMV (375–385k) (Figure 6). Every pair without APMV-APMV-*ori* and APMV-*R.prowazekii* exhibited a significant difference of sequence composition ($p<0.05$, Figure 6). *R. prowazekii* possesses the most similar sequence composition compared with APMV ($p = 0.2528$). The composition similarity of the *R. prowazekii*-APMV pair was higher than those of the *R. prowazekii-H. sapiens* MT and APMV-*H. sapiens* MT pairs (*R. prowazekii-H. sapiens* MT: $p = 0.04685$; APMV-*H. sapiens* MT: $p = 0.00146$), suggesting that the APMV conserves the sequence derived from an ancestor of mitochondria rather than from highly evolved mitochondria in humans. Interestingly, the composition of human mitochondria and *L. lactis* also showed a high similarity ($p = 0.3552$) rather than the *R. prowazekii-H. sapiens* MT pair, suggesting that the mitochondrial genome still conserves a remnant of bacterial characteristics.

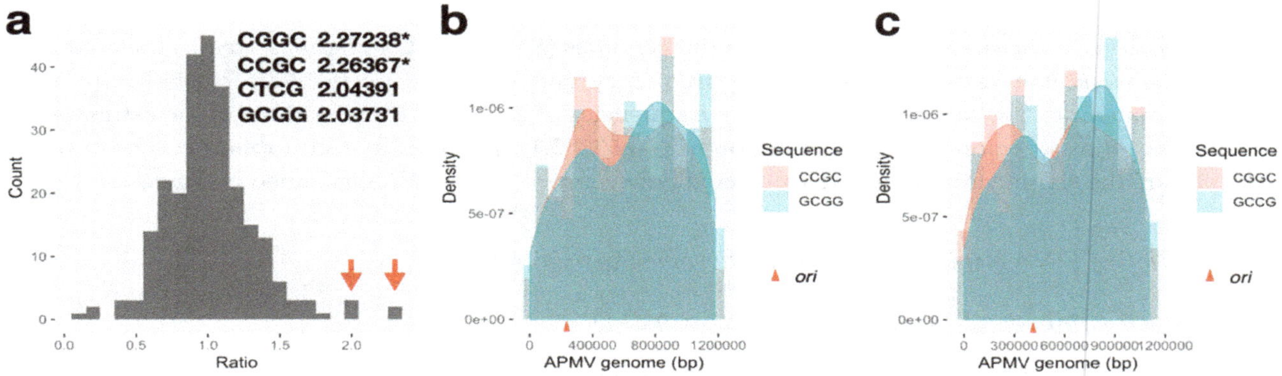

Figure 5. Comparative analysis of the quartet nucleotide composition ratio between the *ori*-containing region and the total genome of APMV. (**a**) Frequency distribution of fold-changes (= [*ori*-containing region]/[whole genome], *ori*-containing region: 375–385 kb of AY653733.1). Red arrows indicate that the fold-changes of the bins are greater than two. Sequences and fold-changes greater than two are listed on the top right of the figure. Asterisks indicate significant outliers from the population calculated by Grubbs test ($p<0.05$). (**b**,**c**) Density graph of the 5′-CCGC-3′/5′-GCGG-3′ and 5′-CGGC-3′/5′-GCCG-3′ on the APMV genome (AY653733.1). Red arrowhead on the scale indicate the estimated location of the *ori* region (382,000 ± 5000 bp).

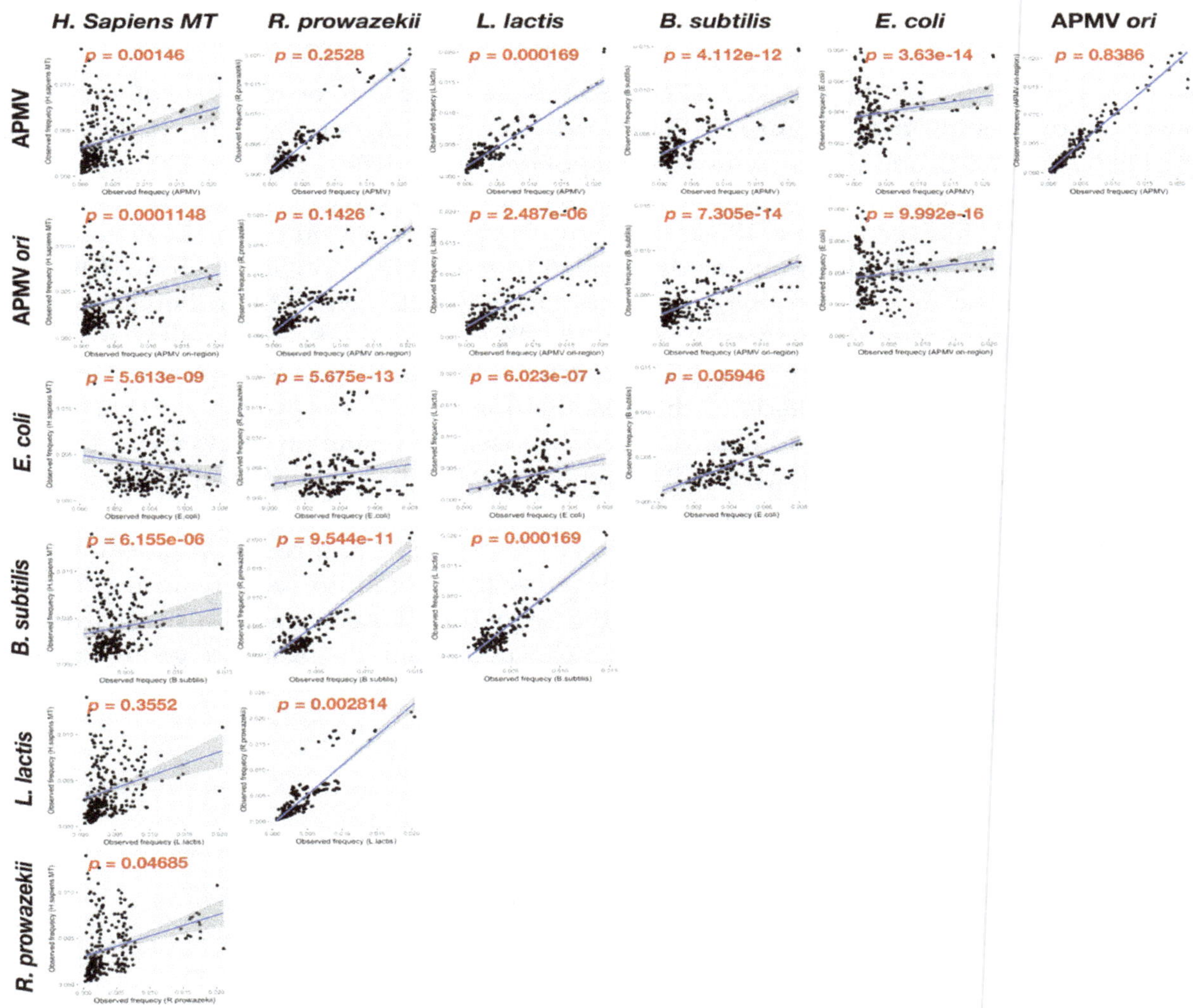

Figure 6. Comparative analysis of quartet nucleotide composition frequencies among APMV, human

mitochondria (*H. sapiens* MT), and four bacterial genomes (*Escherichia Coli, Bacillus subtilis, Lactococcus lactis, Rickettsia prowazekii).* Pairs of every frequency between two species were plotted with an approximate line. "APMV *ori*" indicates the frequencies of sequence compositions in the estimated location of the *ori* region (375–385 kb region of AY653733.1). *p*-values in each top left corner of the graphs are the significant differences calculated between two groups as determined by KS test.

3.3. Termination of DNA Replication and Chromosome Segregation

3.3.1. Sequence Analysis of FtsK-Like Protein in APMV

FtsK mainly conserves the motor domain (alpha/beta domain) and gamma domain, the latter of which is our main target, and recognizes KOPS [5,26]. The DNA binding sites of FtsK has been determined in *P. aeruginosa* [34,41]. We aligned the gamma domain of APMV with *P. aeruginosa* and other bacteria (*L. lactis, B. subtilis,* and *E. coli*), revealing that the DNA binding sites determined in *P. aeruginosa* were conserved to almost the same degree among APMV and bacteria. The phylogenetic analysis of FtsK-gamma domain among these species indicated that the sequence of *L. lactis* was most closely related to the APMV-FtsK gamma domain (Figure 7b). However, in APMV, there was a 10 aa insertion in the region where the DNA binding sites were localized (Figure 7a), suggesting that DNA-binding activities of APMV-FtsK had possibly collapsed. Therefore, we confirmed the three-dimensional structure via homology modeling of the gamma domain in APMV-FtsK, using two models from *P. aeruginosa* as templates [34,41].

As a result, the DNA binding sites of APMV-FtsK were thought to be topologically reconstructed and conserved by this insertion (Figure 8). Estimated DNA binding sites against the DNA backbone were K242, K243, K253, and K256, which corresponded with R770, K771, R778, and R781 in *P. aeruginosa,* respectively, and the binding site for KOPS specifically was N246 at APMV-FtsK, which corresponded to N777 in *P. aeruginosa*. K242, K243, and N246 were placed on the insertion (Figure 7a, Figure 8a) [33]. These four basic amino acids localize in two helix motifs, which bind to the KOPS region and support the recognition of KOPS by asparagine between the two helix motifs (Figure 8a,b) [33]. Next, we confirmed the conservation of the ATP binding sites on the FtsK in APMV. The FtsK conserves the Walker A motif (the ATPase active site), and in *P. aeruginosa,* substitution of amino acid residue in this motif leads to the deactivation of ATPase (K472N) [41]. The sequence alignment showed that the motor domain of APMV conserved lysine on the Walker A motif (K32) as well as other three bacteria (Figure 9a). The other ATP binding sites found in *P. aeruginosa* [41] were partially conserved in the APMV-FtsK (Figure 9a). However, some amino acids were not conserved, even in bacteria (R418 and H675 of *P. aeruginosa*-FtsK, Figure 9a), suggesting that these amino acid residues in binding sites could be replaced with other amino acid residues. Phylogenetic analyses of these motor domains indicated that the sequence of APMV was distinct from bacteria (Figure 9b). Therefore, APMV-FtsK might have speciated from its bacterial group prior to the emergence of bacterial FtsK.

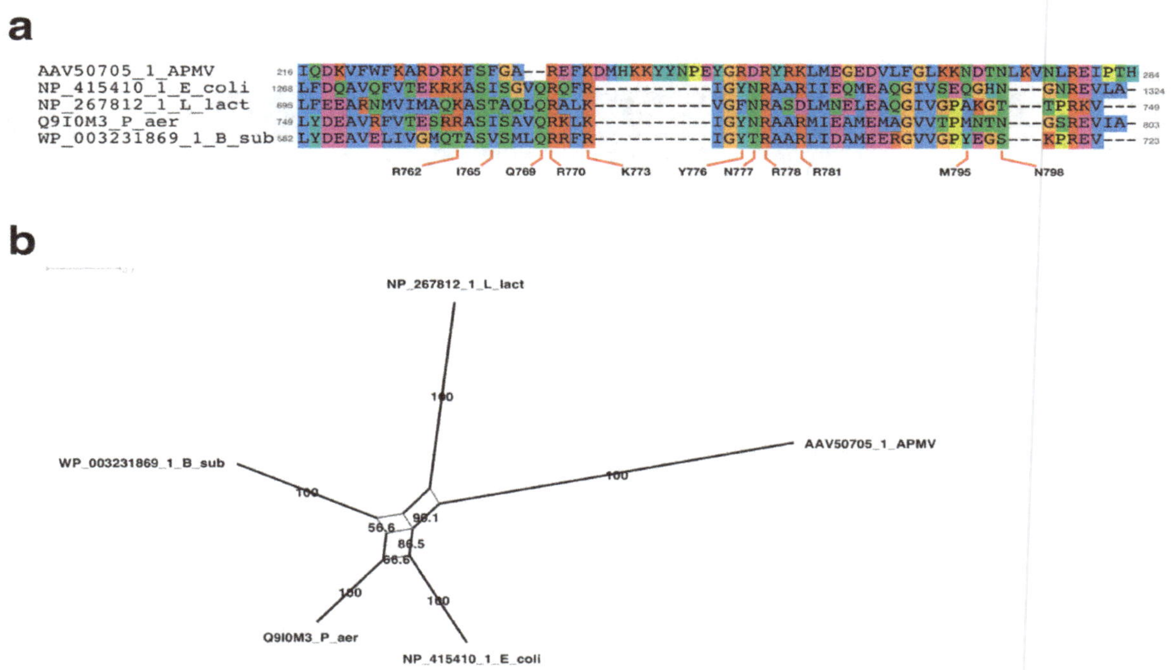

Figure 7. Sequence analyses of APMV-FtsK gamma domain. (**a**) Sequence alignment of FtsK-like protein gamma domain. Accession number and species name are labeled on the left side of each alignment. Grey numbers located on both sides of the alignments indicate the exact positions of each sequence. Amino acids labeled under the alignment indicate the amino acid residues responsible for DNA binding sites of *Pseudomonas aeruginosa* [25]. (**b**) Neighbor-Net network tree of FtsK gamma domains. Accession number and species name corresponding to the sequence alignment are labeled at the end of the branch. Numbers on nearby branches indicate bootstrap test values with 1000 replicates. Scale bar: number of substitutions per site.

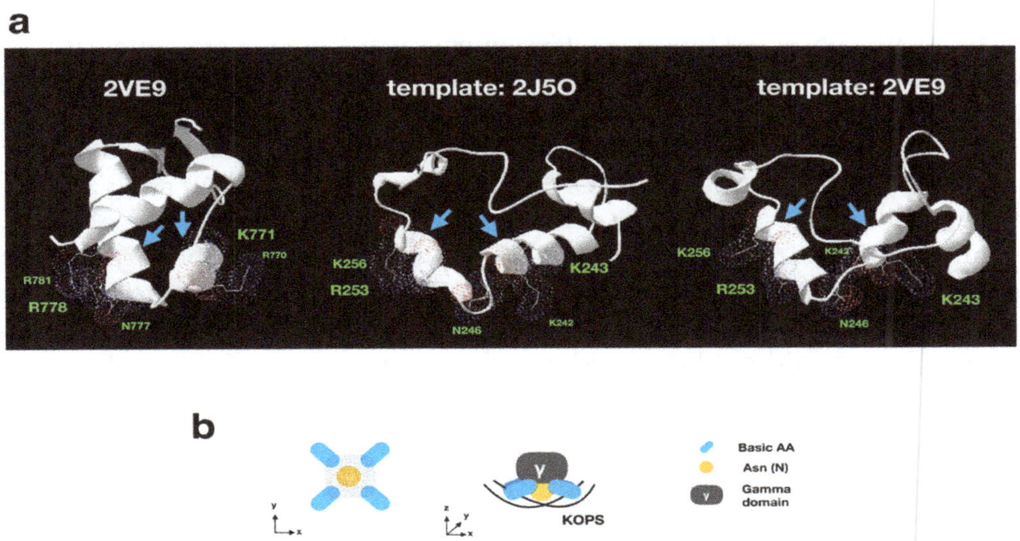

Figure 8. Homology modeling analysis of APMV-FtsK gamma domain. (**a**) Estimated structures of FtsK gamma domain using two different templates of *P. aeruginosa*-FtsK (PDBID; middle: 2J5O; right: 2VE9). The left model is the gamma domain of *P. aeruginosa*-FtsK (PDBID: 2VE9). Side chains on the left model are directly bound to FtsK orienting polar sequences (KOPS), while the other two models are estimated amino acids, which are functionally homologous to 2VE9. Blue arrows indicate two helix motifs harboring KOPS binding residues. (**b**) Model of the gamma domain of APMV-FtsK against KOPS. Four basic amino acids and Asn (N) correspond to the side chain in every model on panel (a).

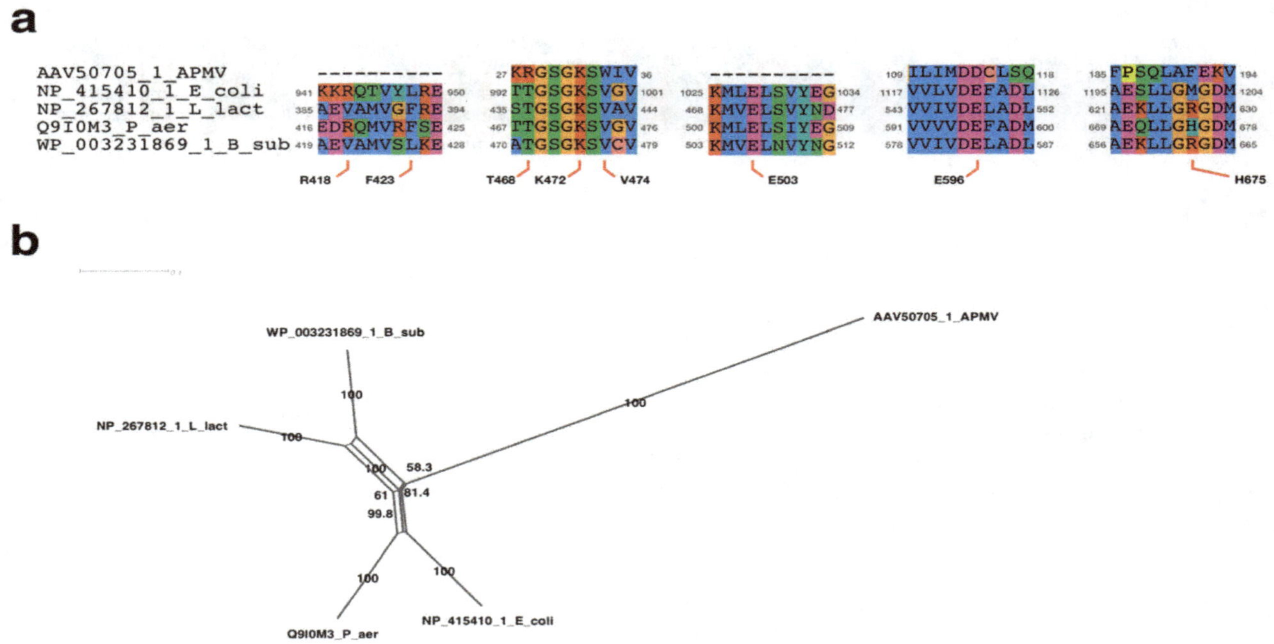

Figure 9. Sequence analyses of APMV-FtsK ATPase domain. (**a**) Motor domain of FtsK-like protein sequence alignment. Accession number and species name are labeled on the left side of alignments. Grey numbers located on both sides of the alignments indicate the exact positions of each sequence. Labels under the alignment indicate the amino acids related to ATPase activities determined in *P. aeruginosa* [41]. (**b**) Neighbor-Net network tree of the gamma domains. Accession number and species name correspond to the sequence alignment labeled at the end of the branch. Numbers on nearby branches indicate bootstrap test values with 1000 replicates. Scale bar: number of substitutions per site.

3.3.2. Distribution Pattern of KOPS on APMV Genome

The most frequently observed type of KOPS sequence was from *L. lactis*, whereas those of *B. subtilis* and *E. coli* were hardly encountered (*L. lactis*, 5′-GAGAAG-3′: 225; *B. subtilis*, 5′-GAGAAGGG-3′: 5; *E. coli*, 5′-GGGNAGGG-3′: 9; Figure 10a). This KOPS distribution pattern was different between the positive and negative strand, and the density graphs showed that these two patterns crossed at the estimated *ori* position (Figure 10b). These KOPS distribution patterns were analyzed in each bacterial genome, and the results suggested that the density of KOPS on the positive/negative strand switched on the exact points of the *ori*/*ter* region (Figure 10c). This is similar to the distribution pattern of *L. lactis*-derived KOPS in the APMV genome and to KOPS in the bacterial genomes, indicating that *L. lactis* KOPS could be one of the commonly used termination sequences in DNA replication in APMV. It should be noted that we did not have evidence that the other two KOPS, 5′-GAGAAGGG-3′ and 5′-GGGNAGGG-3′, found in APMV, were inactive (Figure 10a).

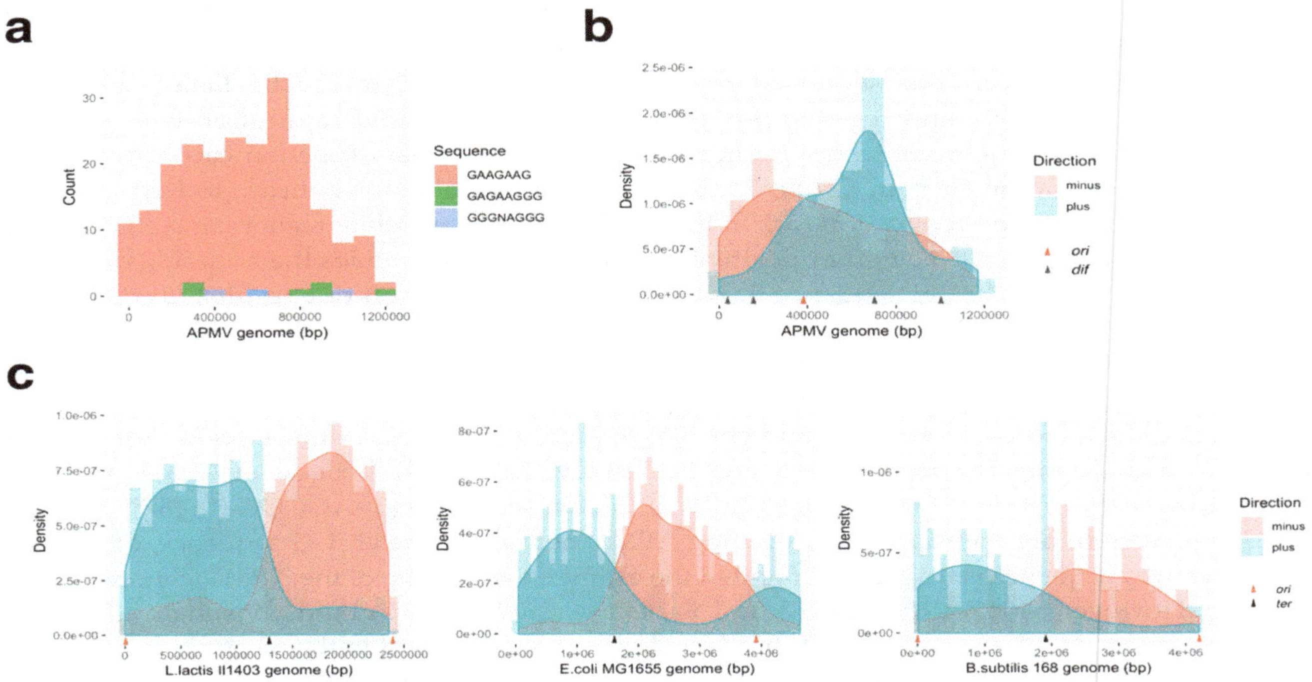

Figure 10. Bacterial KOPS distribution on APMV. (**a**) Frequency distribution of bacterial KOPS distribution in the APMV genome (AY653733.1, *L. lactis*; 5′-GAAGAAG-3′, *B. subtilis*: 5′-GAGAAGGG-3′, *E. coli*: 5′-GGGNAGGG-3′). Complementary sequences were also counted. (**b**) Density graph of the *L. lactis* type KOPS in the APMV genome (AY653733.1). KOPS on the positive and negative strand of the APMV genome were plotted separately. Red arrow: estimated *ori* region; dark gray arrows: *dif* sequence positions (**c**) Bacterial KOPS distributions in bacterial genomes (*L. lactis* Il1403: NC_002662.1, *E. coli* MG1655: NC_000913.3, *B. subtilis* 168: NC_000964.3). Red and black arrows indicate the *ori* and *ter* positions, respectively.

4. Discussion

The GC/AT skew bias is the result of replication bias in bacteria [19], and the replication-transcription conflict causes a high mutation rate in genes, causing genetic transcription and DNA replication to be co-directional [42]. We determined the estimated *ori* region using high-resolution cumulative skew graphs of AT nucleotides and CDSs (382 kb, Figures 1 and 2), which suggests that the transcription and DNA replication of the APMV genome are co-directional. This characteristic is not unique to cellular organisms but is the same in APMV. Both bacteria and viruses replicate faster than eukaryotes. APMV has a 1.2 Mb genome, a size which is similar to that of bacteria. Therefore, its large genomic structure would likely be constructed while facing the selection pressure of DNA replication.

The negative and positive values on both ends of the GC skew plot suggest that the G/C nucleotide composition switches between the 5′ region (<296 kb) and the 3′ region (>882 kb). Both end regions have *dif* sequences, which have been previously estimated [5], suggesting that this nucleotide composition bias is perhaps a key factor for replication termination with directional homologous recombination (Figure S3). Indeed, both regions harbor paralogous genes of ankyrin repeats in opposite directions, indicating that recombination between the 5′ and 3′ ends frequently occurs (Figure 3a). Moreover, the ankyrin repeats are symmetrically placed between *dif* 2 and *dif* 3; thus, these two *dif* sequences might often be used for termination and segregation (Figure 3a). A model for the termination of DNA replication in APMV has been previously described, hypothesizing that the genomic DNA bent symmetrically during DNA replication of the 3′ end [3]. The symmetrical distribution of the sequences and genes in APMV suggest that homologous recombination and/or sequence insertions would occur prior to chromosome segregation. Furthermore, the large, symmetrical inversion between APMV and

Megavirus chillensis (lineage C of *Mimiviridae*) [43] indicates that the topology of the genomic DNA during replication led to the diversification of the viral family *Mimiviridae*.

In bacteria, the *dnaA* box sequence on the *ori* region is recognized by DnaA when DNA replication is initiated [9], and the Rep protein is used when plasmid replication is initiated [44]. Rep is also involved in chromosomal replication, and a lack of Rep function can cause delay in chromosomal replication [11]. In *E. coli*, RecB, which is a homologous helicase to Rep, acts in the reconstruction of a stalled replication fork in the RecBCD pathway, while recognizing chi sequences (5′-GCTGGTGG-3′) [15,16]. The gram-positive bacteria *B. subtilis* encodes the homolog of the RecB, AddA, and recognizes the chi sequence 5′-AGCGG-3′ [12]. Based on our results, UvrD/Rep-like helicase of APMV is more similar to AddA than it is to Rep, and a portion of the gram-positive chi-like sequence (5′-CCGC-3′) is frequently found in the estimated *ori* region of the APMV genome (Figures 4 and 5) This suggests that the initiation of DNA replication is presumably mediated by the interaction between the UvrD/Rep-like helicase and chi-like sequences in APMV. In *B. subtilis*, AddA, Q1155, and I1157, homologous residues of Q922 and I924 in UvrD/Rep-like helicase of APMV are known to recognize the fourth and fifth G nucleotides of the chi sequence (5′-AGCGG-3′) [37]. Therefore, these conservations are appropriate for binding APMV helicase to the GCGG sequence, of which the complementary sequence is frequently found in the estimated *ori* region of the APMV genome (Figures 4b and 5). However, S1015, V1016, S1017, and Y1204 in *B. subtilis* AddA bind phosphate at the 3′ end of the chi sequence [37], suggesting that these residues are not involved in the recognition of a specific chi sequence. Phosphate binding sites were not conserved in UvrD/Rep-like helicase of APMV at the sequence level, however these regions contain polar amino acids similar to *B. subtilis* AddA (Figure 4b), which indicates that the UvrD/Rep-like helicase of APMV could also bind phosphate at the 3′ end of the chi sequence. Furthermore, the other quartet sequence, 5′-CGGC-3′, which was detected on the estimated *ori* region of APMV (Figure 5a,c), is reported to be a part of the DNA replication signal sequence of mitochondria (3′-GGCCG-5′) [38], indicating that the DNA replication of APMV is also initiated by host replication machineries involved in the replication of mitochondrial DNA.

KOPS and FtsK determine the region of DNA replication termination, and FtsK mediates DNA segregation process in bacteria [7,8]. KOPS are different among bacteria [6–8], in that the KOPS recognition mechanism is thought to be defined by the structure of FtsK and the sequence of KOPS itself. Our results showed that, in the APMV genome, the distribution pattern of *L. lactis*-KOPS was similar to those of bacteria (Figure 10b,c). The sequence analysis of the FtsK gamma domain, which interacts with KOPS, also showed a high similarity between *L. Lactis* and APMV (Figure 7), suggesting that the pair of KOPS and the FtsK-gamma domain structure in APMV might be homologous to that of *L. lactis*. Furthermore, we found that there was an insertion in the gamma domain of APMV-FtsK. Interestingly, the three-dimensional structure estimated by homology modeling revealed that this insertion was reconstructed and conserved in the DNA-binding domain (Figure 8). We also showed that the Walker A motif was conserved in the FtsK-motor domain of APMV (Figure 9a). These results suggest that APMV-FtsK might be functionally homologous to that of bacteria. The phylogeny of APMV-FtsK (gamma domain) was also found to be similar to that of *L. lactis* (Figure 7b). Therefore, the mechanisms of interaction between these proteins and specific sequences are presumably homologous between APMV and bacteria such as *L. lactis*, although further molecular biological studies and structural analyses are required to certify this model.

The phylogenetic relations between *Rickettsia* and mitochondria and between mitochondria and Mimivirus were described previously [39,40]. Interestingly, the genome size and GC content of *Rickettsia* are 1.1 Mb and 29%, respectively [39], which is highly similar to APMV (1.2Mb, 28%). Furthermore, the backbone of the Mimivirus genome is reported to be derived from the ancestor of mitochondria [40]. Considering the quartet sequence similarities (Figure 6) and the phylogenetic relation between *Rickettsia* and mitochondria, the ancestor of Mimivirus infected the ancestor of eukaryotic cells (last archaeal common ancestor, LACA) before the endosymbiosis of mitochondrial ancestor to the first eukaryotic common ancestor (FECA), while the ancestor of mitochondria and

Rickettsia also infected to the ancestor of eukaryotic cells. This ancestral virus presumably harbored shorter genomic DNA than the present-day Mimivirus, and it acquired the long genome from ancestor of mitochondria by genomic fusion. The sequence 5′-CGGC-3′ found in the APMV *ori* region might also have been acquired from the ancestor of mitochondria, and is still conserved as a DNA replication signal sequence (3′-GGCCG-5′) in the mitochondrial genome [38]. It has been reported that the comparison between mitochondrial genes and the *Rickettsia* genome shows much higher similarities than that between mitochondrial genes and megaviral genome (poxvirus), while the synteny of these three species are significantly conserved [45]. Thus, according to our results, the genomic remnants of the ancestor of mitochondria may be still conserved in the Mimivirus genome to a greater extent than in the poxvirus genome. Moreover, the co-infection (or preying on) of the LACA cells might have occurred not only in ancestor of mitochondria and Mimivirus but also in other bacteria, and therefore APMV conserves the bacteria-like machineries such as UvrD/Rep-like helicase and APMV-FtsK derived from the co-infected ancestor of gram-positive bacteria by horizontal gene transfer.

5. Conclusions

Here, we presented a proposed model of the initiation and termination of DNA replication and chromosome segregation for APMV. The estimated *ori* region exists at the 382 kb position in the genome, which contains the chi-like sequence recognized by *B. subtilis* AddA, which is homologous to the UvrD/Rep-like helicase of APMV. The other sequence has a homology of a DNA replication signal sequence of mitochondria, indicating that the DNA replication of APMV may initiate with the replication machineries of mitochondria. The KOPS distribution pattern and the structure of APMV-FtsK indicate that the KOPS recognition system by APMV-FtsK is similar to that of *L. lactis*. Consequently, replication initiation, termination and segregation systems of APMV are presumably mediated by DNA repair machineries, similar to that of gram-positive bacteria, such as *L. lactis*. Furthermore, the comparison of quartet sequence compositions shows the similarity between APMV and *Rickettsia*, which may have the closest common ancestor of mitochondria, indicating that Mimivirus has acquired a large bacteria-like genome and its DNA replication machineries from ancestor of mitochondria during the co-infection to the LACA cells. The evolutionary history of APMV remains unclear; however, the further analyses of such a chimeric genome of APMV may illustrate the early stage of evolution of eukaryotic cells and Mimivirus.

Author Contributions: M.A. designed the research and performed the analyses. M.A. and M.T. wrote the manuscript.

Acknowledgments: We would like to thank K. Aoki, S. Fukaya, and M. Kurabayashi from our laboratory for helpful discussions. We would also like to thank Editage (www.editage.jp) for English language editing.

References

1. La Scola, B.; Audic, S.; Robert, C.; Jungang, L.; de Lamballerie, X.; Drancourt, M.; Birtles, R.; Claverie, J.M.; Raoult, D. A giant virus in amoebae. *Science* **2003**, *299*, 2033. [CrossRef] [PubMed]
2. Raoult, D.; Audic, S.; Robert, C.; Abergel, C.; Renesto, P.; Ogata, H.; La Scola, B.; Suzan, M.; Claverie, J.M. The 1.2-Megabase Genome Sequence of Mimivirus. *Science* **2004**, *306*, 1344–1350. [CrossRef] [PubMed]
3. Yoshida, T.; Claverie, J.-M.; Ogata, H. Mimivirus reveals Mre11/Rad50 fusion proteins with a sporadic distribution in eukaryotes, bacteria, viruses and plasmids. *Virol. J.* **2011**, *8*, 427. [CrossRef] [PubMed]
4. Kreuzer, K.N. Recombination-dependent DNA replication in phage T4. *Trends Biochem. Sci.* **2000**, *25*, 165–173. [CrossRef]
5. Chelikani VRanjan, T.; Zade, A.; Shukla, A.; Kondabagil, K. Genome segregation and packaging machinery in *Acanthamoeba polyphaga* mimivirus is reminiscent of bacterial apparatus. *J. Virol.* **2014**, *88*, 6069–6075. [CrossRef] [PubMed]
6. Nolivos, S.; Touzain, F.; Pages, C.; Coddeville, M.; Rousseau, P.; El Karoui, M.; Le Bourgeois, P.; Cornet, F. Co-evolution of segregation guide DNA motifs and the FtsK translocase in bacteria: Identification of the atypical *Lactococcus lactis* KOPS motif. *Nucleic Acids Res.* **2012**, *40*, 5535–5545. [CrossRef]

7. Ptacin, J.L.; Nollmann, M.; Becker, E.C.; Cozzarelli, N.R.; Pogliano, K.; Bustamante, C. Sequence-directed DNA export guides chromosome translocation during sporulation in *Bacillus subtilis*. *Nat. Struct. Mol. Biol.* **2008**, *15*, 485–493. [CrossRef] [PubMed]
8. Bigot, S.; Saleh, O.A.; Lesterlin, C.; Pages, C.; El Karoui, M.; Dennis, C.; Grigoriev, M.; Allemand, J.F.; Barre, F.X.; Cornet, F. KOPS: DNA motifs that control *E. coli* chromosome segregation by orienting the FtsK translocase. *EMBO J.* **2005**, *24*, 3770–3780. [CrossRef]
9. Kaguni, J.M. Replication initiation at the *Escherichia coli* chromosomal origin. *Curr. Opin. Chem. Biol.* **2011**, *15*, 606–613. [CrossRef]
10. Lohman, T.M. *Escherichia coli* DNA helicases: Mechanisms of DNA unwinding. *Mol. Microbiol.* **1992**, *6*, 5–14. [CrossRef]
11. Lane, H.E.; Denhardt, D.T. The *rep* mutation. IV. Slower movement of replication forks in *Escherichia coli rep* strains. *J. Mol. Biol.* **1975**, *97*, 99–112. [CrossRef] [PubMed]
12. Chédin, F.; Noirot, P.; Biaudet, V.; Ehrlich, S.D. A five-nucleotide sequence protects DNA from exonucleolytic degradation by AddAB, the RecBCD analogue of *Bacillus subtilis*. *Mol. Microbiol.* **1998**, *29*, 1369–1377. [CrossRef] [PubMed]
13. Velankar, S.S.; Soultanas, P.; Dillingham, M.S.; Subramanya, H.S.; Wigley, D.B. Crystal structures of complexes of PcrA DNA helicase with a DNA substrate indicate an inchworm mechanism. *Cell* **1999**, *97*, 75–84. [CrossRef]
14. Petit, M.A.; Dervyn, E.; Rose, M.; Entian, K.D.; McGovern, S.; Ehrlich, S.D.; Bruand, C. PcrA is an essential DNA helicase of *Bacillus subtilis* fulfilling functions both in repair and rolling-circle replication. *Mol. Microbiol.* **1998**, *29*, 261–273. [CrossRef] [PubMed]
15. Horiuchi, T.; Fujimura, Y. Recombinational rescue of the stalled DNA replication fork: A model based on analysis of an *Escherichia coli* strain with a chromosome region difficult to replicate. *J. Bacteriol.* **1995**, *177*, 783–791. [CrossRef]
16. Amundsen, S.K.; Sharp, J.W.; Smith, G.R. RecBCD Enzyme "Chi Recognition" Mutants Recognize Chi Recombination Hotspots in the Right DNA Context. *Genetics* **2016**, *204*, 139–152. [CrossRef] [PubMed]
17. Biswas, I.; Maguin, E.; Ehrlich, S.D.; Gruss, A. A 7-base-pair sequence protects DNA from exonucleolytic degradation in *Lactococcus lactis*. *Proc. Natl. Acad. Sci. USA* **1995**, *92*, 2244–2248. [CrossRef]
18. Frank, A.C.; Lobry, J.R. Asymmetric substitution patterns: A review of possible underlying mutational or selective mechanisms. *Gene* **1999**, *238*, 65–77. [CrossRef]
19. Arakawa, K.; Tomita, M. The GC skew index: A measure of genomic compositional asymmetry and the degree of replicational selection. *Evol. Bioinform.* **2007**, *3*, 159–168. [CrossRef]
20. Freeman, J.M. Patterns of Genome Organization in Bacteria. *Science* **1998**, *279*, 1827. [CrossRef]
21. Grigoriev, A. Analyzing genomes with cumulative skew diagrams. *Nucleic Acids Res.* **1998**, *26*, 2286–2290. [CrossRef] [PubMed]
22. Lobry, J.R.; Sueoka, N. Asymmetric directional mutation pressures in bacteria. *Genome Biol.* **2002**, *3*, RESEARCH0058.1. [CrossRef]
23. Lobry, J.R.; Louarn, J.-M. Polarisation of prokaryotic chromosomes. *Curr. Opin. Microbiol.* **2003**, *6*, 101–108. [CrossRef]
24. Krzywinski, M.; Schein, J.; Birol, I.; Connors, J.; Gascoyne, R.; Horsman, D.; Jones, S.J.; Marra, M.A. Circos: An information aesthetic for comparative genomics. *Genome Res.* **2009**, *19*, 1639–1645. [CrossRef] [PubMed]
25. Whelan, S.; Irisarri, I.; Burki, F. PREQUAL: Detecting non-homologous characters in sets of unaligned homologous sequences. *Bioinformatics* **2018**, *34*, 3929–3930. [CrossRef] [PubMed]
26. Katoh, K.; Standley, D.M. MAFFT multiple sequence alignment software version 7: Improvements in performance and usability. *Mol. Biol. Evol.* **2013**, *30*, 772–780. [CrossRef]
27. Capella-Gutiérrez, S.; Silla-Martínez, J.M.; Gabaldón, T. trimAl: A tool for automated alignment trimming in large-scale phylogenetic analyses. *Bioinformatics* **2009**, *25*, 1972–1973. [CrossRef] [PubMed]
28. Nakhleh, L. Evolutionary Phylogenetic Networks: Models and Issues. In *Problem Solving Handbook in Computational Biology and Bioinformatics*; Springer: Boston, MA, USA, 2010; pp. 125–158. [CrossRef]
29. Huson, D.H.; Bryant, D. Application of phylogenetic networks in evolutionary studies. *Mol. Biol. Evol.* **2006**, *23*, 254–267. [CrossRef]
30. Arakawa, K.; Tamaki, S.; Kono, N.; Kido, N.; Ikegami, K.; Ogawa, R.; Tomita, M. Genome Projector: zoomable genome map with multiple views. *BMC Bioinform.* **2009**, *10*, 31. [CrossRef]

31. Rice, P.; Longden, I.; Bleasby, A. EMBOSS: The European Molecular Biology Open Software Suite. *Trends Genet.* **2000**, *16*, 276–277. [CrossRef]
32. Yang, J.; Yan, R.; Roy, A.; Xu, D.; Poisson, J.; Zhang, Y. The I-TASSER Suite: Protein structure and function prediction. *Nat. Methods* **2015**, *12*, 7–8. [CrossRef]
33. Löwe, J.; Ellonen, A.; Allen, M.D.; Atkinson, C.; Sherratt, D.J.; Grainge, I. Molecular mechanism of sequence-directed DNA loading and translocation by FtsK. *Mol. Cell* **2008**, *31*, 498–509. [CrossRef] [PubMed]
34. Sivanathan, V.; Allen, M.D.; de Bekker, C.; Baker, R.; Arciszewska, L.K.; Freund, S.M.; Bycroft, M.; Löwe, J.; Sherratt, D.J. The FtsK gamma domain directs oriented DNA translocation by interacting with KOPS. *Nat. Struct. Mol. Biol.* **2006**, *13*, 965–972. [CrossRef] [PubMed]
35. Arai, N.; Arai, K.; Kornberg, A. Complexes of *Rep* protein with ATP and DNA as a basis for helicase action. *J. Biol. Chem.* **1981**, *256*, 5287–5293.
36. Kooistra, J.; Haijema, B.J.; Hesseling-Meinders, A.; Venema, G. A conserved helicase motif of the AddA subunit of the *Bacillus subtilis* ATP-dependent nuclease (AddAB) is essential for DNA repair and recombination. *Mol. Microbiol.* **1997**, *23*, 137–149. [CrossRef]
37. Krajewski, W.W.; Fu, X.; Wilkinson, M.; Cronin, N.B.; Dillingham, M.S.; Wigley, D.B. Structural basis for translocation by AddAB helicase-nuclease and its arrest at χ sites. *Nature* **2014**, *508*, 416–419. [CrossRef] [PubMed]
38. Hixson, J.E.; Wong, T.W.; Clayton, D.A. Both the conserved stem-loop and divergent 5′-flanking sequences are required for initiation at the human mitochondrial origin of light-strand DNA replication. *J. Biol. Chem.* **1986**, *261*, 2384–2390.
39. Andersson, S.G.; Zomorodipour, A.; Andersson, J.O.; Sicheritz-Pontén, T.; Alsmark, U.C.; Podowski, R.M.; Näslund, A.K.; Eriksson, A.S.; Winkler, H.H.; Kurland, C.G. The genome sequence of *Rickettsia prowazekii* and the origin of mitochondria. *Nature* **1998**, *396*, 133–140. [CrossRef] [PubMed]
40. Seligmann, H. Giant viruses as protein-coated amoeban mitochondria? *Virus Res.* **2018**, *253*, 77–86. [CrossRef]
41. Massey, T.H.; Mercogliano, C.P.; Yates, J.; Sherratt, D.J.; Löwe, J. Double-stranded DNA translocation: Structure and mechanism of hexameric FtsK. *Mol. Cell* **2006**, *23*, 457–469. [CrossRef]
42. Paul, S.; Million-Weaver, S.; Chattopadhyay, S.; Sokurenko, E.; Merrikh, H. Accelerated gene evolution through replication-transcription conflicts. *Nature* **2013**, *495*, 512–515. [CrossRef] [PubMed]
43. Arslan, D.; Legendre, M.; Seltzer, V.; Abergel, C.; Claverie, J.-M. Distant Mimivirus relative with a larger genome highlights the fundamental features of Megaviridae. *Proc. Natl. Acad. Sci. USA* **2011**, *108*, 17486–17491. [CrossRef]
44. Han, M.; Yagura, M.; Itoh, T. Specific interaction between the initiator protein (Rep) and origin of plasmid ColE2-P9. *J. Bacteriol.* **2007**, *189*, 1061–1071. [CrossRef] [PubMed]
45. Seligmann, H. Giant viruses: Spore-like missing links between Rickettsia and mitochondria? *Ann. N. Y. Acad. Sci.* **2019**. [CrossRef]

4

Elongated Flexuous Plant Virus-Derived Nanoparticles Functionalized for Autoantibody Detection

Carmen Yuste-Calvo [1], Mercedes López-Santalla [2,3], Lucía Zurita [1], César F. Cruz-Fernández [1], Flora Sánchez [1], Marina I. Garín [2,3] and Fernando Ponz [1,*]

1. Centro de Biotecnología y Genómica de Plantas, Universidad Politécnica de Madrid-Instituto Nacional de Investigación y Tecnología Agraria y Alimentaria (CBGP, UPM-INIA), Campus Montegancedo, Autopista M-40, km 38. Pozuelo de Alarcón, 28223 Madrid, Spain; carmen.yus.cal@gmail.com (C.Y.-C.); lucia.zurita@inia.es (L.Z.); cesarfcruzf@gmail.com (C.F.C.-F.); florasanchez@telefonica.net (F.S.)
2. Division of Hematopoietic Innovative Therapies, Centro de Investigaciones Energéticas, Medioambientales y Tecnológicas (CIEMAT) and Centro de Investigación Biomédica en Red de Enfermedades Raras (CIBER-ER), 28040 Madrid, Spain; Mercedes.LopezSantalla@externos.ciemat.es (M.L.-S.); marina.garin@ciemat.es (M.I.G.)
3. Advanced Therapy Unit, Instituto de Investigación Sanitaria Fundación Jiménez Díaz (IIS-FJD/UAM), 28040 Madrid, Spain

* Correspondence: fponz@inia.es

Abstract: Nanoparticles derived from the elongated flexuous capsids of *Turnip mosaic virus* (TuMV) have been shown to be efficient tools for antibody sensing with a very high sensitivity if adequately functionalized with the corresponding epitopes. Taking advantage of this possibility, TuMV virus-like particles (VLPs) have been genetically derivatized with a peptide from the chaperonin Hsp60, a protein described to be involved in inflammation processes and autoimmune diseases. Antibodies against the peptide have been previously shown to have a diagnostic value in at least one autoimmune disease, multiple sclerosis. The functionalized Hsp60-VLPs showed their significant increase in sensing potency when compared to monoclonal antibody detection of the peptide in a conventional immunoassay. Additionally, the developed Hsp60-VLPs allowed the detection of autoantibodies against the Hsp60 peptide in an in vivo mouse model of dextran sodium sulfate (DSS)-induced colitis. The detection of minute amounts of the autoantibodies allowed us to perform the analysis of their evolution during the progression of the disease. The anti-Hsp60 autoantibody levels in the sera of the inflamed mice went down during the induction phase of the disease. Increased levels of the anti-HSP60 autoantibodies were detected during the resolution phase of the disease. An extension of a previously proposed model for the involvement of Hsp60 in inflammatory processes is considered, incorporating a role for Hsp60 autoantibodies. This, and related models, can now be experimentally tested thanks to the autoantibody detection hypersensitivity provided by the functionalized VLPs.

Keywords: VNPs; Hsp60; IBD; autoantibody; inflammation; diagnosis

1. Introduction

The use of viral nanoparticles (VNPs) for biomedical applications has become a new tool in theranostics. Specifically, VNPs derived from plant viruses offer advantages in terms of biosafety since they are only plant pathogens and also because of plant virus variability in nature, size, and structure, that allows a specific design of VNPs depending on the application [1–13]. We work with *Turnip mosaic virus* (TuMV), a virion with an elongated and flexuous structure, 700 nm long and 12 nm wide, with *ca.* 2000 copies of the coat protein in each particle, from which multifunctional VNPs can be obtained by genetic fusion and/or chemical conjugation to the coat protein (CP) [14–17]. One potential application is their use as tools for the detection of circulating antibody levels that frequently are too low to be detected by conventional detection methodologies, thus contributing to improve the diagnosis, progression, and/or prognosis of immunity-mediated pathologies, especially at the early stages of the disease when no clear symptoms of the inflammation are evident.

Inflammatory bowel disease (IBD) is a pathology encompassing two chronic inflammatory disorders; Crohn's disease, characterized by transmural inflammation usually affecting the terminal ileus and/or the large intestine; and ulcerative colitis, where inflammation occurs in the lining of the colon mucosa [18,19]. In these pathologies, our research focuses on diagnosis improvement and the development of personalized therapy, in order to achieve improvement in the patients' quality of life. On the other hand, we seek to gain new knowledge about the multifactorial pathogenesis of these disorders, for which there are several representative animal models, capable of mimicking the symptomatology and pathology of the disease. For IBD, a murine model induced by dextran sodium sulfate (DSS) has been developed [20,21]. This compound, when administered in drinking water, induces inflammation of the colon mucosa, as well as ulceration, which leads to severe weight loss and, in extreme cases, lethality.

We have explored the deployment of Heat Shock Protein 60 (Hsp60) as a possible autoantigen in IBD, due to the implication of Hsp60 in inflammatory processes [22–31], including IBD [32–35]. Heat shock proteins are a broad family of molecular chaperons capable of reacting to cellular stress, especially in thermal changes. However, they also respond in other stress situations, such as tissue damage, cellular injury, or heavy metal poisoning. Hsp60, despite being a protein whose main function occurs in the mitochondria, is able to act in the cytoplasm modulating the immune response associated with inflammation. Both the complete protein and the peptides derived therefrom can act as natural regulators of the inflammatory reaction together with other cytokines, chemokines, and autoantibodies, regulating the immune system [30,32,36–39]. The activity of Hsp60 has been associated with various autoimmune and inflammatory pathologies, such as atherosclerosis, diabetes, rheumatoid arthritis, or multiple sclerosis [23,25,29,40–42].

Alterations in the levels of anti-Hsp60 autoantibodies have been described in several inflammatory pathologies [22,23,26,27,37], making the autoantibody detection a novel strategy with potential application for the diagnosis, progression, and/or prognosis of the disease. In these studies, exhaustive analyses by peptide arrays have allowed the identification of certain epitopes for the design of specific VNPs where the election of a peptide with diagnostic value is the key to success. Based on these premises, we have designed functionalized TuMV VNPs with an epitope, described previously in multiple sclerosis [40]. The multimeric presentation of this epitope on TuMV VNPs (Hsp60-virus-like particles (VLPs)) allowed us the quantitative detection of anti-Hsp60 autoantibodies in an in vivo model of intestinal inflammation induced by DSS. The high detection levels of the developed Hsp60-VLPs may represent a novel tool that can be used for the diagnosis, progression, and/or prognosis in inflammation-mediated disorders.

2. Materials and Methods

2.1. *Construction of CP and Hsp60-CP Expression Plasmids*

Constructs corresponding to infectious clones were obtained by designing synthetic genes (GeneArt), based on the p35Tunos-Vec01-Nat1 sequence, as shown in Figure S1, derived from an original TuMV infectious clone [43]. Synthetic DNA consisted of a partial NIb gene, starting with the sequence corresponding to the restriction site *Mlu* I, followed by alanine codon (first CP amino acid) to maintain the protease recognition sequence, followed by the sequence encoding the Hsp60 peptide, which comprises amino acids 301–320 of human Hsp60 (KAPGFGDNRKNQLKDMAIAT, sequence conserved in mice), and the restriction site *Nae* I, corresponding to the first two CP amino acids. The DNA was digested with *Mlu* I and *Nae* I. The resulting fragment was purified. Vector p35Tunos-vec01-Nat1 was also digested with these two restriction enzymes, and the two fragments ligated together to obtain the recombinant vector with the sequence encoding Hsp60 peptide.

For VLP expression plasmids, PCR (GeneAmp®PCR System 9700, Applied Biosystems, CA, United States) was performed, amplifying the whole modified CP with two extra codons—one corresponding to the initial methionine, and a STOP codon. The nucleotide sequence CACC was also added at the 5′ end to allow directional cloning into a pENTR-D-TOPO vector, as shown in Figure S2. Then the fragment was cloned into a pEAQ-HT vector, as shown in Figure S3, by Gateway cloning with LR clonase enzyme.

2.2. Production and Purification of VLPs

For VLP production, the pEAQ construct was transformed into *Agrobacterium tumefaciens* LBA4404 for agroinfiltration mediated transitory expression in *Nicotiana benthamiana* plants [44,45]. The same procedure was followed for non-modified VLPs. Plant growth, *Agrobacterium* culture preparation, agroinfiltration, tissue harvesting, and VLP purification were performed as previously described by us [16,17].

2.3. VLP Characterization

Characterization of assembled purified VLPs was performed by SDS-PAGE, western blot and transmission electron microscopy (TEM, ICTS-CNME, Madrid, Spain).

The conditions for SDS-PAGE and western blot were as described [16]. Anti-Hsp60 D307 antibody was from Invitrogen.

To check structural integrity, TEM was performed. Electron microscopy grids (400 mesh copper, carbon coated) were coated at room temperature for 15 min with a 10 μL drop of VLPs diluted at 0.02 mg/mL final concentration (50 mM borate buffer, pH 8.1), and washed with buffer. Finally, the grids were rinsed with distilled water and stained with 2% uranyl acetate for 2 min. Samples were examined on a transmission electron microscope (JEM JEOL 1010, Tokyo, Japan).

2.4. IBD Murine Model

Adult C57BL/6J (8 weeks-old) mice used in these studies were obtained from the Jackson Laboratory (ref. 000664). The regulations concerning experimental animal welfare (RD 223/1998 and Directive 2010/63/EU protocols) were followed. The ethics committee for animal research of the CIEMAT (Proex. 414/15) and Comunidad de Madrid (based on the RD 53/2013) reviewed and approved all protocols.

The IBD was induced in mice by dextran sodium sulfate (DSS, MP Biochemicals), a sulphated polymer cytotoxic on intestinal epithelial cells and macrophages. In addition, enteral DSS favors Gram-negative anaerobic bacteria increases, which together with the erosive potential on the intestinal barrier and the macrophages' inappropriate response, would lead to the appearance of intestinal lesions [20,21]. This model reproduces the clinical, histopathological, and immune characteristics observed in humans, inducing chronic colitis associated with diarrhea and weight loss.

The chemical compound was administered in drinking water at 1.35% (*w*/*v*) over 7 days *ad libitum* and disease progress was determined by monitoring mouse weight and the number of granulocytes in peripheral blood by an automated blood cell-counter (Abacus, Diatron, Budapest, Hungary). Normal distribution was analyzed by the Shapiro–Wilks test.

Non-parametric techniques (Mann–Whitney U test) were used. Autoantibody levels were determined in mouse sera, obtained at different times after disease induction. All trials were of a "simple blind" type.

2.5. Immunoassays

Indirect enzyme-linked immunosorbent assays (ELISA) were performed in order to evaluate the sensitivity provided by the VLPs, in comparison with the whole Hsp60 protein and free peptide at equal peptide amounts, and also to measure autoantibody levels in peripheral blood. Plates (Nunc

MaxiSorp) were coated with 1 µg purified VLPwt or Hsp60-VLPs, resuspended in 50 mM sodium carbonate buffer, pH 9.6; or equivalent amounts of Hsp60 protein or peptide, resuspended in the same buffer and incubated overnight at 4 °C. Plates were washed intensively, and incubated for 1 h with commercial antibodies [Hsp60 (D307) Antibody 4870S, CellSignal and Anti-Hsp60 antibody (ab46798), Abcam; different dilutions in phosphate-buffered saline, pH 7.4, 0.05% Tween 20 (*v*/*v*), 2% PVP-40 (*w*/*v*)] or overnight at 4 °C for mice sera (diluted 1:100 in the same buffer). Then, secondary antibodies, diluted 1:1500 in the same buffer, were added and incubated for 1 h at room temperature. Color was developed by alkaline phosphatase reaction and detected after addition of *p*-nitrophenylphosphate. Absorbance was measured at 405 nm (TECAN Genios Pro). Differences between both coating-VLPs were calculated, since this would be the measure attributable to autoantibodies against Hsp60. The normalization process was done as follows. A serum pool was made using the sera from healthy mice and this pool was included in all the plates in order to calculate the background of each plate. The normalized values of autoantibodies directed against Hsp60 in the DSS serum samples were obtained by taking into account the mean value of the healthy serum pool in each plate and the mean value of the healthy serum pool from all the plates, in order to minimize the effects of the variation in the background signal. The data obtained were analyzed based on the non-parametric Mann–Whitney U test.

3. Results

3.1. Production and Characterization of Hsp60-VLPs

Our first Hsp60-VNP derivatization approach was the production of virions genetically modified with the Hsp60-peptide fused to the CP N-terminus. However, the insertion of the peptide interfered with the infectivity of the modified virus, so this strategy was discarded. Starting out from the virion construct, a second construction strategy was followed to generate now modified VLPs.

VLPs modified by genetic fusion of the Hsp60-peptide to the CP were made by transient expression of the fusion protein CP-Hsp60 peptide in plants. As shown before [16], the fusion protein will self-assemble into VLPs within plant cells. Once purified nanoparticles were obtained, they were characterized by SDS-PAGE, western-blot assays, and transmission electron microscopy, as shown in Figure 1.

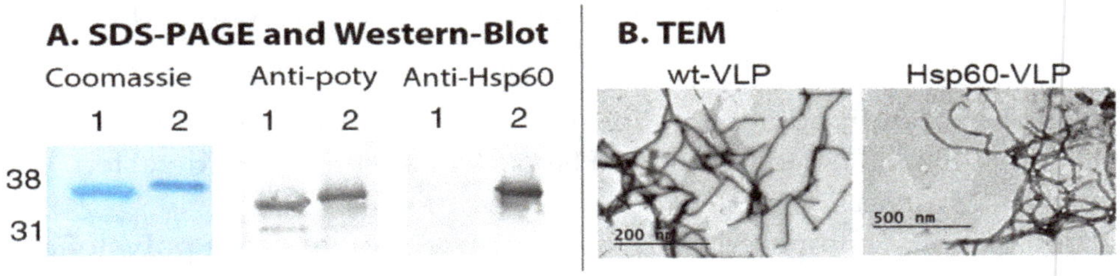

Figure 1. Hsp60-virus-like particles (VLPs) characterization. (**A**) SDS-PAGE and western blot of wild type VLPs (1) and Hsp60-VLPs (2). (**B**) Micrographs of wt-VLPs and Hsp60-VLPs, where particles of about 700 nm are shown.

The results obtained showed a modified electrophoretic mobility for Hsp60-VLPs CP, consistent with a molecular weight increase of approximately 2.5 kDa. This fusion protein was detected specifically not only by antibodies directed against the virus, but also by those directed against the Hsp60 protein. Finally, TEM structural characterization showed elongated and flexuous particles with similar length and thickness than non-modified particles, so no detectable significant changes in VLP structure were found after peptide insertion.

3.2. Increased Sensitivity of Hsp60 Peptide Detection by Specific Hsp60 Antibodies

Multimeric peptide presentation on the surface of TuMV VNPs increases its detectability by specific antibodies in conventional immunoassays like ELISA [15,17,46]. To evaluate this in the Hsp60-VLPs construct, ELISA and dot-blot assays were performed using three types of presentations: Hsp60-VLPs, free Hsp60 peptide, and complete Hsp60 protein, as shown in Figure 2. Assays were performed using equal peptide amounts presented in the three forms. Two different antibodies were used to assess the response, a monoclonal antibody that specifically recognizes an epitope within the peptide 301–320, and a polyclonal antibody against complete Hsp60.

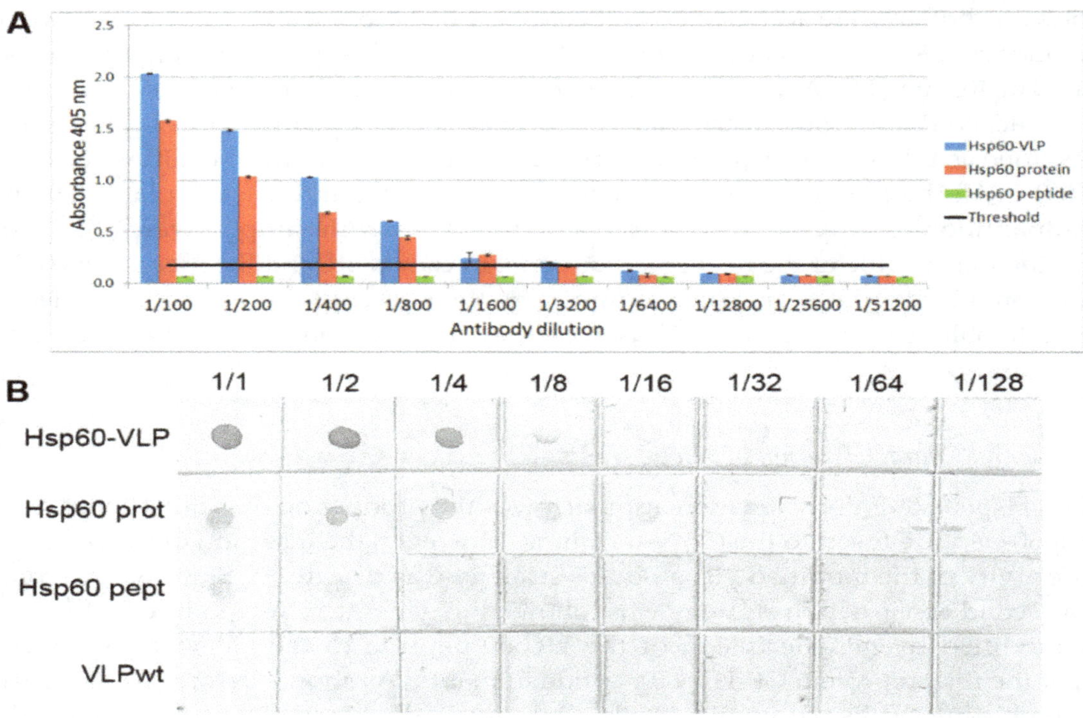

Figure 2. Specificity and sensitivity assays. (**A**) ELISA using monoclonal Hsp60 antibody dilutions, comparing Hsp60-VLPs with free Hsp60 peptide and Hsp60 protein. The black line indicates the positive signal threshold and the standard error is also shown (three replicates). (**B**) Dot blot using the same components, and also VLPs-wt.

Using the monoclonal antibody, it was found that free Hsp60 peptide was hardly recognized, while both the complete Hsp60 protein and the Hsp60-VLPs were easily detected, with Hsp60-VLPs being the most sensitive system.

Hsp60-VLP detectability, compared with the complete protein using polyclonal antibodies, was evaluated, as shown in Figure S4. In this case, complete Hsp60 protein presents greater sensitivity, possibly due to the higher number of epitopes present on its complete surface.

The results obtained show that, although the complete HSP60 protein contains various epitopes, the multimeric presentation in VNPs of a defined epitope allows differential detection by measurement of specific autoantibodies in biological fluids, such as peripheral blood serum. These results highlight the Hsp60-VLPs potential for specific antibody detection, with the epitope election being the key to develop TuMV-VNPs as a diagnostic tool.

3.3. Hsp60-VLPs Autoantibody Detection

Once sensitivity and specificity were shown through different immunoassays, the potential of Hsp60-VLPs as a diagnostic tool of autoimmune pathologies, exemplified by IBD, was tested, measuring autoantibodies against Hsp60 in a pool of control sera, as shown in Figure 3. In these pathologies, autoantibody levels are usually very low at early stages, but early diagnosis by detecting these low levels would improve patients' prognosis.

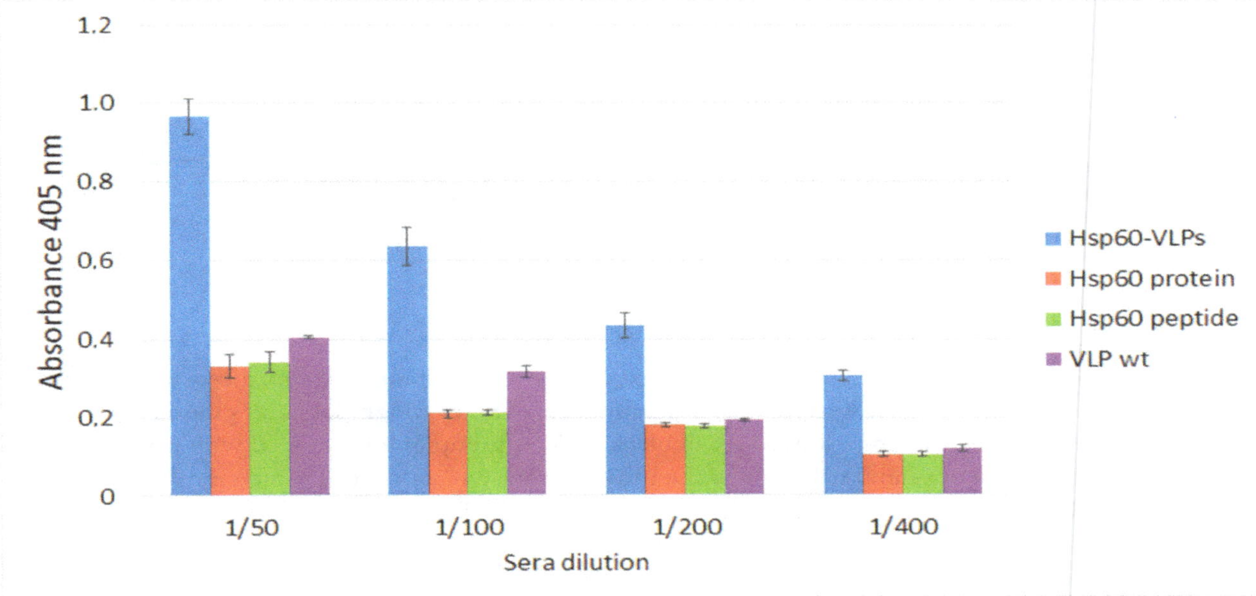

Figure 3. Anti-HSP60 autoantibody detection in the sera of healthy mice by ELISA, using different antigen-presentation platforms; Hsp60-VLP (blue), complete protein Hsp60 (red), Hsp60 peptide (green), and VLPwt (purple) and at different sera dilutions. Standard error bars shown (three replicates).

To compare the data obtained with all detection systems (Hsp60-VLPs, complete Hsp60 protein, and free peptide), an ELISA was developed with a pool of sera from healthy, non-inflamed mice, in order to determine physiological levels of anti-Hsp60 autoantibodies. Unmodified VLPs (VLPwt, not displaying the Hsp60 peptide) were used as a negative control.

The results showed that only with the Hsp60-VLPs could the anti-Hsp60 autoantibodies be detected above the background of the plate and negative controls in healthy mice sera. Therefore, the high ability of Hsp60-VLPs to detect autoantibody levels, undetectable by conventional procedures such as ELISA, was demonstrated. This result indicates that VLPs can be used as a tool to detect variations in the levels of anti-Hsp60 autoantibodies in mice sera when the traditional methodologies based on the complete Hsp60 protein cannot be detected, even at basal autoantibody levels in non-inflamed mice.

3.4. Hsp60-VLPs for Autoantibody Detection in DSS-Induced Colitis

Due to the proven ability of Hsp60-VLPs to detect low levels of autoantibodies in the sera samples, potential diagnostic applications were assessed. Since Hsp60 is deeply involved in inflammatory processes [23,25,47], a murine model of the inflammatory bowel disease (IBD) induced by DSS was selected for further studies [20,21]. DSS-induced colitis is a favorite model for chemically induced colitis because it is a simple one and highly similar to human IBD. DSS induces intestinal inflammation through a yet non-deciphered mechanism, although it probably results from damage to the epithelial monolayer lining the large intestine. This lesion would allow dissemination of proinflammatory intestinal contents (e.g., bacteria and their products) into the tissue underneath [20].

The evolution of Hsp60 autoantibody levels in the inflammatory model induced by DSS was studied. The progress of the pathology, as well as the recovery after removing the chemical agent, was determined by monitoring the body weight and granulocytes levels in peripheral blood, as shown in Figure 4.

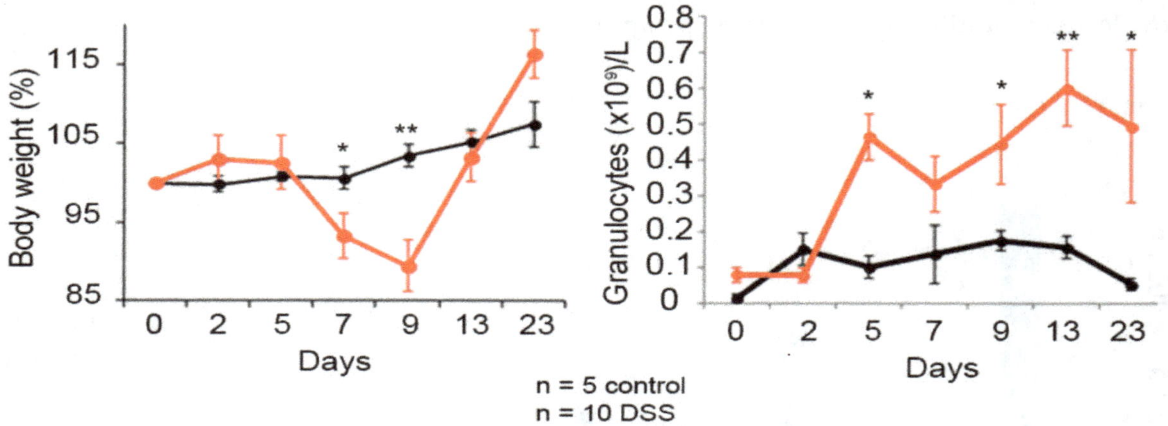

Figure 4. Colitis status of mice treated with dextran sodium sulfate (DSS) in the drinking water. Assessment of the pathology progression by determination of body weight loss and granulocyte levels in peripheral blood. Healthy mice are shown in black and the group of mice treated with DSS is shown in red. The mean and the standard error of the mean are shown, as well as those significant results according to the nonparametric Mann–Whitney U test (* $p \leq 0.05$ and ** $p \leq 0.01$).

The results show a significant decrease in body weight of mice treated with DSS, significant at day 7 and 9. This decrease continues until the DSS was withdrawn, after which the mice recovered weight and reverted to a non-pathological state. Regarding granulocyte levels, there was an increase in blood, in this case significantly at 5 days. In this case the reversal begins later, several days after withdrawing DSS.

To evaluate the potential of Hsp60-VLPs as a novel tool in the knowledge of the inflammatory pathology, an ELISA was performed. Autoantibody levels measured with the complete Hsp60 protein were not detectable, as shown in Figure S5, so only data obtained with Hsp60-VLPs and VLPs-wt were used to assess autoantibody levels in sera of the mice diluted 100-fold. Data were normalized, eliminating the signal attributable to the background which could be originated by the viral component of nanoparticles, and the results were analyzed, as shown in Figure 5.

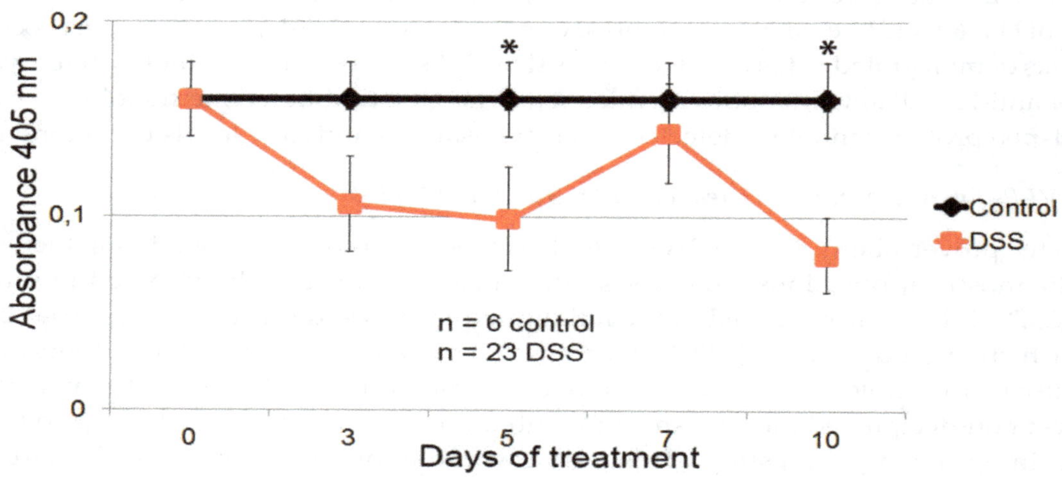

Figure 5. Detection of anti-HSP60 autoantibody in the sera of the healthy and DSS-colitic mice. Healthy mice are represented in black, and in red are mice treated with DSS in the drinking water for 7 days. The levels of anti-HSP60 autoantibodies in the control group has been defined as the average value of all the mice in the control group, including the values of the DSS-treated colitic mice at time 0. Mean and standard error of the mean are shown, as well as those significant results according to the Mann–Whitney U nonparametric test (* $p \leq 0.05$).

The results showed a marked decrease in anti-Hsp60 autoantibody levels following the progress of inflammation at 5 and 10 days (significant according to the Mann–Whitney U nonparametric test), maintaining higher levels in healthy individuals and the basal levels at day 0 in the DSS group. These results demonstrate TuMV functionalized VNPs as a hypersensitive tool in the progression analysis of the inflammation, being able to detect levels of autoantibodies in cases where conventional methodology is not capable.

4. Discussion

In the diagnosis of autoimmune pathologies, current techniques face two main problems, especially when diagnosis is made by detecting specific autoantibodies in serum—background and low autoantibody levels. The high background is due to numerous substances present in serum, and low autoantibody levels occur at early stages of pathology development, a time in which a proper diagnosis would allow early treatment, and a better patient prognosis [48,49]. This is in addition to problems associated with the use of biological samples, such as peripheral blood sera. There is also an added problem when specific epitopes are needed as the basis of a detection system. These epitopes, small peptides, are unable to adhere efficiently to test plates [50], thus leading to very low sensitivity compared to systems deploying high molecular weight proteins in which the handicap is the presence of numerous epitopes, not all of them with diagnostic value. All these inconveniences in conventional methodologies make it increasingly convenient to develop new approaches for a highly sensitive technique, capable of detecting very low levels of autoantibody, or rather small changes, in diluted sera.

To address these challenges, the use of nanoparticles as a multimeric presentation system is an alternative approach, which should provide high sensitivity and solve the epitope–adhesion problem [6,9,51,52]. We have previously shown that functionalized nanoparticles derived from TuMV significantly increased antibody detection with respect to conventional systems [16,17,46]. Although this system has not been used in the diagnosis, progression, and/or prognosis of specific pathologies yet, its potential makes it a good candidate for the design of a hypersensitive analysis system by the detection of serum autoantibodies.

To test TuMV VNPs as a new tool for autoantibody detection, Hsp60 was chosen as an antigen. This protein is involved in numerous inflammatory mechanisms, also related to autoimmune pathologies [27,30,32,33,35,38,40]. Regarding Hsp60 specific commercial antibodies, sensitivity tests were made by comparing the Hsp60-VLPs' detection capacity with respect to the complete recombinant Hsp60 protein. The results showed that, despite not showing a greater sensitivity in polyclonal antibody detection as shown in Figure S1, VLPs sensed epitope-specific monoclonal antibodies with greater sensitivity, as shown in Figure 2. This may be due not only to multimeric presentation in the nanoparticle, but also to a better epitope-exposure on the VLP surface with respect to the complete Hsp60 protein.

According to these results, and because this peptide was chosen due to its diagnostic potential in autoimmune pathologies [49], an evaluation of its efficacy in IBD diagnosis in a DSS-induced colitis murine model [20,21] was decided. In the sera of these mice, nanoparticles exhibited greater detection power not only compared to the free peptide, but also to the complete Hsp60, which was not able to detect autoantibodies under the same conditions. As mentioned above, the enhanced sensitivity of the Hsp60-VLPs could be provided not only because of a multimeric presentation of the chosen epitope, but also because of its exposure on the particle surface since the epitope may have a more internal location in the complete Hsp60 but be exposed on the surface in VLPs. This approach should allow the development of sets of designed nanoparticles for each disease, selecting peptides with diagnostic value present in various autoantigens, and presenting them on the same multifunctionalized nanoparticle [15], improving diagnosis, progression, and prognosis of autoimmune diseases.

Thanks to the high sensitivity shown by Hsp60-VLPs, it was possible to evaluate the evolution of autoantibody levels with disease progress in a murine model. In this case, and against expectations, autoantibody levels decreased concomitantly with the inflammatory process, as shown in Figure 5, suggesting a new view regarding the mechanism of action of Hsp60 (and its autoantibodies) in the inflammatory process. According to these results, anti-Hsp60 autoantibodies could be considered as biomarkers of a non-pathological inflammation state, participating in immune homeostasis as "immunomodulators", recently called "immunculus" ("immune homunculus"), as it was already described for other molecules and autoantibodies [36,39,53]. According to this model, and taking into account studies relating a decrease in autoantibody levels with autoimmune disease development [25,26,32,41,54,55], it seems that autoantibodies in healthy individuals may play a role as a mechanism to regulate inflammation, avoiding a pathological inflammatory response caused by Hsp60. This protein, and peptides derived thereof, have been implicated in pro- and anti-inflammatory processes, and a model has been previously proposed [30]. The results presented in this study allow us to extend tentatively this model, involving now a role for Hsp60 autoantibodies in the progression of the inflammation. A general view of such an extended model is presented in Figure 6. The presence of large amounts of Hsp60 protein (and/or other similar chaperonins present in intestinal microbiota) would be related to an activation of the inflammatory response, triggering the production of autoantibodies. These autoantibodies would act as neutralizers, regulating protein levels, and thereby deactivating the inflammatory response and even activating new anti-inflammatory pathways. The regulatory balance of these pathways in both directions would be essential to maintain a non-pathological physiological state. This proposed process would imply that, when an inflammatory process is induced by external agents, as occurs in experimental colitis induced by DSS, inflammation would be accompanied by a decrease in the levels of regulatory autoantibodies.

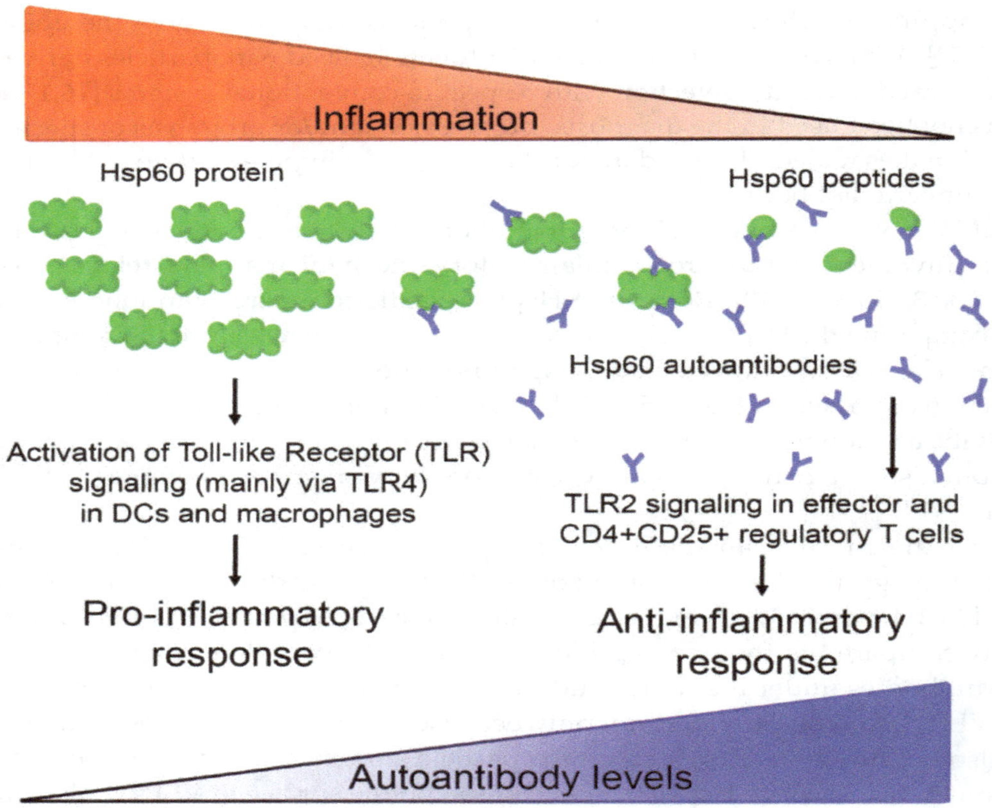

Figure 6. A possible Hsp60 autoantibody regulation system based on a previously described model [30]. High Hsp60 protein levels, related with endogenous proteins and/or similar chaperonins present in microbiota or other pathogens, would activate inflammation pathways, which in turn would induce the production of autoantibodies against Hsp60, neutralizing the antigen, and generating different peptides that activate an anti-inflammatory response.

All the ideas raised would place Hsp60 as a study target. In this case, TuMV VLPs allowed the identification of possible markers in inflammatory pathologies, being able to generate diagnostic systems of high sensitivity and specificity, useful to identify new diagnostic and therapeutic targets. In addition, considering cases such as Hsp60, where autoantibodies can act as regulatory agents, functionalized VLPs could be used not only as a diagnostic tool, but also in therapy for immunization, where the great immunogenic power of VNPs from TuMV has been demonstrated [17]. This would be related to other studies connecting autoantibodies and inflammation, in which immunization with the protein in patients induced symptomatic amelioration [25,41,56]. This work demonstrates that VNPs are a good alternative in the biomedical field, acting as a functionalizable platform for different applications, from diagnosis to treatment. In this context, exploring the possibility of using several epitopes from Hsp60 (or even from other new autoantigens) in multifunctional nanoparticles, would open the door to new theranostic tools to improve prognosis of patients with autoimmune pathologies.

5. Conclusions

This study lays the groundwork for the development of new theranostic tools, based on viral nanoparticles functionalized with specific epitopes involved in autoimmune pathologies, allowing the creation of new systems for autoantibody-based specialized diagnosis and for targeted treatments.

Author Contributions: Conceptualization, F.P., F.S., and M.I.G.; methodology, C.Y.-C., L.Z., C.F.C.-F., M.L.-S., and F.S.; software, C.Y.-C. and M.L.-S.; validation, C.Y.-C. and M.L.-S; formal analysis, C.Y.-C., M.L.-S., F.S., M.I.G., and F.P.; writing—original draft preparation, C.Y.-C.; writing—review and editing, M.L.-S, M.I.G., and F.P.

Acknowledgments: We thank George Lomonossoff for providing the pEAQ vectors.

References

1. Steele, J.F.C.; Peyret, H.; Saunders, K.; Castells-Graells, R.; Marsian, J.; Meshcheriakova, Y.; Lomonossoff, G.P. Synthetic plant virology for nanobiotechnology and nanomedicine. *Wiley Interdiscip. Rev. Nanomed. Nanobiotechnol.* **2017**, *9*, e1447. [CrossRef] [PubMed]
2. Rohovie, M.J.; Nagasawa, M.; Swartz, J.R. Virus-like particles: Next-generation nanoparticles for targeted therapeutic delivery. *Bioeng. Transl. Med.* **2017**, *2*, 43–57. [CrossRef] [PubMed]
3. Lee, K.L.; Twyman, R.M.; Fiering, S.; Steinmetz, N.F. Virus-based nanoparticles as platform technologies for modern vaccines. *Wiley Interdiscip. Rev. Nanomed. Nanobiotechnol.* **2016**, *8*, 554–578. [CrossRef] [PubMed]
4. Schwarz, B.; Douglas, T. Development of virus-like particles for diagnostic and prophylactic biomedical applications. *Wiley Interdiscip. Rev. Nanomed. Nanobiotechnol.* **2015**, *7*, 722–735. [CrossRef] [PubMed]
5. Lomonossoff, G.P.; Evans, D.J. Applications of plant viruses in bionanotechnology. *Curr. Top. Microbiol. Immunol.* **2014**, *375*, 61–87. [CrossRef] [PubMed]
6. Glasgow, J.; Tullman-Ercek, D. Production and applications of engineered viral capsids. *Appl. Microbiol. Biotechnol.* **2014**, *98*, 5847–5858. [CrossRef] [PubMed]
7. Chen, Q.; Lai, H. Plant-derived virus-like particles as vaccines. *Hum. Vaccines Immunother.* **2013**, *9*, 26–49. [CrossRef] [PubMed]
8. Alonso, J.M.; Gorzny, M.L.; Bittner, A.M. The physics of tobacco mosaic virus and virus-based devices in biotechnology. *Trends Biotechnol.* **2013**, *31*, 530–538. [CrossRef]
9. Yildiz, I.; Shukla, S.; Steinmetz, N.F. Applications of viral nanoparticles in medicine. *Curr. Opin. Biotechnol.* **2011**, *22*, 901–908. [CrossRef]
10. Pokorski, J.K.; Steinmetz, N.F. The art of engineering viral nanoparticles. *Mol. Pharm.* **2011**, *8*, 29–43. [CrossRef]
11. Plummer, E.M.; Manchester, M. Viral nanoparticles and virus-like particles: Platforms for contemporary vaccine design. *Wiley Interdiscip. Rev. Nanomed. Nanobiotechnol.* **2011**, *3*, 174–196. [CrossRef] [PubMed]
12. Ludwig, C.; Wagner, R. Virus-like particles-universal molecular toolboxes. *Curr. Opin. Biotechnol.* **2007**, *18*, 537–545. [CrossRef] [PubMed]
13. Grgacic, E.V.; Anderson, D.A. Virus-like particles: Passport to immune recognition. *Methods* **2006**, *40*, 60–65. [CrossRef] [PubMed]

14. Cuenca, S.; Mansilla, C.; Aguado, M.; Yuste-Calvo, C.; Sánchez, F.; Sánchez-Montero, J.M.; Ponz, F. Nanonets derived from *Turnip mosaic virus* as scaffolds for increased enzymatic activity of immobilized *Candida antarctica* lipase B. *Front. Plant Sci.* **2016**, *7*, 464. [CrossRef] [PubMed]
15. Yuste-Calvo, C.; González-Gamboa, I.; Pacios, L.F.; Sánchez, F.; Ponz, F. Structure-based multifunctionalization of flexuous elongated viral nanoparticles. *ACS Omega* **2019**, *4*, 5019–5028. [CrossRef]
16. González-Gamboa, I.; Manrique, P.; Sánchez, F.; Ponz, F. Plant-made potyvirus-like particles used for log-increasing antibody sensing capacity. *J. Biotechnol.* **2017**, *254*, 17–24. [CrossRef] [PubMed]
17. Sánchez, F.; Sáez, M.; Lunello, P.; Ponz, F. Plant viral elongated nanoparticles modified for log-increases of foreign peptide immunogenicity and specific antibody detection. *J. Biotechnol.* **2013**, *168*, 409–415. [CrossRef] [PubMed]
18. Lemmens, B.; De Hertogh, G.; Sagaert, X. Inflammatory Bowel Diseases. In *Pathobiology of Human Disease*; McManus, L.M., Mitchell, R.N., Eds.; Academic Press: San Diego, CA, USA, 2014; Volume 19, pp. 1297–1304.
19. Denmark, V.K.; Mayer, L. Chapter 60—Inflammatory Bowel Diseases. In *The Autoimmune Diseases*, 5th ed.; Rose, N.R., Mackay, I.R., Eds.; Academic Press: Boston, MA, USA, 2014; pp. 873–888.
20. Eichele, D.D.; Kharbanda, K.K. Dextran sodium sulfate colitis murine model: An indispensable tool for advancing our understanding of inflammatory bowel diseases pathogenesis. *World J. Gastroenterol.* **2017**, *23*, 6016–6029. [CrossRef]
21. Chassaing, B.; Aitken, J.D.; Malleshappa, M.; Vijay-Kumar, M. Dextran sulfate sodium (DSS)-induced colitis in mice. *Curr. Protoc. Immunol.* **2014**, *104*, 15–25. [CrossRef]
22. Van Eden, W.; Jansen, M.A.A.; Ludwig, I.S.; Leufkens, P.; van der Goes, M.C.; van Laar, J.M.; Broere, F. Heat shock proteins can be surrogate autoantigens for induction of antigen specific therapeutic tolerance in rheumatoid arthritis. *Front. Immunol.* **2019**, *10*, 279. [CrossRef]
23. Mantej, J.; Polasik, K.; Piotrowska, E.; Tukaj, S. Autoantibodies to heat shock proteins 60, 70, and 90 in patients with rheumatoid arthritis. *Cell Stress Chaperones* **2019**, *24*, 283–287. [CrossRef] [PubMed]
24. Gonciarz, W.; Matusiak, A.; Rudnicka, K.; Rechcinski, T.; Chalubinski, M.; Czkwianianc, E.; Broncel, M.; Gajewski, A.; Chmiela, M. Autoantibodies to a specific peptide epitope of human Hsp60 (ATVLA) with homology to *Helicobacter pylori* HspB in *H. pylori*-infected patients. *APMIS* **2019**, *127*, 139–149. [CrossRef] [PubMed]
25. Ulmansky, R.; Naparstek, Y. Protective antibodies against HSP60 for autoimmune inflammatory diseases. *Clin. Immunol.* **2018**, *186*, 63. [CrossRef] [PubMed]
26. Ponnusamy, T.; Venkatachala, S.K.; Ramanjappa, M.; Kakkar, V.V.; Mundkur, L.A. Inverse association of ApoB and HSP60 antibodies with coronary artery disease in Indian population. *Heart Asia* **2018**, *10*, e011018. [CrossRef] [PubMed]
27. Selli, M.E.; Wick, G.; Wraith, D.C.; Newby, A.C. Autoimmunity to HSP60 during diet induced obesity in mice. *Int. J. Obes. (Lond.)* **2017**, *41*, 348–351. [CrossRef] [PubMed]
28. Jang, E.J.; Jung, K.Y.; Hwang, E.; Jang, Y.J. Characterization of human anti-heat shock protein 60 monoclonal autoantibody Fab fragments in atherosclerosis: Genetic and functional analysis. *Mol. Immunol.* **2013**, *54*, 338–346. [CrossRef] [PubMed]
29. Kimura, A.; Sakurai, T.; Yamada, M.; Koumura, A.; Hayashi, Y.; Tanaka, Y.; Hozumi, I.; Takemura, M.; Seishima, M.; Inuzuka, T. Elevated anti-heat shock protein 60 antibody titer is related to white matter hyperintensities. *J. Stroke Cerebrovasc. Dis.* **2012**, *21*, 305–309. [CrossRef]
30. Quintana, F.J.; Cohen, I.R. The HSP60 immune system network. *Trends Immunol.* **2011**, *32*, 89–95. [CrossRef]
31. Khalil, A.A.; Kabapy, N.F.; Deraz, S.F.; Smith, C. Heat shock proteins in oncology: Diagnostic biomarkers or therapeutic targets? *Biochim. Biophys. Acta* **2011**, *1816*, 89–104. [CrossRef]
32. Cappello, F.; Mazzola, M.; Jurjus, A.; Zeenny, M.N.; Jurjus, R.; Carini, F.; Leone, A.; Bonaventura, G.; Tomasello, G.; Bucchieri, F.; et al. Hsp60 as a novel target in IBD management: A prospect. *Front. Pharmacol.* **2019**, *10*, 26. [CrossRef]
33. Tomasello, G.; Sciume, C.; Rappa, F.; Rodolico, V.; Zerilli, M.; Martorana, A.; Cicero, G.; De Luca, R.; Damiani, P.; Accardo, F.M.; et al. Hsp10, Hsp70, and Hsp90 immunohistochemical levels change in ulcerative colitis after therapy. *Eur. J. Histochem.* **2011**, *55*, e38. [CrossRef] [PubMed]
34. Tomasello, G.; Rodolico, V.; Zerilli, M.; Martorana, A.; Bucchieri, F.; Pitruzzella, A.; Marino Gammazza, A.; David, S.; Rappa, F.; Zummo, G.; et al. Changes in immunohistochemical levels and subcellular localization after therapy and correlation and colocalization with CD68 suggest a pathogenetic role of Hsp60 in ulcerative colitis. *Appl. Immunohistochem. Mol. Morphol.* **2011**, *19*, 552–561. [CrossRef] [PubMed]

35. Rodolico, V.; Tomasello, G.; Zerilli, M.; Martorana, A.; Pitruzzella, A.; Gammazza, A.M.; David, S.; Zummo, G.; Damiani, P.; Accomando, S.; et al. Hsp60 and Hsp10 increase in colon mucosa of Crohn's disease and ulcerative colitis. *Cell Stress Chaperones* **2010**, *15*, 877–884. [CrossRef] [PubMed]
36. Kolinski, T.; Marek-Trzonkowska, N.; Trzonkowski, P.; Siebert, J. Heat shock proteins (HSPs) in the homeostasis of regulatory T cells (Tregs). *Cent. Eur. J. Immunol.* **2016**, *41*, 317–323. [CrossRef] [PubMed]
37. Alard, J.E.; Pers, J.O.; Youinou, P.; Jamin, C. Chapter 41—Heat shock protein autoantibodies. In *Autoantibodies*, 3rd ed.; Shoenfeld, Y., Meroni, P.L., Gershwin, M.E., Eds.; Elsevier: San Diego, CA, USA, 2014; pp. 343–348.
38. Grundtman, C.; Kreutmayer, S.B.; Almanzar, G.; Wick, M.C.; Wick, G. Heat shock protein 60 and immune inflammatory responses in atherosclerosis. *Arterioscler. Thromb. Vasc. Biol.* **2011**, *31*, 960–968. [CrossRef] [PubMed]
39. Habich, C.; Burkart, V. Heat shock protein 60: Regulatory role on innate immune cells. *Cell. Mol. Life Sci.* **2007**, *64*, 742–751. [CrossRef] [PubMed]
40. Pinar, O.; Ozden, Y.A.; Omur, E.; Muhtesem, G. Heat shock proteins in multiple sclerosis. In *Multiple Sclerosis: Bench to Bedside: Global Perspectives on a Silent Killer*; Asea, A.A.A., Geraci, F., Kaur, P., Eds.; Springer International Publishing Ag: Champaign, IL, USA, 2017; Volume 958, pp. 29–42.
41. Ulmansky, R.; Landstein, D.; Moallem, E.; Loeb, V.; Levin, A.; Meyuhas, R.; Katzavian, G.; Yair, S.; Naparstek, Y. A humanized monoclonal antibody against Heat Shock Protein 60 suppresses murine arthritis and colitis and skews the cytokine balance toward an anti-inflammatory response. *J. Immunol.* **2015**, *194*, 5103–5109. [CrossRef] [PubMed]
42. Wu, C.T.; Ou, L.S.; Yeh, K.W.; Lee, W.I.; Huang, J.L. Serum heat shock protein 60 can predict remission of flare-up in juvenile idiopathic arthritis. *Clin. Rheumatol.* **2011**, *30*, 959–965. [CrossRef] [PubMed]
43. Touriño, A.; Sánchez, F.; Fereres, A.; Ponz, F. High expression of foreign proteins from a biosafe viral vector derived from *Turnip mosaic virus*. *Span. J. Agric. Res.* **2008**, *6*, 48–58. [CrossRef]
44. D'Aoust, M.A.; Lavoie, P.O.; Couture, M.M.; Trepanier, S.; Guay, J.M.; Dargis, M.; Mongrand, S.; Landry, N.; Ward, B.J.; Vezina, L.P. Influenza virus-like particles produced by transient expression in *Nicotiana benthamiana* induce a protective immune response against a lethal viral challenge in mice. *Plant Biotechnol. J.* **2008**, *6*, 930–940. [CrossRef] [PubMed]
45. Sainsbury, F.; Liu, L.; Lomonossoff, G.P. Cowpea mosaic virus-based systems for the expression of antigens and antibodies in plants. *Methods Mol. Biol.* **2009**, *483*, 25–39. [CrossRef] [PubMed]
46. Duval, F.; Cruz-Vega, D.E.; González-Gamboa, I.; González-Garza, M.T.; Ponz, F.; Sánchez, F.; Alarcon-Galvan, G.; Moreno-Cuevas, J.E. Detection of autoantibodies to vascular endothelial growth factor receptor-3 in bile duct ligated rats and correlations with a panel of traditional markers of liver diseases. *Dis. Markers* **2016**, *2016*. [CrossRef] [PubMed]
47. Van Eden, W.; van Herwijnen, M.; Wagenaar, J.; van Kooten, P.; Broere, F.; van der Zee, R. Stress proteins are used by the immune system for cognate interactions with anti-inflammatory regulatory T cells. *FEBS Lett.* **2013**, *587*, 1951–1958. [CrossRef] [PubMed]
48. Gecse, K.B.; Vermeire, S. Differential diagnosis of inflammatory bowel disease: Imitations and complications. *Lancet Gastroenterol. Hepatol.* **2018**, *3*, 644–653. [CrossRef]
49. Quintana, F.J.; Farez, M.F.; Viglietta, V.; Iglesias, A.H.; Merbl, Y.; Izquierdo, G.; Lucas, M.; Basso, A.S.; Khoury, S.J.; Lucchinetti, C.F.; et al. Antigen microarrays identify unique serum autoantibody signatures in clinical and pathologic subtypes of multiple sclerosis. *Proc. Natl. Acad. Sci. USA* **2008**, *105*, 18889–18894. [CrossRef] [PubMed]
50. Antimisiaris, G.S. Protein/peptide drug delivery: Advanced drug delivery systems (DDSs). *Biotech Int.* **2010**, *22*, 10–13.
51. Lico, C.; Benvenuto, E.; Baschieri, S. The two-faced *Potato virus X*: From plant pathogen to smart nanoparticle. *Front. Plant Sci.* **2015**, *6*, 1009. [CrossRef]
52. Higa, A.M.; Mambrini, G.P.; Ierich, J.C.M.; Garcia, P.S.; Scramin, J.A.; Peroni, L.A.; Okuda-Shinagawa, N.M.; Machini, M.T.; Trivinho-Strixino, F.; Leite, F.L. Peptide-conjugated silver nanoparticle for autoantibody recognition. *J. Nanosci. Nanotechnol.* **2019**, *19*, 7564–7573. [CrossRef]
53. Poletaev, A.; Osipenko, L. General network of natural autoantibodies as immunological homunculus (Immunculus). *Autoimmun. Rev.* **2003**, *2*, 264–271. [CrossRef]
54. Cohen, I.R. Autoantibody repertoires, natural biomarkers, and system controllers. *Trends Immunol.* **2013**, *34*, 620–625. [CrossRef]

55. Papuc, E.; Krupski, W.; Kurys-Denis, E.; Rejdak, K. Antibodies against small heat-shock proteins in Alzheimer's disease as a part of natural human immune repertoire or activation of humoral response? *J. Neural Transm. (Vienna)* **2016**, *123*, 455–461. [CrossRef] [PubMed]
56. Zhong, Y.; Tang, H.; Wang, X.; Zeng, Q.; Liu, Y.; Zhao, X.I.; Yu, K.; Shi, H.; Zhu, R.; Mao, X. Intranasal immunization with heat shock protein 60 induces CD4(+) CD25(+) GARP(+) and type 1 regulatory T cells and inhibits early atherosclerosis. *Clin. Exp. Immunol.* **2016**, *183*, 452–468. [CrossRef] [PubMed]

5

Evaluation of Three Morphologically Distinct Virus-Like Particles as Nanocarriers for Convection-Enhanced Drug Delivery to Glioblastoma

Joel A. Finbloom [1,†], Ioana L. Aanei [1,†], Jenna M. Bernard [1,†], Sarah H. Klass [1], Susanna K. Elledge [1], Kenneth Han [1], Tomoko Ozawa [2], Theodore P. Nicolaides [3], Mitchel S. Berger [2] and Matthew B. Francis [1,4,*]

1 Department of Chemistry, University of California, Berkeley, CA 94720, USA; jaf@berkeley.edu (J.A.F.); aanei.ioana@gmail.com (I.L.A.); bernarje@gmail.com (J.M.B.); sklass@berkeley.edu (S.H.K.); susannaelle@berkeley.edu (S.K.E.); kennethhan@berkeley.edu (K.H.)

2 Department of Neurological Surgery, University of California, San Francisco, CA 94158, USA; tomoko.ozawa@ucsf.edu (T.O.); Mitchel.Berger@ucsf.edu (M.S.B.)

3 Department of Pediatrics, NYU Langone Medical Center, New York, NY 10016, USA; Theodore.nicolaides@nyumc.org

4 Materials Sciences Division, Lawrence Berkeley National Laboratories, Berkeley, CA 94720, USA

* Correspondence: mbfrancis@berkeley.edu

† These authors contributed equally to this work.

Abstract: Glioblastoma is a particularly challenging cancer, as there are currently limited options for treatment. New delivery routes are being explored, including direct intratumoral injection via convection-enhanced delivery (CED). While promising, convection-enhanced delivery of traditional chemotherapeutics such as doxorubicin (DOX) has seen limited success. Several studies have demonstrated that attaching a drug to polymeric nanoscale materials can improve drug delivery efficacy via CED. We therefore set out to evaluate a panel of morphologically distinct protein nanoparticles for their potential as CED drug delivery vehicles for glioblastoma treatment. The panel consisted of three different virus-like particles (VLPs), MS2 spheres, tobacco mosaic virus (TMV) disks and nanophage filamentous rods modified with DOX. While all three VLPs displayed adequate drug delivery and cell uptake in vitro, increased survival rates were only observed for glioma-bearing mice that were treated via CED with TMV disks and MS2 spheres conjugated to doxorubicin, with TMV-treated mice showing the best response. Importantly, these improved survival rates were observed after only a single VLP–DOX CED injection several orders of magnitude smaller than traditional IV doses. Overall, this study underscores the potential of nanoscale chemotherapeutic CED using virus-like particles and illustrates the need for further studies into how the overall morphology of VLPs influences their drug delivery properties.

Keywords: virus-like particles; glioblastoma; convection-enhanced delivery; tobacco mosaic virus; bioconjugation; doxorubicin; drug delivery; protein-based nanomaterials; viral capsid

---

1. Introduction

Protein-based nanomaterials are a promising class of nanocarriers for drug delivery and diagnostic applications. These materials are formed through the self-assembly of protein monomers into larger nanoscale scaffolds of varying morphologies and properties [1–4]. Protein-based nanomaterials and specifically virus-like particles (VLPs) based on naturally occurring viruses, have demonstrated effectiveness in drug delivery and imaging applications [1–12]. Unlike other synthetic delivery vehicles, such as polymeric micelles and liposomes, VLPs are homogenous in their size distribution and are produced by inexpensive recombinant expression [1,2]. Further, VLPs are degradable in the body and have demonstrated few toxicity issues [1,5]. These protein-based nanomaterials also allow for site-selective modification through amino acid mutagenesis of natural or noncanonical amino acids into the protein backbone [6,13–15]. This site-selective conjugation allows for greater control over the location and amount of cargo loaded onto the VLPs, which can have significant effects on cancer targeting and delivery efficiencies.

Glioblastoma multiforme (GBM) is one of the deadliest and hardest to treat cancers, with over 17,000 new diagnoses per year [16]. While traditional small molecule chemotherapeutic approaches have largely failed to treat GBM, many nanomaterial-based approaches are being developed to enhance GBM treatment [16–20], although the reports of such approaches with virus-like particles is limited [21]. Alternative delivery routes are also being investigated for GBM treatment. Particularly, intratumoral injection of chemotherapeutics via convection-enhanced delivery has seen success in enhancing GBM treatment [22–24].

A constant pressure is maintained during the injection through a microfluidic pump that creates a fluid convection to facilitate a homogenous diffusion of the drug molecule throughout the targeted area [22]. After CED infusion, it is important for the chemotherapeutic agent to remain in the tumor tissue. We therefore hypothesized that VLP nanocarriers might be retained better in the tumor tissue after CED, as has been observed for polymeric constructs [22,25]. Additionally, the differing morphologies of VLPs could influence their tumor treatment efficacies in complex ways. This may be particularly important when developing drug delivery systems for glioblastoma treatment, as gliomas are known to form nanodimensional (50–200 nm) pores in their associated vasculature [26,27]. These pores may critically influence nanoparticle extravasation in a morphologically-dependent manner. Intriguingly, one report has noted differences in the pore sizes of malignant and benign glioma [27].

To determine the efficacies of VLPs as nanocarriers for GBM treatment, a panel of distinct VLPs was evaluated for their drug delivery potential. This panel consisted of a 27 nm MS2 sphere, an 18 nm tobacco mosaic virus (TMV) disk and a 50 nm nanophage filamentous rod (Figure 1). MS2 VLPs have been widely used by our lab and others as drug delivery and diagnostic nanocarriers, as MS2 has a series of 2 nm pores along the surface that allow for interior capsid modification and subsequent drug release [2,6,13,28,29]. While MS2 has been extensively studied, it has not been tested for glioblastoma drug delivery. We recently reported a stable nanodisk composed of a double-arginine mutant of the tobacco mosaic virus [30]. These TMV disks maintained their structural assembly within all biologically-relevant conditions tested, which is uncommon for recombinantly expressed TMV mutants that lack their genomic material [31,32].

The TMV disks were further functionalized with chemotherapeutic cargo to showcase their potential as drug delivery vehicles, as disk-shaped nanomaterials have shown promise in other studies for their enhanced tumor accumulation and cell penetration [33,34]. The last member of the VLP panel was a nanoscale variation of the filamentous fd phage. The nanophage (NP) VLP was recently reported as a new potential nanocarrier [35] but has not previously been functionalized and evaluated for its drug delivery potential. This panel therefore encompasses disparate VLP morphologies, while maintaining relatively consistent dimensions of 15–50 nm. All three of the VLPs presented herein were functionalized with chemotherapeutic cargo to evaluate the efficacies of our VLP panel for drug delivery in glioblastoma models.

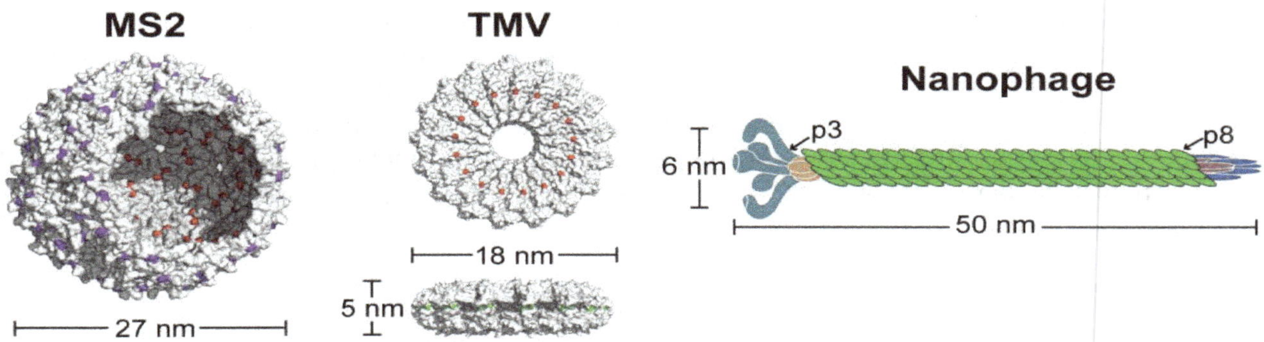

Figure 1. Panel of virus-like particles under evaluation. Three different nanocarriers composed of MS2 spheres, TMV disks and nanophage filamentous rods were tested to determine drug delivery efficacies. Each VLP contains reactive handles for bioconjugation, such as cysteines (**red**), reactive amines (**green**), or noncanonical p-aminophenylalanine moieties (**purple**).

2. Materials and Methods

2.1. Reagents and Instruments

Unless otherwise noted, reagents were purchased from Sigma (St. Louis, MO, USA) and used without further purification. Sulfo-LC-SPDP was purchased from Thermo (Waltham, MA, USA). Water was deionized using the NANOpure purification system (Thermo). *N*-methylpyridinium-4-carboxaldehyde benzenesulfonate hydrate (Rapoport's salt, RS) was obtained from Alfa Aesar (Ward Hill, MA, USA). NAP desalting columns were purchased from GE Healthcare (Marlborough, MA, USA). Spin concentrators with different molecular weight cutoffs (MWCO) were from Millipore (Billerica, MA, USA). Aminophenol-PEG$_{5k}$-OMe [14], alkoxyamine-PEG$_{5k}$-OMe [36] and DOX-EMCH [37] were synthesized as reported previously. EMEM media for cell culture was purchased from ATCC (Manassas, VA, USA). MTS Assay ((3-(4,5-dimethylthiazol-2-yl)-5-(3-carboxymethoxyphenyl)-2-(4-sulfophenyl)-2*H*-tetrazolium)) was purchased from Promega (Madison, WI, USA). Doxorubicin was purchased from Pfizer as a 2 mg/mL stock solution in saline. Liposomal Dox was purchased from SunPharma (Mumbai, India) as a 2 mg/mL solution.

Liquid chromatography mass spectrometry (LC/MS) analysis was performed using acetonitrile (Optima grade, 99.9%, Thermo Fisher), formic acid (99+%, Pierce, Rockford, IL, USA) and dd-H_2O as mobile phase solvents. Electrospray ionization mass spectrometry (ESI-MS) of proteins was performed using an Agilent 1260 series liquid chromatograph outfitted with an Agilent 6224 time-of-flight (TOF) LC/MS system (Santa Clara, CA, USA). The LC was equipped with a Poroshell 300SB-C18 (5 μm particles, 1.0 mm × 75 mm, Agilent) analytical column. Data were collected and analyzed using Agilent MassHunter Qualitative Analysis B.05.00.

Sodium dodecyl sulfate-polyacrylamide gel electrophoresis (SDS-PAGE) was carried out on TRIS gels in a Mini-Protean apparatus from Bio-Rad (Hercules, CA, USA) or on Bis-TRIS gels in a Mini Gel Tank apparatus (Thermo), following the protocol from the manufacturer. The protein electrophoresis samples were heated for 10 min at 95 °C in the presence of β-mercaptoethanol to ensure reduction of any disulfide bonds. Gels were run for 35–60 min at 150–200 V in 2-(*N*-morpholino)ethanesulfonic acid (MES)—SDS buffer to allow good separation of the bands. Commercially available markers (Bio-Rad) were applied to at least one lane of each gel for assignment of apparent molecular masses. Visualization of protein bands was accomplished by staining with Coomassie Brilliant Blue R-250 (Bio-Rad). Quantification of the degree of modification was obtained by imaging on a Gel Doc™ EZ imager (Bio-Rad) and subsequent optical densitometry using the ImageJ (National Institutes of Health, Bethesda, MD, USA) or Image Lab (Bio-Rad) software.

UV-Vis spectrophotometer readings were carried out using a Cary 50 Bio Spectrophotometer (Agilent, Santa Clara, CA, USA) or a NanoDrop 1000 (Thermo Scientific). Analytical size exclusion was performed on an Agilent 1100 series HPLC equipped with a PolySep-GFC-P 5000 column (Phenomenex, Torrance, CA, USA), at a flow rate of 1 mL/min. Incucyte live cell imaging (Essen Bioscience, Ann Arbor, MI, USA) was used to monitor the cellular uptake of DOX-protein conjugates. An IVIS 50 Lumina imaging system (Perkin Elmer, Waltham, MA, USA) was used to measure in vivo and ex vivo bioluminescence.

2.2. Protein Purification and Expression

MS2 and TMV proteins were expressed and purified using previously published methods [15,28,30]. Nanophage was expressed and purified using an adapted method from a previously published report [35]. All proteins were purified using anion exchange chromatography with a diethylaminoethanol (DEAE) Sepharose column followed by size exclusion chromatography. Purity was confirmed by SDS-PAGE, LC/MS and HPLC-SEC. Detailed protocols of nanophage expression and purification are available in the Supporting Information.

2.3. Protein Modification

Protein modification reactions varied depending on the protein and the synthetic cargo. Full modification protocols of each protein and each modification are available in the Supporting Information. Typical modifications involved the use of 10 equiv maleimide, 40 equiv of isothiocyanate, 10–20 equiv of PEG and 1–2 equiv of DOX-EMCH. In the case of nanophage, the p8 monomer of the coat protein was the primary target for bioconjugation. All protein conjugates were purified via elution through a Nap desalting column followed by multiple rounds of spin concentration with a designated molecular weight cut-off (MWCO) that would allow for small-molecule flow-through but prevent protein flow-through. Protein conjugates were characterized by a combination of LC/MS, gel electrophoresis and HPLC-SEC (Supporting Figures S1–S4).

2.4. Cell Culture

U87-MG human glioblastoma cells were obtained from the UC Berkeley Cell Culture Facility. U87-MG human glioblastoma cells bearing the luciferase reporter (U87-Luc) were acquired from UCSF. Cells were cultured in DMEM containing phenol red (ATCC, Manassas, VA, USA) or DMEM without phenol red (Thermo, Waltham, MA, USA) with 10% fetal bovine serum (Omega Scientific, Tarzana, CA, USA) and 1% penicillin/streptomycin (Thermo) at 37 °C and 5% CO_2.

2.5. Cell Viability Assays

U87-MG cells were trypsinized and diluted to a density of 50,000 cells/mL. An aliquot of 100 μL of cell stock was placed in each well of a 96 well plate (Corning, Corning, NY, USA) for a density of 5000 cells/well. The plate was incubated at 37 °C, 5% CO_2 overnight. Following this, media was removed from the plate and 100 μL of appropriate sample stocks media was added. The cells were incubated at 37 °C, 5% CO_2 for 3 d. The media containing the sample was removed from the well and 100 μL of MTS media (20% MTS in media) was added to each well and incubated for 1–3 h. An Infinite 200 Pro plate reader (Tecan, Switzerland) was used to measure the MTS absorbance at 490 nm. Cell viability was calculated as an absorbance percent relative to the untreated cell control. The experiment was performed in triplicate.

2.6. Cell Uptake Studies

Cell uptake studies were performed as previously described [30]. U87-MG cells were trypsinized and diluted to a density of 50,000 cells/mL. An aliquot of 200 μL of cell stock was placed in a well of a 96 well plate (Corning) for a density of 10,000 cells/well. The plate was incubated at 37 °C, 5% CO_2 for

48 h. Following this, media was removed from the plate and 200 μL of appropriate sample stocks (standardized to 1 μM DOX) in phenol free media was added. The cells were incubated at 37 °C, 5% CO_2 for 48 h. Incucyte Zoom Live-Cell Analysis System (EssenBio, Ann Arbor, MI, USA) was used to collect images every hour post incubation. Phase and green fluorescence images at 20× magnification were collected, capitalizing on the intrinsic fluorescence of doxorubicin. The images were processed using Incucyte Zoom proprietary software v.2016A and the Top-Hat background subtraction algorithm (radius 10 μm, threshold 0.5% of green calibration dye signal, GCU) was used to define the boundaries of the cells (green objects). The mean fluorescence intensities of 4 images taken in each well were plotted against the incubation time.

2.7. Tumor Growth and Survival Studies in Glioblastoma Models

All animal procedures were performed according to a protocol approved by the University of California San Francisco Institutional Animal Care and Use Committee (IACUC). Five-week old athymic (nude) female mice weighing 18–23 g were purchased from Simonsen Labs (Gilroy, CA, USA). For tumor inoculation, 3×10^5 U87-Luc glioblastoma cells with luciferase reporter gene were implanted intracranially [38].

Mice bearing U87-Luc intracranial tumors in sets of 9 animals per study group were injected via convection enhanced delivery (CED) with 10 μL of PBS, DOX, Lipo-Dox, or VLP-DOX conjugates in sterile saline, at day 11 post-implantation. CED infusion cannula were made with silica tubing (Polymicro Technologies, Phoenix, AZ, USA) fused to a 0.1 mL syringe (Plastic One, Roanoke, VA, USA) with a 0.5 mm stepped-tip needle that protruded from the silica guide base. Syringes were loaded with the agents and attached to a microinfusion pump (Bioanalytical Systems, Lafayette, IN, USA). The silica cannula attached with a infusion pump was lowered to a 4 mm depth through a skull hole, which was made by skull puncture with a coordination of 3 mm to the right from bregma and just on top of the coronal suture (the same region in the caudate putamen at which tumor cells were injected). The agents were infused at a rate of 1 μL/min until a volume of 10 μL had been delivered. Cannulae were removed 5 min post completion of infusion.

Tumor size was monitored using the luciferase reporter system and measuring bioluminescence on an IVIS 50 Lumina system (Supporting Figure S5). Mice were euthanized when tumor burden reached levels determined by IACUC guidelines. Kaplan Meier survival curves were plotted based on these survival points. Log rank tests were performed on the Kaplan Meier curves to gauge statistical significance. When dividing mice into small and large tumor cohorts, a Bioluminescence Intensity (BLI) of 10^7 was taken as the cutoff, as this was the standard midpoint across groups when assessing tumor size prior to CED injection. Statistical comparison between small and large tumor cohorts was performed using log rank tests of the Kaplan Meier survival curves for each cohort. A full listing of tumor size in all mice that were sacrificed due to tumor burden is available in Supporting Figure S6. Detailed side-by-side survival curves of key cohorts are available in Supporting Figure S7.

3. Results and Discussion

3.1. Chemical Modification of VLPs for Drug Delivery

In order to investigate the potential of our VLP panel as nanocarriers for glioblastoma treatment, we synthesized a series of conjugates with the chemotherapeutic molecule doxorubicin (DOX). Doxorubicin is an anthracycline antibiotic commonly used for the treatment of many types of cancer. However, its systemic toxicity (especially cardiac toxicity) and poor penetration through the blood-brain barrier limit its use in the treatment of brain tumors [16,39]. Several different formulations of doxorubicin have been developed, including PEGylated liposomal doxorubicin (Doxil) [40]. Unfortunately, none of these agents showed activity in clinical trials investigating its use to treat brain cancer [16]. We hypothesized that VLP carriers could enhance the delivery of DOX for glioblastoma treatment. By attaching doxorubicin to the protein scaffolds through acid-labile linkers, the release and distribution of DOX can be controlled to achieve more effective drug release and tumor treatment.

The ketone moiety on DOX was modified via a condensation reaction with *N*-ε-maleimidocaproic acid hydrazide (EMCH), as previously described (Figure S1) [30,37]. The maleimide moiety of DOX–EMCH was reacted with thiol residues on MS2, TMV and Nanophage VLPs to produce VLP–DOX conjugates (Figure 2). This modification was shown to retain the cytotoxic activity of doxorubicin upon hydrolysis of the hydrazone linkage between EMCH and DOX, which releases DOX in its native state. It is anticipated that the DOX will release upon endocytosis of the VLP–DOX conjugates, as the late endosome and subsequent lysosome have internal pH in the range of pH 4.5–5.5, which is sufficient to cleave the hydrazone linkage. For the nanophage, native Lys residues were first converted to thiols using Traut's reagent [41]. The VLPs were further modified with PEG_{5k}, as this is a standard modification for improving biodistribution and minimizing the immune response against VLPs [3,4,28]. Additionally, VLP PEGylation helped improve the solubility of the protein constructs, as DOX is a relatively hydrophobic molecule.

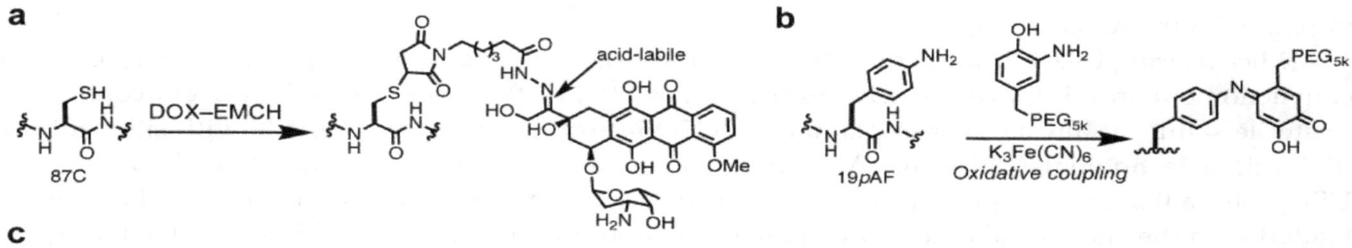

c

Virus-like Particle	DOX attachment site	DOX modification %	PEG attachment site	PEG modification %
MS2 spheres	87C (interior)	35% (63 per capsid)	19pAF (exterior)	70% (126 per capsid)
TMV disks	123C (exterior face)	30% (10 per capsid)	N terminus (exterior edge)	50% (17 per capsid)
Nanophage rods	Lys (exterior)	20% (20 per capsid)	Lys (exterior)	15% (15 per capsid)

Figure 2. Doxorubicin (DOX) and PEG conjugation to virus-like particles for drug delivery studies. (**a**) Bioconjugation scheme for the modification of MS2 with (**a**) DOX–EMCH and (**b**) PEG. DOX–VLP conjugates contain an acid-labile hydrazone linkage, which is anticipated to cleave upon VLP endocytosis and degradation. Modification schemes for TMV and nanophage are available in the Supporting Information. (**c**) Each VLP was modified with DOX–EMCH using maleimide addition to cysteine residues (MS2, TMV) or to thiols synthetically installed onto lysine residues (nanophage) as shown in Figure S1. All VLPs were modified with PEG using either oxidative couplings (MS2, TMV) or NHS ester reactions (nanophage) to improve biodistribution.

3.2. Evaluation of VLP–DOX Conjugates In Vitro

After chemical modification, the VLP–DOX conjugates were evaluated for their effects on cell viability. U87-MG glioblastoma cells were incubated with the agents at different concentrations for 72 h to allow sufficient time for DOX release from the VLP conjugates upon endocytosis and a cell viability assay was performed (Figure 3). All nanocarriers showed a similar dose-response curve and had comparable efficiency compared to the free drug, suggesting that the drug was released and was able to reach the nucleus and induce cell death.

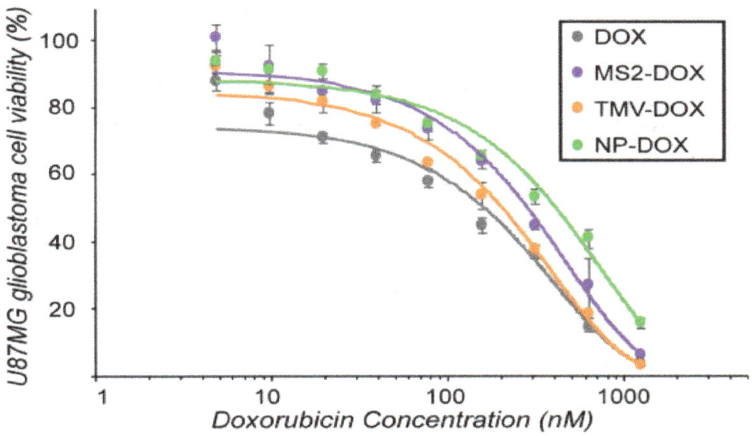

Figure 3. VLP delivery of doxorubicin to U87-MG glioblastoma cells. U87-MG glioblastoma cells were treated with varying amounts of VLP–DOX conjugates for 72 h. Significant cell death was observed for all VLPs tested. Treatment with the VLPs alone displayed no toxicity (data not shown).

In order to understand the uptake kinetics of DOX conjugates in glioblastoma cells, cell uptake was monitored using live cell imaging. Cells were incubated with VLPs bearing a standard dose of 1 μM DOX and images were collected every hour for 48 h to assess the rate of VLP–DOX accumulation. Both MS2–DOX and TMV–DOX displayed fast uptake into cells, while the nanophage–DOX conjugates displayed slower uptake in comparison (Figure 4). This may be due to different cellular uptake efficiencies between nanoparticle morphologies, as has been observed previously [33,42–45]. Overall, despite subtle differences in uptake kinetics and drug delivery efficacies, all three VLPs were amenable to DOX modification and were able to deliver their chemotherapeutic payload in cell culture effectively.

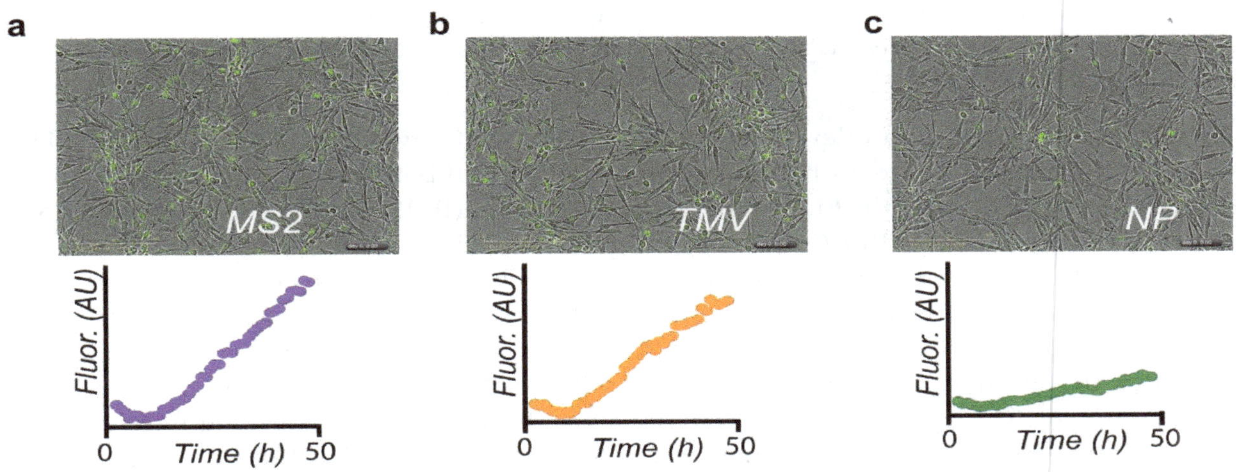

Figure 4. VLP-DOX uptake into U87-MG cells at 48 h. Uptake kinetics of VLP-DOX conjugates into U87-MG glioblastoma cells were monitored in the green channel to detect DOX fluorescence. Data are presented with consistent scaling. Initial intensities are artificially high due to the presence of autofluorescent compounds in the cell media. Upon photobleaching, the FLI drops to more accurate baselines. Both (**a**) MS2 and (**b**) TMV demonstrated significant cellular uptake, while the uptake of (**c**) nanophage (NP) was markedly slower. The TMV data appeared previously in reference [30].

3.3. Convection-Enhanced Delivery of VLP–DOX Conjugates

Each of the three VLP–DOX conjugates was evaluated for its ability to treat glioblastoma after CED injection. VLP-DOX conjugates were compared to PBS, DOX and liposomal DOX (Lipo–DOX) treatment via CED injection 11 days after intracranial tumor implantation (Figure 5). While the payload of DOX per nanocarrier was variable, each glioma-bearing mouse was injected with a standardized DOX payload of 20 μg/kg and tumor size was monitored via bioluminescence intensity (BLI, Figure 5a). The tumor growth of MS2 and TMV-treated mice demonstrated some growth inhibition, although the large variability in tumor reduction within groups led to no statistical difference between treatment groups. Kaplan Meier survival analysis of treated mice suggested improved outcomes for mice treated with TMV–DOX conjugates, as three out of eight mice survived past 40 days, whereas for the PBS-treated group, no mouse survived past 34 days (Figure 5b). This is particularly encouraging, as the CED injected dose of DOX was orders of magnitude lower than the milligrams per kilogram of doxorubicin typically injected intravenously [18,46]. However, log rank analysis of survival curves of TMV and PBS-treated mice did not demonstrate statistical significance between the groups for the full set of animals ($p = 0.052$).

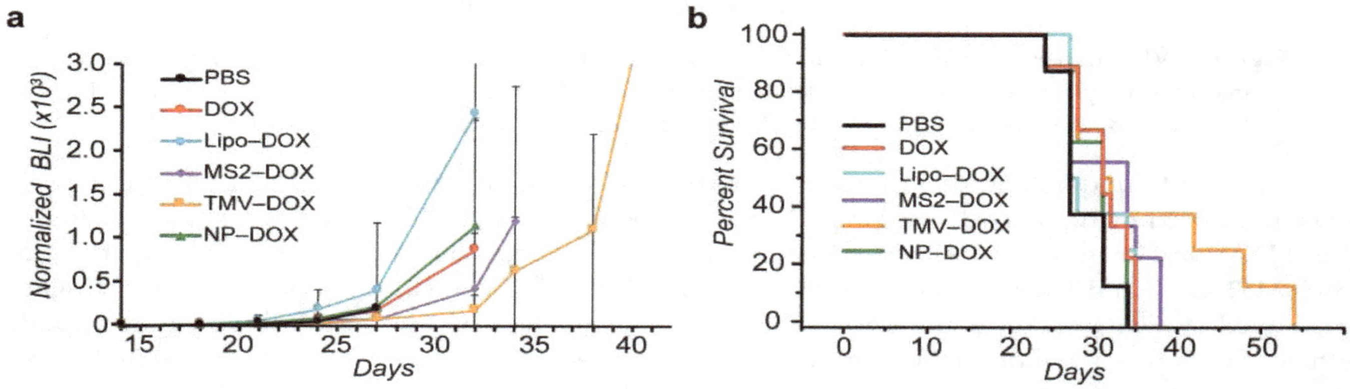

Figure 5. Evaluation of VLP panel for glioblastoma treatment. MS2 spheres, TMV disks and nanophage rods bearing DOX payloads of 20 μg/kg were injected via convection-enhanced delivery infusion (CED) into U87-MG glioma-bearing mice. (**a**) Tumor growth analysis suggested modest tumor growth inhibition from MS2 and TMV-treated groups. (**b**) Kaplan-Meier survival curves of mice with U87-MG glioma. The survival curve results suggest improved efficacy with TMV–DOX treatment.

The glioma-bearing mice under study could be divided into two cohorts: one with small tumors prior to CED injection and one with larger tumors. The cutoff between large and small tumor size was determined by analyzing the bioluminescence intensities (BLI) of treatment groups and arriving at a general midpoint of 10^7 BLI (Supporting Figure S6). This cutoff allowed for relatively even distributions between large and small tumors within each treatment group, with the exception of Lipo–DOX, which only had two large tumors within that group. Analysis of survival times between large and small tumors within treatment groups revealed significant differences for MS2–DOX and TMV–DOX treated groups (Supporting Figure S7), with smaller tumors responding better to the VLP treatment (Figure 6). This analysis reveals that the size of the glioblastoma could significantly influence the drug delivery efficacy of VLPs. This may be due to differences in VLP diffusion within large versus small tumors, or due to vasculature differences between large and small tumors that could influence drug delivery efficacies [47–49]. It is also possible that larger tumors would require higher injected doses to reduce the tumor burden and increase survival times. Further analysis of these tumor size cohorts revealed that both MS2–DOX and TMV–DOX treatments significantly increased the survival times of small tumor-bearing mice, when compared to PBS controls, with TMV–DOX treatment displaying the highest improved survival (Figure 6). Lastly, mice treated with MS2–DOX displayed significantly improved survival rates when compared to DOX treatment alone. Taken together, these findings underscore the potential of VLPs as nanocarriers for GBM treatment via CED, especially for the treatment of smaller tumors and illustrate some of the morphological effects that are at play in VLP drug delivery.

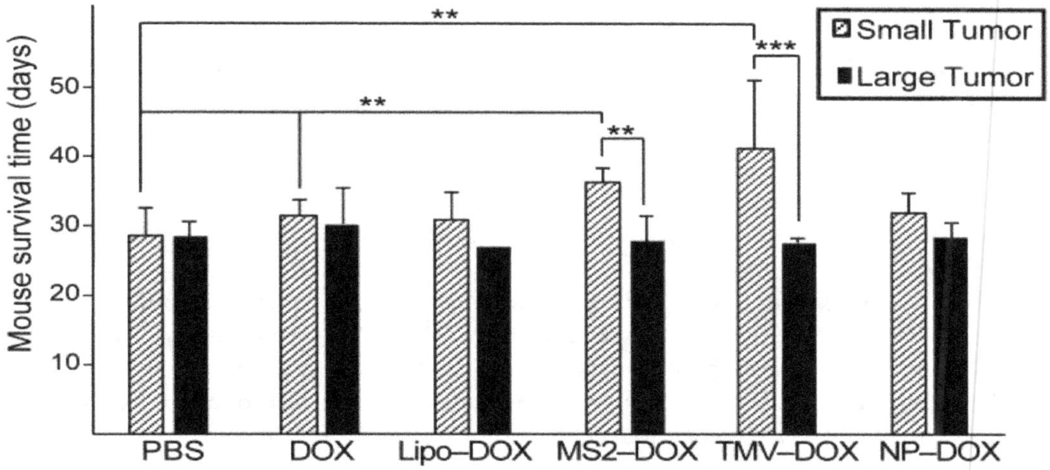

Figure 6. Tumor size dependence of mouse response to VLP treatment. Treatment groups could be divided into two categories based on tumor size as measured by bioluminescence intensity (BLI). Small tumors (as defined by a BLI < 107 AU) showed significantly increased response to CED treatment in the case of TMV and MS2 treated mice when compared to larger tumors (BLI > 107 AU). This BLI cutoff led to a relatively even distribution of mice within each treatment group with the exception of Lipo–DOX, where only two mice had large tumors. Both MS2–DOX and TMV–DOX treatment of small tumors led to significantly enhanced survival when compared to PBS treatment. ** $p < 0.05$. *** $p < 0.01$ as measured by log rank test of survival points. See Supporting Information Figures S5–S7 for the tumor size and survival data for individual animals.

4. Conclusions

We present here an analysis of three morphologically distinct virus-like particles on their ability to deliver drugs to glioblastoma. A panel of virus-like particles representing spheres, disks and filamentous rods was modified with chemotherapeutic cargo and PEGylated via orthogonal bioconjugation strategies. These VLPs were taken up into cancer cells and effectively released their chemotherapeutic payload in cell culture. The VLPs were further tested in glioma-bearing mouse models and showed significant differences in their efficacies for glioblastoma treatment via CED injection. Importantly, glioblastoma-bearing mice treated with TMV-DOX demonstrated improved survival. This is especially encouraging given the low dose of DOX delivered to the mice. Lastly, tumor size dependence of mouse survival was observed for VLP-treated mice, indicating that the tumor environment could have significant implications for GBM treatment via protein-based nanomaterials. Future studies will focus on the evaluation of these VLPs in other tumor models to determine if different tumor types require different VLPs for effective delivery of a chemotherapeutic payload. These studies will help expand our understanding of the effects of virus-like particle morphology on drug delivery, which may have significant implications in the growing field of nanomedicine and help lead to the development of functional protein-based nanomaterials for cancer treatment.

Supplementary Materials: , Figure S1: Modification Scheme for TMV and Nanophage Conjugation, Figure S2: LC/MS characterization of DOX-modified VLPs, Figure S3: HPLC-SEC of Nanophage-PEG_{5k}-DOX, Figure S4: Protein gel characterization of VLP-PEG_{5k}-DOX conjugates, Figure S5: Individual tumor growth plots for each treatment group, Figure S6: Tumor size distribution of glioblastoma-bearing mice and survival points, Figure S7: Kaplan Meier Survival Curves of small and large tumor cohorts.

Author Contributions: J.A.F., I.L.A. and J.M.B. contributed equally toward this work. Conceptualization, J.A.F., I.L.A., J.M.B., S.H.K. and M.B.F.; Formal analysis, J.A.F., I.L.A., J.M.B., S.H.K., T.O., T.P.N. and M.B.F.; Funding acquisition, T.P.N., M.S.B. and M.B.F.; Methodology, J.A.F., I.L.A., J.M.B., S.H.K., S.K.E., K.H. and T.O.; Supervision, T.P.N., M.S.B. and M.B.F.; Writing—original draft, J.A.F., I.L.A., J.M.B. and M.B.F.

Acknowledgments: We thank Youngho Seo for scientific guidance. We thank Raquel Santos and Edgar Lopez Lepe for technical assistance.

References

1. Wen, A.M.; Steinmetz, N.F. Design of virus-based nanomaterials for medicine, biotechnology, and energy. *Chem. Soc. Rev.* **2016**, *45*, 4074–4126. [CrossRef] [PubMed]
2. Dedeo, M.T.; Finley, D.T.; Francis, M.B. Viral capsids as self-assembling templates for new materials. *Prog. Mol. Biol. Transl. Sci.* **2011**, *103*, 353–392. [PubMed]
3. Shukla, S.; Ablack, A.L.; Wen, A.M.; Lee, K.L.; Lewis, J.D.; Steinmetz, N.F. Increased tumor homing and tissue penetration of the filamentous plant viral nanoparticle potato virus X. *Mol. Pharm.* **2013**, *10*, 33–42. [CrossRef] [PubMed]
4. Shukla, S.; Eber, F.J.; Nagarajan, A.S.; DiFranco, N.A.; Schmidt, N.; Wen, A.M.; Eiben, S.; Twyman, R.M.; Wege, C.; Steinmetz, N.F. The Impact of Aspect Ratio on the Biodistribution and Tumor Homing of Rigid Soft-Matter Nanorods. *Adv. Healthc. Mater.* **2015**, *4*, 874–882. [CrossRef] [PubMed]
5. Czapar, A.E.; Steinmetz, N.F. Plant viruses and bacteriophages for drug delivery in medicine and biotechnology. *Curr. Opin. Chem. Biol.* **2017**, *38*, 108–116. [CrossRef] [PubMed]
6. Wu, W.; Hsiao, S.C.; Carrico, Z.M.; Francis, M.B. Genome-free viral capsids as multivalent carriers for taxol delivery. *Angew. Chem. Int. Ed.* **2009**, *48*, 9493–9497. [CrossRef]
7. Qazi, S.; Liepold, L.O.; Abedin, M.J.; Johnson, B.; Prevelige, P.; Frank, J.A.; Douglas, T. P22 Viral Capsids as Nanocomposite High-Relaxivity MRI Contrast Agents. *Mol. Pharm.* **2013**, *10*, 11–17. [CrossRef] [PubMed]
8. Rhee, J.-K.; Baksh, M.; Nycholat, C.; Paulson, J.C.; Kitagishi, H.; Finn, M.G. Glycan-targeted virus-like nanoparticles for photodynamic therapy. *Biomacromolecules* **2012**, *13*, 2333–2338. [CrossRef]
9. Zhao, X.; Chen, L.; Luckanagul, J.A.; Zhang, X.; Lin, Y.; Wang, Q. Enhancing Antibody Response against Small Molecular Hapten with Tobacco Mosaic Virus as a Polyvalent Carrier. *ChemBioChem* **2015**, *16*, 1279–1283. [CrossRef]
10. Czapar, A.E.; Zheng, Y.-R.; Riddell, I.A.; Shukla, S.; Awuah, S.G.; Lippard, S.J.; Steinmetz, N.F. Tobacco mosaic virus delivery of phenanthriplatin for cancer therapy. *ACS Nano* **2016**, *10*, 4119–4126. [CrossRef]
11. Chariou, P.L.; Lee, K.L.; Pokorski, J.K.; Saidel, G.M.; Steinmetz, N.F. Diffusion and Uptake of Tobacco Mosaic Virus as Therapeutic Carrier in Tumor Tissue: Effect of Nanoparticle Aspect Ratio. *J. Phys. Chem. B* **2016**, *120*, 6120–6129. [CrossRef] [PubMed]
12. Douglas, T.; Young, M. Host–guest encapsulation of materials by assembled virus protein cages. *Nature* **1998**, *393*, 152–155. [CrossRef]
13. Witus, L.S.; Francis, M.B. Using Synthetically Modified Proteins to Make New Materials. *Acc. Chem. Res.* **2011**, *44*, 774–783. [CrossRef] [PubMed]
14. Obermeyer, A.C.; Jarman, J.B.; Francis, M.B. N-terminal modification of proteins with o-aminophenols. *J. Am. Chem. Soc.* **2014**, *136*, 9572–9579. [CrossRef] [PubMed]
15. Obermeyer, A.C.; Capehart, S.L.; Jarman, J.B.; Francis, M.B. Multivalent viral capsids with internal cargo for fibrin imaging. *PLoS ONE* **2014**, *9*, e100678. [CrossRef] [PubMed]
16. Omuro, A.; DeAngelis, L.M. Glioblastoma and Other Malignant Gliomas. *JAMA* **2013**, *310*, 1842–1850. [CrossRef]
17. Jensen, S.A.; Day, E.S.; Ko, C.H.; Hurley, L.A.; Luciano, J.P.; Kouri, F.M.; Merkel, T.J.; Luthi, A.J.; Patel, P.C.; Cutler, J.I.; et al. Spherical nucleic acid nanoparticle conjugates as an RNAi-based therapy for glioblastoma. *Sci. Transl. Med.* **2013**, *5*, 209ra152. [CrossRef]
18. Wohlfart, S.; Gelperina, S.; Kreuter, J. Transport of drugs across the blood–brain barrier by nanoparticles. *J. Control. Release* **2012**, *161*, 264–273. [CrossRef]
19. Shatsberg, Z.; Zhang, X.; Ofek, P.; Malhotra, S.; Krivitsky, A.; Scomparin, A.; Tiram, G.; Calderón, M.; Haag, R.; Satchi-Fainaro, R. Functionalized nanogels carrying an anticancer microRNA for glioblastoma therapy. *J. Control. Release* **2016**, *239*, 159–168. [CrossRef]
20. Chen, W.; Zou, Y.; Zhong, Z.; Haag, R. Cyclo(RGD)-Decorated Reduction-Responsive Nanogels Mediate Targeted Chemotherapy of Integrin Overexpressing Human Glioblastoma In Vivo. *Small* **2017**, *13*, 1601997. [CrossRef]
21. Chao, C.-N.; Yang, Y.-H.; Wu, M.-S.; Chou, M.-C.; Fang, C.-Y.; Lin, M.-C.; Tai, C.-K.; Shen, C.-H.; Chen, P.-L.; Chang, D.; et al. Gene therapy for human glioblastoma using neurotropic JC virus-like particles as a gene delivery vector. *Sci. Rep.* **2018**, *8*, 2213. [CrossRef]

22. Allard, E.; Passirani, C.; Benoit, J.-P. Convection-enhanced delivery of nanocarriers for the treatment of brain tumors. *Biomaterials* **2009**, *30*, 2302–2318. [CrossRef] [PubMed]
23. Xi, G.; Robinson, E.; Mania-Farnell, B.; Vanin, E.F.; Shim, K.-W.; Takao, T.; Allender, E.V.; Mayanil, C.S.; Soares, M.B.; Ho, D.; et al. Convection-enhanced delivery of nanodiamond drug delivery platforms for intracranial tumor treatment. *Nanomed. Nanotechnol. Biol. Med.* **2014**, *10*, 381–391. [CrossRef] [PubMed]
24. Yamashita, Y.; Krauze, M.T.; Kawaguchi, T.; Noble, C.O.; Drummond, D.C.; Park, J.W.; Bankiewicz, K.S. Convection-enhanced delivery of a topoisomerase I inhibitor (nanoliposomal topotecan) and a topoisomerase II inhibitor (pegylated liposomal doxorubicin) in intracranial brain tumor xenografts. *Neuro. Oncol.* **2007**, *9*, 20–28. [CrossRef] [PubMed]
25. Zhan, W.; Wang, C.-H. Convection enhanced delivery of liposome encapsulated doxorubicin for brain tumour therapy. *J. Control. Release* **2018**, *285*, 212–229. [CrossRef] [PubMed]
26. Hashizume, H.; Baluk, P.; Morikawa, S.; McLean, J.W.; Thurston, G.; Roberge, S.; Jain, R.K.; McDonald, D.M. Openings between Defective Endothelial Cells Explain Tumor Vessel Leakiness. *Am. J. Pathol.* **2000**, *156*, 1363–1380. [CrossRef]
27. Mittapalli, R.K.; Adkins, C.E.; Bohn, K.A.; Mohammad, A.S.; Lockman, J.A.; Lockman, P.R. Quantitative Fluorescence Microscopy Measures Vascular Pore Size in Primary and Metastatic Brain Tumors. *Cancer Res.* **2017**, *77*, 238–246. [CrossRef]
28. Farkas, M.E.; Aanei, I.L.; Behrens, C.R.; Tong, G.J.; Murphy, S.T.; O'Neil, J.P.; Francis, M.B. PET Imaging and biodistribution of chemically modified bacteriophage MS2. *Mol. Pharm.* **2013**, *10*, 69–76. [CrossRef]
29. Aanei, I.L.; ElSohly, A.M.; Farkas, M.E.; Netirojjanakul, C.; Regan, M.; Taylor Murphy, S.; O'Neil, J.P.; Seo, Y.; Francis, M.B. Biodistribution of Antibody-MS2 Viral Capsid Conjugates in Breast Cancer Models. *Mol. Pharm.* **2016**, *13*, 3764–3772. [CrossRef]
30. Finbloom, J.A.; Han, K.; Aanei, I.L.; Hartman, E.C.; Finley, D.T.; Dedeo, M.T.; Fishman, M.; Downing, K.H.; Francis, M.B. Stable Disk Assemblies of a Tobacco Mosaic Virus Mutant as Nanoscale Scaffolds for Applications in Drug Delivery. *Bioconjug. Chem.* **2016**, *27*, 2480–2485. [CrossRef]
31. Miller, R.A.; Presley, A.D.; Francis, M.B. Self-assembling light-harvesting systems from synthetically modified tobacco mosaic virus coat proteins. *J. Am. Chem. Soc.* **2007**, *129*, 3104–3109. [CrossRef] [PubMed]
32. Dedeo, M.T.; Duderstadt, K.E.; Berger, J.M.; Francis, M.B. Nanoscale protein assemblies from a circular permutant of the tobacco mosaic virus. *Nano Lett.* **2010**, *10*, 181–186. [CrossRef] [PubMed]
33. Gratton, S.E.; Ropp, P.; Pohlhaus, P.D.; Luft, J.C.; Madden, V.J.; Napier, M.E.; DeSimone, J.M. The effect of particle design on cellular internalization pathways. *Proc. Natl. Acad. Sci. USA* **2008**, *105*, 11613–11618. [CrossRef] [PubMed]
34. Adriani, G.; de Tullio, M.D.; Ferrari, M.; Hussain, F.; Pascazio, G.; Liu, X.; Decuzzi, P. The preferential targeting of the diseased microvasculature by disk-like particles. *Biomaterials* **2012**, *33*, 5504–5513. [CrossRef] [PubMed]
35. Sattar, S.; Bennett, N.J.; Wen, W.X.; Guthrie, J.M.; Blackwell, L.F.; Conway, J.F.; Rakonjac, J. Ff-nano, short functionalized nanorods derived from Ff (f1, fd, or M13) filamentous bacteriophage. *Front. Microbiol.* **2015**, *6*, 316. [CrossRef] [PubMed]
36. Schlick, T.L.; Ding, Z.; Kovacs, E.W.; Francis, M.B. Dual-Surface Modification of the Tobacco Mosaic Virus. *J. Am. Chem. Soc.* **2005**, *127*, 3718–3723. [CrossRef] [PubMed]
37. Toita, R.; Murata, M.; Abe, K.; Narahara, S.; Piao, J.S.; Kang, J.-H.; Hashizume, M. A nanocarrier based on a genetically engineered protein cage to deliver doxorubicin to human hepatocellular carcinoma cells. *Chem. Commun.* **2013**, *49*, 7442–7444. [CrossRef]
38. Ozawa, T.; James, C.D. Establishing Intracranial Brain Tumor Xenografts With Subsequent Analysis of Tumor Growth and Response to Therapy using Bioluminescence Imaging. *J. Vis. Exp.* **2010**, e1986. [CrossRef]
39. Pardridge, W.M. The blood-brain barrier: Bottleneck in brain drug development. *NeuroRx* **2005**, *2*, 3–14. [CrossRef]
40. Gabizon, A.; Shmeeda, H.; Barenholz, Y. Pharmacokinetics of Pegylated Liposomal Doxorubicin. *Clin. Pharmacokinet.* **2003**, *42*, 419–436. [CrossRef]
41. Jue, R.; Lambert, J.M.; Pierce, L.R.; Traut, R.R. Addition of sulfhydryl groups of Escherichia coli ribosomes by protein modification with 2-iminothiolane (methyl 4-mercaptobutyrimidate). *Biochemistry* **1978**, *17*, 5399–5406. [CrossRef] [PubMed]

42. Agarwal, R.; Singh, V.; Jurney, P.; Shi, L.; Sreenivasan, S.V.; Roy, K. Mammalian cells preferentially internalize hydrogel nanodiscs over nanorods and use shape-specific uptake mechanisms. *Proc. Natl. Acad. Sci. USA* **2013**, *110*, 17247–17252. [CrossRef] [PubMed]
43. Chithrani, B.D.; Ghazani, A.A.; Chan, W.C. Determining the Size and Shape Dependence of Gold Nanoparticle Uptake into Mammalian Cells. *Nano Lett.* **2006**, *6*, 662–668. [CrossRef] [PubMed]
44. He, C.; Hu, Y.; Yin, L.; Tang, C.; Yin, C. Effects of particle size and surface charge on cellular uptake and biodistribution of polymeric nanoparticles. *Biomaterials* **2010**, *31*, 3657–3666. [CrossRef] [PubMed]
45. Jiang, W.; Kim, B.Y.S.; Rutka, J.T.; Chan, W.C.W. Nanoparticle-mediated cellular response is size-dependent. *Nat. Nanotechnol.* **2008**, *3*, 145–150. [CrossRef] [PubMed]
46. Ulbrich, K.; Etrych, T.; Chytil, P.; Pechar, M.; Jelinkova, M.; Rihova, B. Polymeric anticancer drugs with pH-controlled activation. *Int. J. Pharm.* **2004**, *277*, 63–72. [CrossRef] [PubMed]
47. Holash, J.; Wiegand, S.J.; Yancopoulos, G.D. New model of tumor angiogenesis: Dynamic balance between vessel regression and growth mediated by angiopoietins and VEGF. *Oncogene* **1999**, *18*, 5356–5362. [CrossRef] [PubMed]
48. Alfonso, J.C.L.; Köhn-Luque, A.; Stylianopoulos, T.; Feuerhake, F.; Deutsch, A.; Hatzikirou, H. Why one-size-fits-all vaso-modulatory interventions fail to control glioma invasion: In silico insights. *Sci. Rep.* **2016**, *6*, 37283. [CrossRef] [PubMed]
49. Gevertz, J.L.; Torquato, S. Modeling the effects of vasculature evolution on early brain tumor growth. *J. Theor. Biol.* **2006**, *243*, 517–531. [CrossRef]

6

Calcium Phosphate Nanoparticle-Based Vaccines as a Platform for Improvement of HIV-1 Env Antibody Responses by Intrastructural Help

Dominik Damm [1], Leonardo Rojas-Sánchez [2], Hannah Theobald [1], Viktoriya Sokolova [2], Richard T. Wyatt [3], Klaus Überla [1], Matthias Epple [2] and Vladimir Temchura [1,*]

1 Institute of Clinical and Molecular Virology, Friedrich-Alexander University Erlangen-Nürnberg, 91054 Erlangen, Germany; dominik.damm@uk-erlangen.de (D.D.); hannah.theobald@fau.de (H.T.); klaus.ueberla@fau.de (K.Ü.)

2 Inorganic Chemistry and Center for Nanointegration Duisburg-Essen (CeNIDE), University of Duisburg-Essen, 45141 Essen, Germany; leonardo.rojas-sanchez@uni-due.de (L.R.-S.); viktoriya.sokolova@uni-due.de (V.S.); matthias.epple@uni-due.de (M.E.)

3 Department of Immunology and Microbial Science, The Scripps Research Institute, La Jolla, CA 92037, USA; wyatt@scripps.edu

* Correspondence: vladimir.temchura@fau.de

Abstract: Incorporation of immunodominant T-helper epitopes of licensed vaccines into virus-like particles (VLP) allows to harness T-helper cells induced by the licensed vaccines to provide intrastructural help (ISH) for B-cell responses against the surface proteins of the VLPs. To explore whether ISH could also improve antibody responses to calcium phosphate (CaP) nanoparticle vaccines we loaded the nanoparticle core with a universal T-helper epitope of Tetanus toxoid (p30) and functionalized the surface of CaP nanoparticles with stabilized trimers of the HIV-1 envelope (Env) resulting in Env-CaP-p30 nanoparticles. In contrast to soluble Env trimers, Env containing CaP nanoparticles induced activation of naïve Env-specific B-cells in vitro. Mice previously vaccinated against Tetanus raised stronger humoral immune responses against Env after immunization with Env-CaP-p30 than mice not vaccinated against Tetanus. The enhancing effect of ISH on anti-Env antibody levels was not attended with increased Env-specific IFN-γ CD4 T-cell responses that otherwise may potentially influence the susceptibility to HIV-1 infection. Thus, CaP nanoparticles functionalized with stabilized HIV-1 Env trimers and heterologous T-helper epitopes are able to recruit heterologous T-helper cells induced by a licensed vaccine and improve anti-Env antibody responses by intrastructural help.

Keywords: nano-vaccines; HIV-1 Env trimers; B-cell targeting; intrastructural help

---

1. Introduction

The unique characteristic of HIV-1 to infect activated CD4 T-cells requires alternative vaccination strategies that differ from the classical approaches used [1]. Polyfunctional serum antibodies against envelope glycoprotein (Env) of HIV-1 correlate with spontaneous HIV-1 control by elite controllers [2] and appear to be crucial for vaccine-mediated protection against HIV-1 [3–5]. However, limited breadth, poor persistence of antibody responses to Env, and potential enhancement of susceptibility to HIV infection by vaccine-induced HIV-specific T-helper cell responses compromise HIV-1 vaccines previously designed for clinical trials [6,7]. Recently we demonstrated how T-helper cells induced by a licensed vaccine against Tetanus toxoid (TT) can be harnessed to provide help for Env-specific B-cells in mice immunized with HIV-1 virus-like particles (VLP) containing T-helper epitopes of TT [8]. This "intrastructural help" [9] (ISH) can be explained by uptake of entire VLPs by Env-specific B-cells and subsequent presentation of epitopes from all proteins of the VLPs on their major histocompatibility complex class II (MHC-II) molecules to harness corresponding T-cell help [8,10].

Biodegradable calcium phosphate (CaP) nanoparticles have several advantages compared to biological and polymer-based nanoparticles and have been used for experimental vaccination during the past decade [11]. Previously we designed CaP nanoparticles covered with a model antigen in order to achieve efficient targeting and activation of cognate B-cells in vitro [12] as well as induction of humoral immune responses in vivo [13]. In animal models, where CD4 T-cells were genetically non-reactive to the surface antigen of the CaP nanoparticles, incorporation of the p30 peptide, a promiscuous (universal) T-helper epitope from TT [14], overcame the lack of functionally active CD4 T-cell epitopes [13]. These data indicate that CaP nanoparticle-based vaccines might be potentially used for improvement of HIV-Env antibody responses by the ISH approach.

A gp160 precursor of HIV-1 Env is proteolytically cleaved into non-covalently linked gp120 and gp41 subunits, which assemble into a trimer of heterodimers [15]. Historically, the initial candidate vaccines for immunization with Env-based antigens used monomeric gp120 [16]. Recent advances in design and manufacturing of soluble HIV-1 Env trimers stabilized in the closed pre-fusion state provide an antigenic form of Env that induces Env antibodies recognizing the quaternary conformation associated with neutralizing activity against autologous HIV-1 [15,16].

In this study, we therefore designed CaP nanoparticles with stabilized HIV-1 Env trimers coupled to the surface and universal T-helper peptides of TT in the core. These nanoparticles were used to demonstrate effects of ISH in mice previously immunized with a licensed vaccine against Tetanus toxoid.

2. Materials and Methods

2.1. Mice, Ethical Statement

Six- to eight-week-old female wild-type (wt) C57bl/6NRj (Bl6) mice (Janvier, Le Genest-Saint-Isle, France), as well as mice with transgenic B-cell receptors (BCR) specific for HIV-1 Env (PGT-121 mice [17], in-house breeding, kindly provided by Dr. M. Nussenzweig, The Rockefeller University, New York, NY, USA) were used in this study. Mice were accommodated in the animal facility of the Faculty of Medicine, FAU (Erlangen, Germany), in accordance with the national law and were handled according to instructions of the Federation of European Laboratory Animal Science Associations. All animal experiments were approved by an external ethics committee of the North Rhine-Westphalian Ministry for Nature, Environment and Consumer Protection (license AZ 84-02.04.2014. A191) and confirmed by the Government of Lower Franconia (license 55.2-2532-2-96).

2.2. Plasmids and VLP Production

The plasmid encoding for BG505 NFL2P gp140 [18] was used for expression and purification of stabilized Env trimers (see 2.3). We introduced recombinant transmembrane (TM) and cytoplasmic (CD) domains of the surface protein G from the vesicular stomatitis virus (VSV-G; GenBank Accession Number: GU177825.1) downstream of the sequence encoding for the soluble Env protein to elicit pseudo-typed, membrane-bound Env trimers (BG505 NFL2P gp140-GTMCD). We further introduced the nucleotide sequence ttcaacaacttcaccgttagcttctggctgcgcgttccgaaagtttctgcttcccacctggaa, that encodes for the TT-derived peptide p30 (FNNFTVSFWLRVPKVSASHLE) into the Gag open reading frame of the plasmid Hgpsyn [19] (Hgpsyn-TTp30) that contains the codon-optimized genetic information for the HIV-1 structural precursor protein (Gag) and viral enzymes (Pol). The p30 sequence was inserted between the p17 matrix protein and the p24 capsid protein. VLPs lacking Env (Δ-VLP), VLPs bearing BG505 NFL2P gp140-GTMCD Env trimers on the surface (Env-VLP) and Env-VLP with the Gag-p30 fusion protein inside (Env-VLP-p30) were produced by co-transfection of 293T cells with equal amounts of the corresponding plasmids and purified as described previously [20].

2.3. Production and Analyses of HIV-1 Env Trimers

2.3.1. Production of HIV-1 Env Trimers

The HIV-1 subtype A envelope protein BG505 NFL2P gp140 (Env) was used as the primary antigen. In brief, this antigen consists of three heterodimeric gp120-$gp41_{ecto}$ subunits and was stabilized for the soluble, trimeric conformation by truncation of the transmembrane and cytoplasmic domains, introduction of a flexible linker (2 x G4S) between the globular gp120 subunit and the gp41 ectodomain as well as a point mutation at amino acid position 559 (I559P) [18]. Env trimers were produced by transfection of 293F cells at a density of 1.0×10^6 cells per mL with the plasmid encoding for BG505 NFL2P gp140 (1 μg/mL DNA). Linear polyethylenimine (PEI, Polysciences Inc., Warrington, PA, USA) was used as transfection reagent in a threefold excess compared to the mass of DNA. Cell culture medium was changed 6 h after the transfection. The supernatant was harvested after three days, sterile-filtered and run over a lectin affinity column (Agarose-bound *Galanthus nivalis* lectin, Vector Laboratories Inc., Burlingame, CA, USA). Env trimers were eluted using 1M Methyl-α-D-mannopyranoside and concentrated with a 10 kDa Amicon cutoff filter (Sigma-Aldrich, St. Louis, MO, USA). Between the centrifugation steps, buffer changes were performed by refilling with DPBS without bivalent cations (Thermo Fisher Scientific, Waltham, MA, USA). The final Env concentration was determined by photometric measurement with a nanodrop device (Thermo Fisher Scientific, Waltham, MA, USA). As a control, we introduced a stop codon downstream of the flexible linker and deleted the nucleotide sequence for the gp41 ectodomain in order to produce gp120 monomers. These proteins were purified in the same way as described above for the trimers. All purified proteins were analyzed by NativePAGE and western blot (WB). For the UV-Vis spectroscopy (see 2.4.4) Env trimers were labelled with AlexaFluor®-488 fluorescent dye using a protein labeling kit (Thermo Fisher Scientific, Waltham, MA, USA) following the manufacturer's instructions.

2.3.2. NativePAGE Analysis of HIV-1 Env Trimers

The native conformation of Env trimers and monomers was addressed using the NativePAGE system (Thermo Fisher Scientific, Waltham, MA, USA) following the manufacturer's guidelines. In brief, 1 μg of each protein was mixed with G-250 additive and loaded onto a 4–16% Native Page Bis-Tris gel. The proteins were separated according to their native size by gel electrophoresis. Afterwards, excessive Coomassie stain was removed from the native gel by overnight fixation in 10% acetic acid and 30% ethanol in ultra-pure H_2O. Faint protein bands on the gel were then developed by staining with silver nitrate using a Silver Stain Kit (Pierce Biotechnology Inc., Rockford, IL, USA).

2.3.3. Western Blot Analysis of HIV-1 Env Trimers

For the antigen-specific immunoblotting of purified Env, 1 μg of each soluble protein sample or 300 ng Env on VLPs were mixed with house-made, reducing SDS sample buffer, boiled and then loaded onto a 12% SDS gel. After gel electrophoresis, proteins were transferred onto a nitrocellulose membrane which was subsequently blocked with 5% skimmed milk in DPBS supplemented with 0.1% Tween20 (PBS-T). The blocked membranes were incubated with polyclonal goat anti-gp120 (Acris Antibodies GmbH, Herford, Germany) and horseradish peroxidase-coupled secondary anti-goat IgG antibody (Dianova, Hamburg, Germany) with multiple washing steps in between. The membranes were finally developed with house-made ECL solution and protein bands were imaged using an Advanced Fluorescence Imager (Intas, Göttingen, Germany).

2.4. Production of Calcium Phosphate Nanoparticles

2.4.1. Instruments

Dynamic light scattering (DLS) and zeta potential were measured with a Zetasizer Nano ZS instrument (laser wavelength $\lambda = 633$ nm, Malvern Instruments, Malvern, UK) with the Smoluchowski approximation. Data obtained from Malvern software were used without further treatment; the particle size results refer to the z-average. Scanning electron microscopy (SEM) was performed with an ESEM Quanta 400 instrument (FEI Co., Hillsboro, OR, USA) and gold/palladium-sputtered samples. Calcium concentrations were measured by atomic absorption spectroscopy (AAS) with an M-Series AA spectrometer (Thermo Electron Corporation, Schwerte, Germany). UV-Vis absorption spectra were measured with a DS-11 FX+ spectrophotometer ("Nanodrop", DeNovix, Wilmington, DE, USA) and a Cary 300 Bio spectrophotometer (Agilent Technologies, Santa Clara, CA, USA). Ultracentrifugation was done at 20 °C with a Sorvall WX Ultra Series centrifuge (Thermo Electron Corporation, Schwerte, Germany). Freeze-drying (lyophilization) was carried out with a Christ Alpha 2-4 LSC instrument (Martin Christ GmbH, Osterode am Harz, Germany). The endotoxin concentration was measured with an Endosafe Nexgen-PTS handheld spectrophotometer (Charles River, Boston, MA USA). Ultrapure water (Purelab, ELGA LabWater, Celle, Germany) was used for all preparations. All nanoparticles were prepared and analyzed at room temperature.

2.4.2. Synthesis of Calcium Phosphate Nanoparticles

The CaP nanoparticle synthesis was performed according to our previously described method [21]. In brief, aqueous solutions of calcium lactate (18 mM, pH = 10, p.a., Sigma-Aldrich Corp., St. Louis, MO, USA), diammonium hydrogen phosphate (10.8 mM, pH = 10, p.a., VWR Life-Sciences) and branched polyethyleneimine (PEI, Mw = 25 kDa, Sigma-Aldrich Corp., St. Louis, MO, USA) were simultaneously pumped at a volume ratio of 5:5:7 mL during one minute into a stirred vessel with 20 mL of ultrapure water. The dispersion was stirred for 20 min. and the CaP/PEI nanoparticle dispersion was used immediately for the following steps. For the synthesis of adjuvant-containing nanoparticles, 1 mL CaP/PEI nanoparticle dispersion was mixed with aqueous solutions of either 60 μL p30 peptide (1 mg/mL) or 40 μL CpG (1 mg/mL) under stirring, followed by 30 min. stirring at room temperature (RT). The adjuvant loading was determined by measuring the residual concentration in the supernatant by UV microvolume spectroscopy ("nanodrop"). For further surface modifications a silica shell was added to the nanoparticles. To this end, 1 mL of either CaP/PEI or of adjuvant-containing nanoparticle dispersion was added to a mixture of 4 mL ethanol (Fisher Chemicals, Hampton, NH, USA), 5 μL tetraethylorthosilicate (TEOS, Sigma-Aldrich Corp., St. Louis, MO, USA) and 10 μL aqueous ammonia solution (7.8 wt.%) and stirred for 16 h. After this time, the nanoparticles were isolated by ultracentrifugation (66,000× *g*, 30 min, 20 °C) and redispersed with 1 mL ultrapure water followed by ultrasonication (UP50H, Hielscher, Teltow, Germany, sonotrode MS7m cycle 0.8, amplitude 70%, 4 s). Silica-terminated calcium phosphate nanoparticles (CaP/PEI/SiO_2) were obtained. To prepare the thiol-terminated calcium phosphate nanoparticles a surface modification with (3-mercaptopropyl)trimethoxysilane (MPS, Sigma-Aldrich Corp., St. Louis, MO, USA) was performed. For this, 1 mL of CaP/PEI/SiO_2 nanoparticles was added to a mixture of 4 mL ethanol and 50 μL (3-mercaptopropyl)trimethoxysilane and stirred for 8 h. The nanoparticles were isolated by ultracentrifugation (66,000× *g*, 30 min, 20 °C), redispersed with 1 mL H_2O and ultrasonicated. After this step, thiol-terminated calcium phosphate nanoparticles (CaP/PEI/SiO_2-SH) were obtained.

2.4.3. Functionalization of CaP Nanoparticles with Env Trimers

The Env trimers (Mw = 140 kDa) were coupled to the nanoparticle surface via a sulfo-SMCC cross-linker (sulfosuccinimidyl-*trans*-4-(N-maleimidomethyl)cyclohexane-1-carboxylate, Merck, Darmstadt, Germany). One end of the linker reacts with the primary amines in Env and the other end with the thiol groups on the nanoparticle surface. For this, 300 μL aqueous sulfo-SMCC solution (1.78 mg/mL) were given to 600 μL Env in DPBS (1 mg/mL) and incubated for 2 h at 4 °C. After incubation, the protein was purified from the unreacted cross-linker with a 3 kDa Amicon ultracentrifuge filter (Merck, Darmstadt, Germany), following the manufacturer recommendations. To attach the Env trimers to the nanoparticle surface, 330 μL activated protein (1 mg/mL) were given to 4 mL CaP/PEI/SiO_2-SH nanoparticle dispersion and incubated for 24 h at 4 °C. After this time, the particles were isolated by centrifugation at 21,000× *g* and 8 °C and washed once with 1 mL H_2O. Finally, the nanoparticles were redispersed in 4 mL H_2O and ultrasonicated (UP50H, Hielscher, sonotrode MS7m cycle 0.8, amplitude 70%, 4 s). The different syntheses were carried out with sterile-filtered solutions.

2.4.4. CaP nanoparticle Characterization and Storage

For the final nanoparticle dispersion, an endotoxin quantification assay was performed with an Endosafe Nexgen-PTS device. No endotoxin was detected in all synthesized nanoparticles (<0.1 EU/mL). To calculate the number of nanoparticles per volume unit, the $Ca2^+$ concentration was measured by AAS and then tentatively expressed as the most common calcium phosphate, i.e., hydroxyapatite, $Ca_{10}(PO4)_6(OH)_2$. The nanoparticle concentration in 1 mL dispersion is calculated from the measured calcium phosphate concentration and the density of hydroxyapatite (3140 kg/m^3) assuming a complete spherical morphology. Additionally, the number of Env units on the nanoparticle surface was determined by UV-Vis spectroscopy with an AlexaFluor®-488 labelled Env protein and UV microvolume spectroscopy ("Nanodrop") (see [22] for typical calculation steps to obtain these data). For storage and transportation, the nanoparticle dispersion was lyophilized according to our previously reported protocol [23]. 20 mg D-(+)-trehalose dihydrate (Sigma-Aldrich Corp., St. Louis, MO, USA) were added to 1 mL of the nanoparticle dispersions as cryoprotectant followed by shock-freezing with liquid nitrogen and lyophilization for 72 h at 0.31 mBar and −10 °C. Immediately before the application, the nanoparticles were redispersed in 1 mL ultrapure water and gently sonicated with an ultrasonication bath.

2.5. B-cell Activation In Vitro

B-cells from wt Bl6 mice or PGT-121 BCR-transgenic mice were isolated from the spleen by magnetic cell separation (Miltenyi Biotec, Bergisch Gladbach, Germany, #130-090-862,). 2.0 × 10^5 cells were incubated with different concentrations of Env-coupled nanoparticles in U-bottom 96-well plates. As controls, we additionally incubated B cells with soluble Env trimers, with Env-VLP (0.2 µg of Env/mL) and with 2 µg/mL of LPS (Sigma-Aldrich, Corp., St. Louis, MO, USA). After 18 h incubation at 37 °C and 5% CO_2, cells were stained with Fixable Viability Dye (Thermo Fisher Scientific, Waltham, MA, USA) and with antibodies against the B-cell surface antigen CD19 (Thermo Fisher Scientific, Waltham, MA, USA) and the early activation marker CD69 (Thermo Fisher Scientific, Waltham, MA, USA). B-cell activation in living B-cells was subsequently measured on a benchtop flow cytometer BD™ LSR II (BD Biosciences, Franklin Lakes, NJ, USA) and analyzed with the FlowJo software (BD Biosciences, Franklin Lakes, NJ, USA).

2.6. Analyses of In Vivo Induced Immune Responses

2.6.1. Immunization, Collection of Blood and Organ Samples

All immunizations were performed intramuscularly in both hind legs in Bl6 mice that were at the age of 6–8 weeks by the time of the first injection. For the induction of intrastructural help, mice were immunized on day 0 and day 28 with Tetanus toxoid vaccine (Tetanol®pur, GSK Vaccines GmbH, Marburg, Germany) diluted 1:10 in sterile DPBS to induce CD4 T-cell responses against the TT peptide p30 (ISH group) or with DPBS alone (control group). All mice were then boosted on day 56, day 84 and day 112 with one of the following particle types: Env trimer-coupled CaP nanoparticles (Env-CaP); Env-CaP with encapsulated p30 peptide (Env-CaP-p30); or Env-VLP-p30 with the membrane-bound form of the Env trimer on the surface and the Gag-p30 fusion protein inside. To evaluate the influence of encapsulated CpG as an adjuvant, mice were boosted thrice with Env-CaP-CpG particles in the same immunization protocol as described above. The injection doses were normalized to 10 µg of Env delivered with CaP nanoparticles and 300 ng of Env delivered with VLPs per mouse. To investigate humoral immune responses against Env, blood samples were collected 2 weeks after each particle immunization. To investigate humoral immune responses against TT, blood samples were collected on day 49. All blood samples were taken under isoflurane anesthesia from the retrobulbar venous plexus with non-heparinized, single-use capillaries (minicaps®, Hirschmann, Eberstadt, Germany). Collected blood samples were centrifuged for 5 min at 5000 rpm. The upper serum fractions were isolated and stored at −20 °C. Mice were sacrificed on day 126 and spleens were isolated for further assessment of the cellular immune response against both Env and p30.

2.6.2. Analyses of Humoral Immune Responses

Serum ELISAs to address the humoral immune response against Env were performed as described previously [8]. Briefly, white opaque MaxiSorp 96-well plates (Greiner Bio One, Frickenhausen, Germany) were coated with 100 ng Env trimer in coating buffer (0.1 M Na_2CO_3, 0.1 M $NaHCO_3$ in H_2O, pH 9.6) per well at 4 °C overnight. After 1 h blocking with 5% skimmed milk (diluted in DPBS containing 0.05% Tween-20), the wells were incubated with serum from immunized mice diluted 1:1000 in 2% skimmed milk for 1 h at RT. Plates were washed and then incubated for 1 h at RT with 1:4000 dilutions of HRP-coupled secondary antibodies specific for different murine IgG subtypes as well as total IgG (Southern Biotech, Birmingham, AL, USA). After thorough washing, serum binding was detected by addition of ECL solution and measurement of the relative light units per second (RLU/s) with an Orion microplate luminometer (Berthold Detection Systems GmbH, Pforzheim, Germany).

2.6.3. Analyses of Cellular Immune Responses

For the analysis of the cytokine profiles of both Env- and p30-specific CD4 T-cells after immunizations, we performed in vitro intracellular cytokine staining (ICS) of CD4 T-cells for interferon gamma (IFN-γ), tumor necrosis factor alpha (TNF-α) and interleukin-2 (IL-2), as well as cytokine ELISA for IL-5. Isolated spleens were dissociated by using a gentleMACS™ Dissociator (Miltenyi Biotech, Bergisch Gladbach, Germany) following the manufacturer's guidelines. The cell suspensions were run through 70 µm cell strainers. We removed erythrocytes by incubation with ACK lysis buffer (150 mM NH_4Cl, 10 mM $KHCO_3$, 0.1 mM EDTA in H_2O, pH 7.2) for 8 min at RT; the lysis reaction was stopped by the addition of R10 medium (RPMI 1640, 10% FCS, 1% Penicillin-Streptomycin, 10 mM HEPES, 2 mM L-glutamine, 50 µM 2-mercaptoethanol). The isolated splenocytes were washed twice by centrifugation and resuspension in R10 and subsequently counted using a Countess Automated Cell Counter (Thermo Fisher Scientific, Waltham, MA, USA). Isolated splenocytes were seeded in 96-well U-bottom plates (1.0×10^6 cells per well). For antigen-driven cytokine production by p30-specific CD4 T-cells, splenocytes were re-stimulated with 5 µg/mL of p30 peptide. Since there is no identified immunodominant MHC-II restricted Env peptides for Bl6 mice, we established a re-stimulation protocol to induce Env-specific cellular responses. To this end, we purified PGT121 B-cells (see above) and incubated them for 3 h at 37 °C with Env-VLPs containing 0.2 µg/mL Env or with the same amount of Δ-VLP. Thereafter, the B-cells, that were supposed to take up the Env-VLP in a BCR-dependent manner and subsequently present an array of different Env peptides to the CD4 T-cells, were washed with R10 and added to splenocyte suspensions (1.5×10^5 PGT121 B-cells per well).

Intracellular Cytokine Staining

Antigen re-stimulated and unstimulated (control) splenocytes were incubated in the presence of 2 µg/mL anti-mouse CD28 antibody and 3 µg/mL Brefeldin A (eBioscience, San Diego, CA, USA) for 6 h at 37 °C and 5% CO_2. After incubation, cells were washed with FACS buffer (1% FCS, 1 mM EDTA in DPBS without bivalent cations) and stained with anti-mouse CD4 antibody (eBioscience, San Diego, CA, USA) and Fixable Viability Dye (Thermo Fisher Scientific, Waltham, MA, USA). Then, the cells were fixated with 2% paraformaldehyde in DPBS and subsequently permeabilized using 0.5% Saponin in FACS buffer and intracellularly stained with anti-mouse IL-2, IFN-γ and TNF-α (eBioscience, San Diego, CA, USA). The cells were washed twice with permeabilization buffer and twice with FACS buffer. The cytokine accumulation in the CD4 T-cells was then analyzed by flow cytometry. The background values of unstimulated cultures (Δ-VLP for Env) were subtracted for each individual mouse.

Cytokine ELISA

Antigen re-stimulated and control splenocytes were incubated in the presence of 2 μg/mL anti-mouse CD28 antibody for 60 h at 37 °C and 5% CO_2. The culture media were harvested and the IL-5 cytokine secretion was analyzed using a Mouse IL-5 ELISA kit (Invitrogen, Carlsbad, CA, USA) following the manufacturer's instructions. The background values of unstimulated cultures were subtracted for each individual mouse.

2.7. Statistical Analysis

Statistical analyses were performed as indicated in the figure legends with the GraphPad Prism 7 software (Graphpad Software Inc., San Diego, CA, USA).

3. Results and Discussion

3.1. Production and Characterization of Soluble HIV-1 Env Trimers and Env-VLP-p30

For the coupling onto the surface of the CaP nanoparticles we needed faithful mimetics of the HIV-1 Env spike. During the last decade, there were great efforts to create soluble Env trimers that are stabilized in a closed pre-fusion conformation. This was achieved by truncation of the protein at amino acid position 664 within the gp41 subunit as well as by the introduction of various stabilizing mutations i.e., I559P [24].

In this study, we used an Env trimer that is derived from the membrane-embedded native form (Figure 1A; left) of a subtype A HIV-1 isolate from an infected baby (BG505). In addition to the modifications mentioned above, a flexible linker (2xG4S) was more recently introduced between the gp120 subunit and the gp41 ectodomain to overcome the need for proper cleavage of the precursor protein in order to receive stabilized well-folded proteins [18].

We expressed these Env trimers (BG505 NFL2P gp140) (Figure 1A, middle left) as well as monomeric gp120 subunits (BG505 NFL2P gp120) (Figure 1A, middle right) in 293F cells and purified them by Lectin affinity chromatography. The purified and concentrated proteins were analyzed on a native gel followed by a silver staining. A prominent band at approximately 700 kDa represented the globular native form of an Env trimer that is formed by three non-covalently assembled gp120-$gp41_{ecto}$ (gp140) heterodimers (Figure 1B). Two minor bands showed a low percentage of gp140 dimers and monomers. The purified gp120 resulted in two major bands on the native gel: a monomeric protein at 200 kDa and a gp120 dimer at 480 kDa that is formed by aberrant intermolecular di-sulfide bridges [25] as well as a large aggregate fraction (Figure 1B). As expected, the expression of gp140 (but not of gp120) results in the production of Env trimers, since the trimerization domain is located in the gp41 ectodomain and stabilized by the I559P mutation.

To compare intrastructural help for CaP nanoparticles side by side with that for HIV-1 VLPs, we generated VLPs with a membrane-bound form of the stabilized Env trimer on the surface and the TT-derived peptide p30 inside (Env-VLP-p30). In order to array stabilized BG505 NFL2P gp140 trimers on HIV-1 VLPs, the transmembrane and cytoplasmic domains of the vesicular stomatitis virus G-protein (VSV-G) were fused to the open reading frame encoding Env. Figure 1C represents the BG505 NFL2P gp140-GTMCD expression construct and the corresponding protein (Figure 1A; right) has an estimated size between 150 and 160 kDa. To produce HIV-1 VLPs that contain the p30 peptide of TT, we inserted the coding sequence for the peptide in frame between the open-reading frame for matrix and capsid of the HIV-1 structural Gag protein. Figure 1D represents the Hgpsyn-TTp30 construct. Env-VLPs-p30 particles were produced by co-transfection of 293T cells with both the BG505 NFL2P gp140-GTMCD and the Hgpsyn-TTp30 construct.

The purified Env proteins as well as the produced Env-VLP-p30 particles were analyzed by Western Blot under reducing conditions (Figure 1E). gp120 and gp140 were represented by prominent bands in the respective sizes. The pseudotyped BG505 NFL2P gp140-GTMCD protein was approximately 150 to 160 kDa in size and, therefore, runs slightly higher than the soluble trimer (Figure 1E; upper panel). In addition, three major Gag bands represent different states of the capsid maturation of the HIV-VLPs (Figure 1E; lower panel).

These characterized Env trimers were further used for the coupling onto the surface of the CaP nanoparticles.

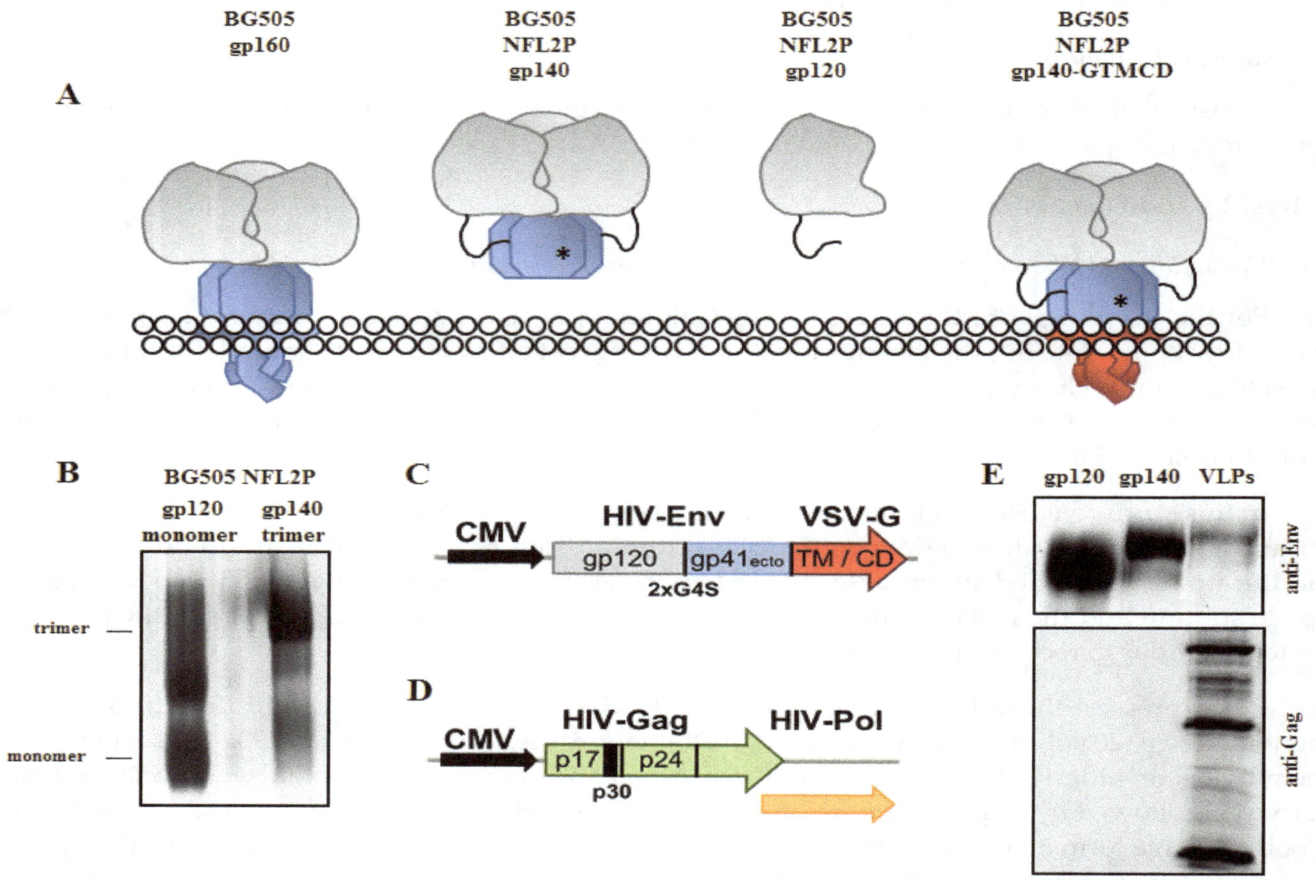

Figure 1. Production and characterization of soluble HIV-1 Env constructs and Env-p30-VLPs. (**A**) Overview of different membrane-embedded and soluble Env constructs. The native BG505 gp160 (left), that is composed of three gp120 (grey) – gp41 (blue) heterodimers, has been used previously to create a soluble stabilized recombinant trimer (BG505 NFL2P gp140; middle left) by truncation at amino acid 664 and introduction of both a point mutation (I559P) indicated with an asterisk (*) and a flexible linker (2xG4S). As a purification control, we produced a BG505 NFL2P gp120 monomer (middle right) by truncation of the protein downstream of the flexible linker. A recombinant VSV-G transmembrane and cytoplasmic domain (TM/CD; red) was introduced downstream of the gp41 ectodomain to create a membrane-bound form of the stabilized trimer for production of Env-VLP-p30 particles (BG505 NFL2P gp140-GTMCD; right). (**B**) Native PAGE of purified gp140 and gp120. Three gp140 molecules formed a globular trimer molecule represented by a major band at 700 kDa. The gp120 subunit was expressed as monomeric or dimeric proteins, but did not form trimers. (**C**,**D**) Schematic overview of the plasmids used for Env-VLP-p30 production. (**C**) BG505 NFL2P gp140-GTMCD construct. Colors are matched to their respective protein domains in Figure 1A. (**D**) Hgpsyn-TTp30 construct: the nucleotide sequence for the p30 peptide was introduced between the HIV-1 p17 matrix protein and the spacer protein 1 followed by the capsid protein p24. (**E**) Reducing SDS-PAGE and Western Blot of purified soluble Env proteins and Env-VLP-p30.

3.2. *Design, Production and Characterization of CaP Nanoparticles Functionalized with Soluble HIV-1 Env Trimers*

Stable CaP nanoparticles were synthesized and the surface modified with soluble stabilized HIV-1 Env trimer proteins (Figure 1). The obtained nanoparticles had an average solid core diameter between 38-57 nm (as determined by SEM) (Figure 2) with a positively charged surface near +27 mV, due to the stabilizing PEI polymer, and a hydrodynamic diameter between 300–400 nm. These parameters make the nanoparticles suitable for cellular uptake [26]. The complete characterization data are shown in Supplementary Table S1. For nanoparticles loaded with the T-helper peptide p30 or with CpG, the load was absorbed onto the nanoparticle core before forming the external silica shell around it. This coating provides protection to the internal loading but also permits a covalent surface modification with Env [21].

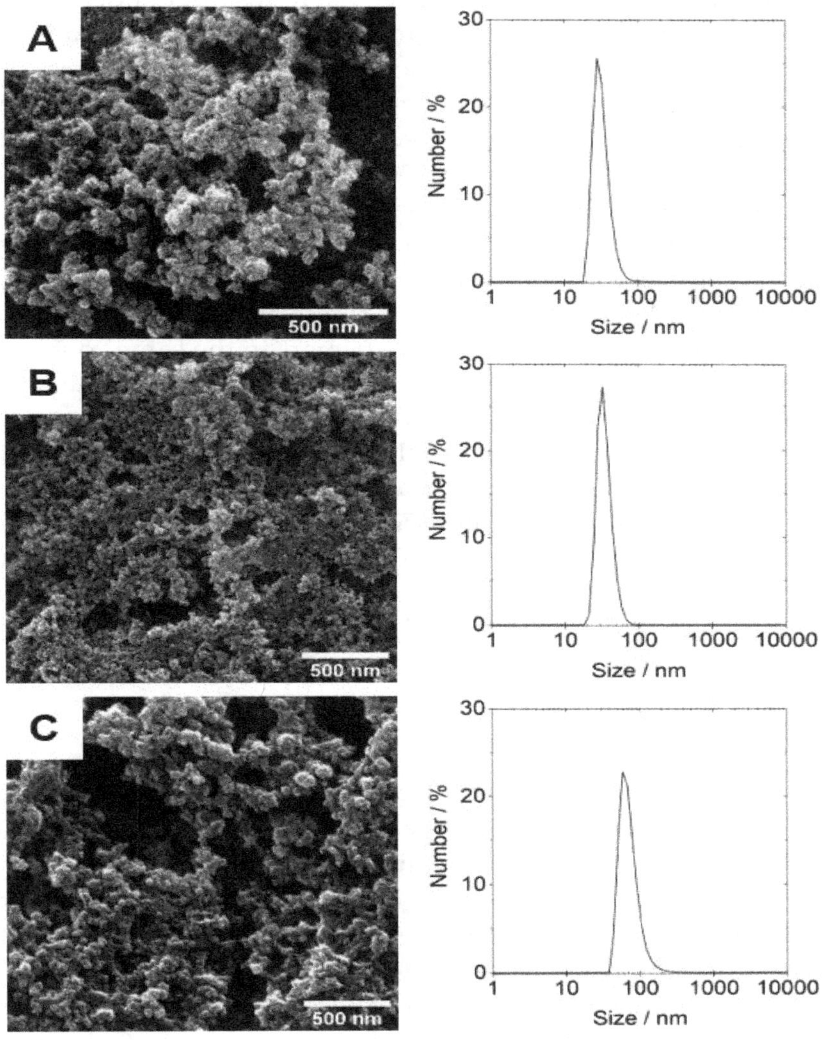

Figure 2. Scanning electron micrographs and dynamic light scattering (DLS) particle size distribution of the prepared nanoparticles: (**A**) Env-CaP, (**B**) Env-CaP-p30, (**C**) Env-CaP-CpG.

To make the protein reactive to the thiol-terminated calcium phosphate nanoparticles Env was activated with the crosslinker sulfo-SMCC by which the primary amines from the protein react with the N-hydroxysuccimide group. In the protein, 33 lysine residues are expected to have the side chain exposed to the solvent and can react with the NHS ester [27,28]. After the addition of the nanoparticles the maleimide group at the other end of the crosslinker reacts with the thiol groups on the nanoparticle surface to form a stable covalent bond between the nanoparticle and the protein.

To determine the coupling efficiency of the Env proteins to the nanoparticles, an AlexaFluor®-488 labelled Env protein was used for the surface modification. After attaching this labelled protein, the reaction yield was determined to 85% by UV-Vis spectroscopy. This factor was assumed to calculate the concentration of Env after attaching the unlabelled protein under equivalent reaction conditions. Additionally, a comparison for the Env concentration determined by nanodrop was performed. The measured concentrations are shown in Supplementary Table S1. The Env concentration determined by UV-Vis spectroscopy was used to calculate Env concentrations for all subsequent nanoparticle preparations.

3.3. *Antigen-Specific Activation of Naïve B-Cells with Env-CaP Nanoparticles In Vitro*

Previously we demonstrated in vitro that CaP nanoparticles functionalized with a monovalent model antigen activated naïve antigen-specific B-cells in a dose-dependent manner more efficiently than the dissolved antigen alone [12]. In contrast to monomeric proteins, each stabilized soluble Env trimer exposes up to 3 identical epitopes per molecule [29].

To investigate how the surface functionalization of CaP nanoparticle with stabilized HIV-1 Env trimers influences the activation of naïve Env-specific B cells, we incubated B-cells from transgenic mice that express the human PGT121 antibody [30] as B-cell receptors (Env B-cells) with Env-CaP and soluble Env trimers (Figure 3A). Env-VLPs served as a positive control for BCR-specific B-cell activation [31]. The stimulation with LPS showed the total polyclonal capability of the B-cells to be activated [32]. Within a concentration range of 8 ng to 200 ng of Env per 1 mL of culture medium, Env trimers arrayed on the CaP nanoparticle surface were able to activate Env-specific B cells more efficiently in a dose-dependent manner than soluble trimers in the same concentrations (Figure 3A).

Incubation of wt B-cells in the presence of Env-CaP nanoparticles, soluble Env trimers or Env-VLPs did not reveal Env-specific early B-cell activation (Figure 3B). The polyclonal activation of wt B-cells with LPS (comparable to those of Env-specific B-cells) indicated the functional activity of the wt B-cells (Figure 3) and, therefore, clearly demonstrates that all experimental Env preparations used were free of polyclonal activators. Our results are consistent with the data reported by Ingale et al., who showed that HIV-1 trimer-conjugated liposomes activated Env-specific B-cells better than soluble trimers [33].

Thus, the surface functionalization of CaP nanoparticles with soluble HIV-1 Env trimers improved the dose-dependent activation of naïve Env-specific B-cells in vitro in comparison to soluble HIV-1 Env trimers alone.

3.4. *Improvement of HIV-1 Env Antibody Responses by Intrastructural Help*

To analyze the induction of anti-Env antibody responses in vivo, we applied Env-CaP nanoparticles intramuscularly, as it was demonstrated to be the most appropriate way for the delivery of B- and T-cell antigens with CaP nanoparticles into draining lymphoid organs [13]. In contrast to the model antigen [13] in a highly reactive C3H mouse strain (as discussed in Reference [34]), Env-CaP nanoparticles at a dose of 10 μg of Env protein per immunization induced poor anti-Env antibody responses in Bl6 mice (Figure 4A, Env-CaP vs. naive).

Although Env-CaP nanoparticles are able to directly activate Env-specific B-cells in vitro (Figure 3), the primary anti-Env antibody response in vivo is also dependent on CD4 T-cell help [35]. The low magnitude of Env-specific IgG antibody responses induced with Env-CaP might be due to (i) a suboptimal MHC-II restricted T-cell help [14] elicited by HIV-1 BG505 Env protein in Bl6 mice [36], or (ii) suboptimal immunogenic capacities of Env-CaP (as discussed in [13]).

We have already demonstrated that the functionalization of CaP-HEL nanoparticles with the p30 peptide of TT (a universal T-helper epitope [14]) overcomes the lack of functionally active HEL-derived CD4 T-cell epitopes in Bl6 mice [13]. The functionalization of Env-CaP nanoparticles with the p30 peptide increased the magnitude of the anti-Env IgG1 antibody response (Figure 4A, Env-CaP vs. Env-CaP-p30), indicating an initially suboptimal MHC-II restricted T-cell help in mice immunized with

Env-CaP. The anti-Env IgG2c antibody response, however, remained unaffected (Figure 4B, Env-CaP vs. Env-CaP-p30).

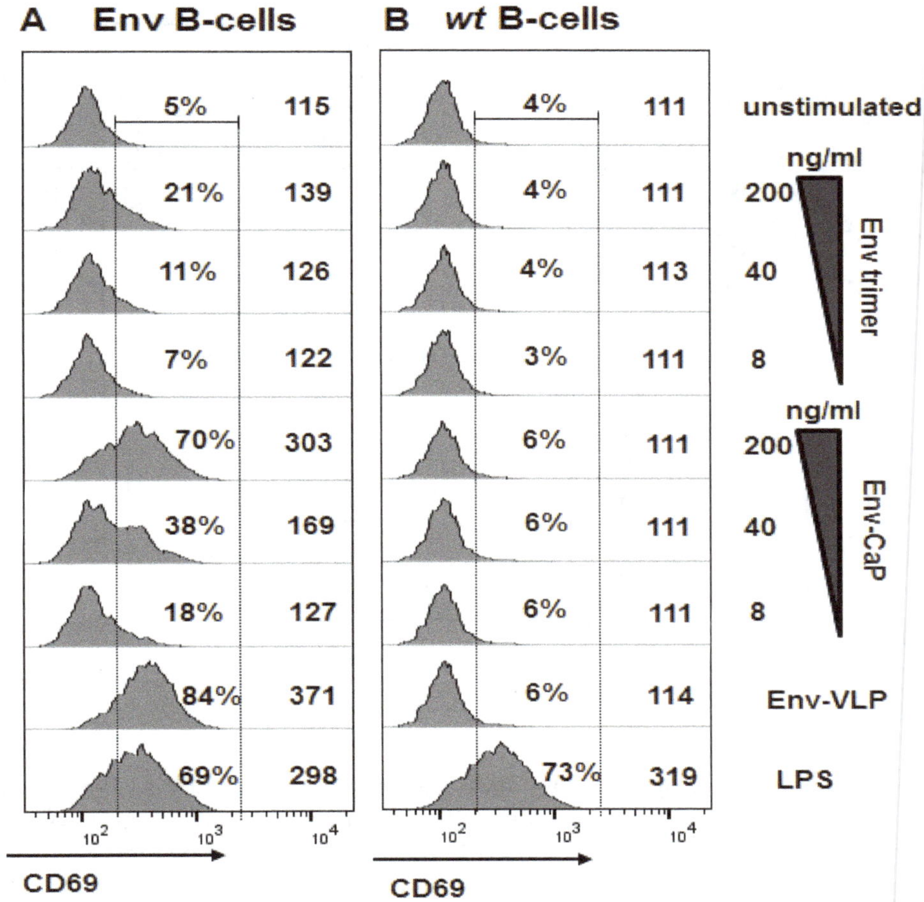

Figure 3. Activation of naïve B-cells in vitro. Naïve B-cells were isolated from PGT121 (**A**) or wt (**B**) mice and stimulated with Env trimers (8, 40 or 200 ng of Env/mL), Env-CaP nanoparticles (8, 40 or 200 ng of Env/mL), Env-VLPs (200 ng of Env/mL), or LPS (2 μg/mL). After 18h incubation, the cells were stained with a viability dye and with anti-CD19 and anti-CD69 antibodies. The histograms represent the expression of CD69 on the surface of viable CD19-positive B-cells. The numbers on the histograms indicate the percentage of the gated CD69-positive cells and medians of the total CD69 fluorescence intensity.

To further improve the Env-CaP-p30 induced anti-Env humoral immune responses, we recruited pre-existing T-helper cells generated by the licensed Tetanol pur vaccine to provide ISH for Env-specific B-cells. Mice previously vaccinated against Tetanus demonstrated significantly higher anti-Env antibody responses after Env-CaP-p30 nanoparticle immunization, than the control animals injected with DPBS instead of Tetanol pur (Figure 4A,B; Env-CaP-p30 vs. Env-CaP-p30 (ISH)). The in vitro reactivation of p30-specific CD4 T-cells from the spleens of differently immunized mice revealed an increase of potential heterologous T-cell help in the ISH group (Figure 4C–E). Intracellular cytokine staining for the Th1 cytokines [37] IFN-γ (Figure 4C) and TNF-α (Figure 4D) showed a trend of being more pronounced in the ISH group. The secretion of the Th2 cytokine [37] IL-5 after p30 reactivation was significantly higher in the ISH group (Figure 4E). CD4 T-cells that produce either Th1 or Th2 cytokines support the generation of IgG2a (IgG2c in Bl6 mice) and IgG1 antibody subclasses in mice, respectively [37,38]. The prominent Th2 cytokine profile of p30-specific CD4 T-cells is, therefore, consistent with the predominant induction of the anti-Env IgG1 antibody subclass after immunizations with Env-CaP-p30 nanoparticles (Figure 4A,B).

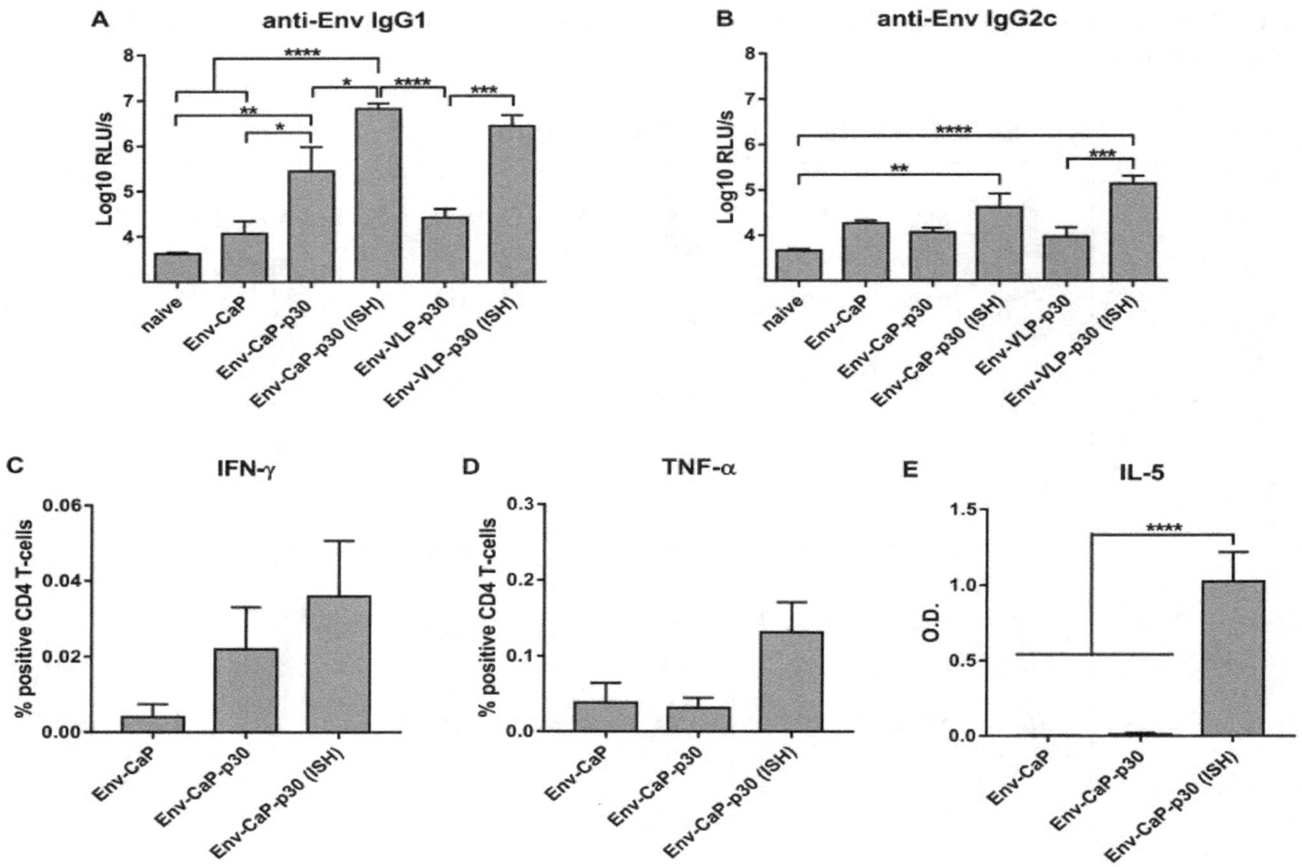

Figure 4. Improvement of Env-specific IgG subtype responses after Env-CaP-p30 immunizations by intrastructural help from Tetanus toxoid-specific T-helper cells. (**A**,**B**) IgG1 (**A**) and IgG2c (**B**) Env-specific antibody responses were measured in sera from wt mice primed twice with DPBS or Tetanol (ISH) and boosted 3 times with different CaP nanoparticles or Env-VLP-p30. (**C**,**D**) Percentages of CD4 T-cells producing IFN-γ (**C**) and TNF-α (**D**) from differently immunized wt animals after in vitro re-stimulation with p30 peptide were measured by intracellular cytokine staining. (**E**) p30-specific IL-5 cytokine secretion as determined by ELISA. The columns represent the mean values of six animals ± SEM. * $p < 0.05$; ** $p < 0.001$; *** $p < 0.0005$; **** $p < 0.0001$; one-way ANOVA with Tukey multiple comparison post-hoc test.

Previously, we demonstrated ISH in mice immunized with Tetanol pur prior to administration of HIV-1 derived VLPs containing different T helper epitopes of TT [8]. For direct comparison of the ISH effects on the anti-Env antibody induction between the CaP nanoparticle platform and the HIV-1 VLP system, we produced VLPs containing the p30 epitope of TT within Gag (Figure 1D) and presenting stabilized BG505 Env trimers in a membrane-anchored form on their surface (Figure 1A,C). The VLP antigen dosage was selected based on our previous studies [8,35] and was 0.3 μg of Env protein per immunization. For the Env-VLP-p30 particles we observed intrastructural help (Figure 4A,B; Env-VLP-p30 vs. Env-VLP-p30 (ISH)) in a magnitude comparable to the CaP nanoparticle platform (Figure 4A,B; Env-VLP-p30 (ISH) vs. Env-CaP-p30 (ISH)).

Altogether, the incorporation of T-helper cell epitopes of non-HIV proteins into CaP nanoparticles functionalized with HIV-1 Env trimers may allow the Env-specific B-cells to get T-cell help from non-HIV specific CD4 T-cells via the ISH mechanism.

3.5. Distinct Effects of ISH and CpG-Adjuvants on the Induction of Env-Specific CD4 T-Cell Responses

Along with engaging the heterologous universal p30 T-helper epitope, functionalization of the B-cell targeting CaP-nanoparticles with TLR-ligands significantly increased IgG antibody responses against a model antigen in mice [13]. The TLR9 ligand CpG demonstrated a number of advantages to other TLR-ligands tested in the study (as discussed in Reference [13]). In addition, the CpG-based adjuvant 1018 ISS [39] is already approved for clinical use.

To compare the improvement of antibody responses by CpG and ISH side-by-side, animals previously immunized against Tetanus received either Env-CaP-p30 or Env-Cap-CpG nanoparticles. Both Env-CaP-p30 and Env-Cap-CpG induced comparable total anti-Env IgG immune responses (Figure 5A), although the anti-Env IgG subtype distribution varied between the groups (Figure 5B).

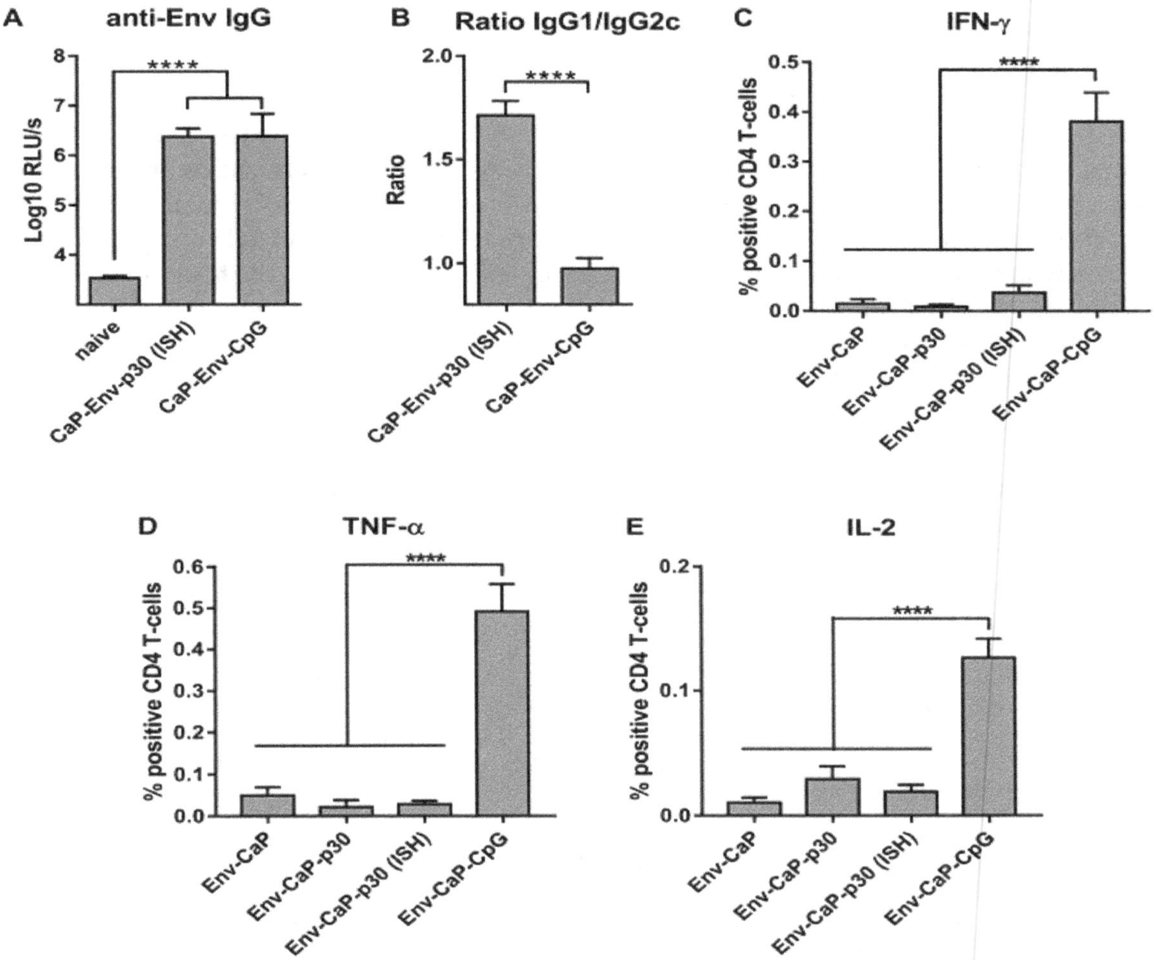

Figure 5. Characterization of anti-Env immune responses after heterologous ISH and CpG adjuvantation. (**A**) Env-specific antibody responses were measured in sera from wt mice primed twice with Tetanol and boosted 3 times with either Env-CaP-p30 or Env-CaP-CpG. The columns represent the mean values of six animals ± SEM. **** $p < 0.0001$; one-way ANOVA with Tukey multiple comparison post-hoc test. (**B**) Env-specific IgG1/IgG2c ratios of individual mice. The columns represent the mean values of six animals ± SEM. **** $p < 0.0001$; unpaired Student t-test. (**C–E**) Percentages of CD4 T-cells producing IFN-γ (**C**), TNF-α (**D**) and IL-2 (**E**) from differently immunized wt animals after in vitro re-stimulation with Env-VLPs were measured by intracellular cytokine staining. The columns represent the mean values of six animals ± SEM. **** $p < 0.0001$; one-way ANOVA with Tukey multiple comparison post-hoc test.

The efficacy of an HIV-1 vaccine, however, may not only depend on the strength of a protective antibody response induced, but also on the magnitude of vaccine-induced immune mechanisms increasing the susceptibility to infection. In particular, HIV-specific CD4 T-cell responses may increase the susceptibility to infection by expanding the number of activated CD4 T-cells as targets for the HIV-1 infection [1,6,7,40]. We therefore also compared CD4 T-cell responses in both vaccine groups.

In contrast to Env-CaP-p30 nanoparticles that recruit and support heterologous (non-HIV-1 specific) T-cell help (Figure 4), Env-CaP-CpG nanoparticles might induce increased Env-specific CD4 T-cell response as a result of TLR9 ligation in dendritic cells [41,42]. Indeed, the percentage of Env-specific CD4 T-cells in the spleens of immunized animals that can be reactivated in vitro after Env-VLP exposure was significantly higher in the Env-CaP-CpG immunized animals (Figure 5B–D). Reactivation of Env-specific CD4 T-cells producing Th1 pro-inflammatory [43] cytokines IFN-γ (Figure 5B) as well as TNF-α (Figure 5C) and IL-2 (Figure 5E), a potent mitogen and growth factor for CD4 T-cells, in mice from the Env-CaP-p30 (ISH) group was comparable to those in mice that received Env-CaP or Env-CaP-p30 after DPBS injections. These results clearly demonstrated that ISH does not increase expansion of Env-specific CD4 T-cells.

In non-human primate models of HIV/SIV infection, vaccinated animals with high numbers of vaccine-induced IFN-γ secreting T-cells were more susceptible to acquisition of challenge virus infection than poor T-cell responders [7,40], indicating that CpG functionalization of Env-CaP vaccines may potentially enhance the susceptibility to acquisition of HIV-1 infection in vaccinees.

Thus, the incorporation of T-helper epitopes of heterologous (non-HIV) proteins in Env-CaP nanoparticles was able to increase the magnitude of anti-Env specific IgG antibody responses by ISH without significant induction of Env-specific IFN-γ producing CD4 T-helper cells. This suggests that the ISH strategy for nanoparticle-based HIV-1 vaccines may allow to avoid induction of HIV-1 specific CD4 T-cell responses suspected to enhance the susceptibility to HIV-1 infection.

4. Conclusions

We demonstrated that the incorporation of T-helper cell epitopes of non-HIV proteins into CaP nanoparticles functionalized with HIV-1 Env trimers allows the Env-specific B-cells to receive T-cell help from non-HIV specific CD4 T-cells via the ISH mechanism. In a mouse model, the Env-CaP-p30 nanoparticle-based vaccine was able to improve HIV-1 Env-specific antibody responses without additional induction of HIV-specific CD4 T-cells suspected to increase the susceptibility to infection.

Author Contributions: Conceptualization, V.T., K.Ü. and M.E.; methodology, V.T., M.E. and R.T.W.; validation, V.T. and M.E.; formal analysis, D.D., L.R.-S., V.S. and V.T.; investigation, D.D., L.R.-S. and H.T.; resources, K.Ü., R.T.W. and M.E.; writing—original draft preparation, V.T., D.D. and L.R.-S.; writing—review and editing, V.T.; visualization, D.D., L.R.-S. and V.T.; supervision, V.T., V.S., M.E. and K.Ü.; project administration, V.T. and V.S. All authors read and approved the final manuscript.

Acknowledgments: We thank Ghulam Nabi (Ruhr University Bochum, Germany) for his contribution to the generation of the Hgpsyn-TTp30 expression plasmid, and Michel Nussenzweig (The Rockefeller University, New York, NY, USA) for PGT-121 mice donation. We acknowledge support by Deutsche Forschungsgemeinschaft and Friedrich-Alexander-Universität Erlangen-Nürnberg (FAU) within the funding programme Open Access Publishing. L.R.-S. thanks the Ministry of Science, Technology and Telecommunications of Costa Rica (MICITT, Costa Rica) for a Ph.D. scholarship.

References

1. Temchura, V.; Tenbusch, M. The two faces of vaccine-induced immune response: Protection or increased risk of HIV infection?! *Virol. Sin.* **2014**, *29*, 7–9. [CrossRef] [PubMed]
2. Chung, A.W.; Ghebremichael, M.; Robinson, H.; Brown, E.; Choi, I.; Lane, S.; Dugast, A.S.; Schoen, M.K.; Rolland, M.; Suscovich, T.J.; et al. Polyfunctional Fc-Effector Profiles Mediated by IgG Subclass Selection Distinguish RV144 and VAX003 Vaccines. *Sci. Transl. Med.* **2014**, *6*, 228ra38. [CrossRef] [PubMed]
3. Kim, J.H.; Excler, J.-L.; Michael, N.L. Lessons from the RV144 Thai Phase III HIV-1 Vaccine Trial and the Search for Correlates of Protection. *Annu. Rev. Med.* **2015**, *66*, 423–437. [CrossRef] [PubMed]
4. Excler, J.-L.; Ake, J.; Robb, M.L.; Kim, J.H.; Plotkin, S.A. Nonneutralizing Functional Antibodies: A New "Old" Paradigm for HIV Vaccines. *Clin. Vaccine Immunol.* **2014**, *21*, 1023–1036. [CrossRef] [PubMed]
5. Ackerman, M.E.; Mikhailova, A.; Brown, E.P.; Dowell, K.G.; Walker, B.D.; Bailey-Kellogg, C.; Suscovich, T.J.; Alter, G. Polyfunctional HIV-Specific Antibody Responses Are Associated with Spontaneous HIV Control. *PLoS Pathog.* **2016**, *12*, e1005315. [CrossRef] [PubMed]
6. Lewis, G.K.; DeVico, A.L.; Gallo, R.C. Antibody persistence and T-cell balance: Two key factors confronting HIV vaccine development. *Proc. Natl. Acad. Sci. USA* **2014**, *111*, 15614–15621. [CrossRef] [PubMed]

7. Fouts, T.R.; Bagley, K.; Prado, I.J.; Bobb, K.L.; Schwartz, J.A.; Xu, R.; Zagursky, R.J.; Egan, M.A.; Eldridge, J.H.; LaBranche, C.C.; et al. Balance of cellular and humoral immunity determines the level of protection by HIV vaccines in rhesus macaque models of HIV infection. *Proc. Natl. Acad. Sci. USA* **2015**, *112*, E992–E999. [CrossRef]
8. Elsayed, H.; Nabi, G.; McKinstry, W.J.; Khoo, K.K.; Mak, J.; Salazar, A.M.; Tenbusch, M.; Temchura, V.; Überla, K. Intrastructural Help: Harnessing T Helper Cells Induced by Licensed Vaccines for Improvement of HIV Env Antibody Responses to Virus-Like Particle Vaccines. *J. Virol.* **2018**, *92*, e00141-18. [CrossRef]
9. Russell, S.M.; Liew, F.Y. T cells primed by influenza virion internal components can cooperate in the antibody response to haemagglutinin. *Nature* **1979**, *280*, 147–148. [CrossRef]
10. Temchura, V.; Überla, K. Intrastructural help. *Curr. Opin. HIV AIDS* **2017**, *12*, 272–277. [CrossRef]
11. Sokolova, V.; Westendorf, A.M.; Buer, J.; Überla, K.; Epple, M. The potential of nanoparticles for the immunization against viral infections. *J. Mater. Chem. B* **2015**, *3*, 4767–4779. [CrossRef]
12. Temchura, V.V.; Kozlova, D.; Sokolova, V.; Überla, K.; Epple, M. Targeting and activation of antigen-specific B-cells by calcium phosphate nanoparticles loaded with protein antigen. *Biomaterials* **2014**, *35*, 6098–6105. [CrossRef] [PubMed]
13. Zilker, C.; Kozlova, D.; Sokolova, V.; Yan, H.; Epple, M.; Überla, K.; Temchura, V. Nanoparticle-based B-cell targeting vaccines: Tailoring of humoral immune responses by functionalization with different TLR-ligands. *Nanomed. Nanotechnol. Biol. Med.* **2017**, *13*, 173–182. [CrossRef] [PubMed]
14. Oscherwitz, J.; Yu, F.; Cease, K.B. A heterologous helper T-cell epitope enhances the immunogenicity of a multiple-antigenic-peptide vaccine targeting the cryptic loop-neutralizing determinant of Bacillus anthracis protective antigen. *Infect. Immun.* **2009**, *77*, 5509–5518. [CrossRef] [PubMed]
15. De Taeye, S.W.; Moore, J.P.; Sanders, R.W. HIV-1 Envelope Trimer Design and Immunization Strategies to Induce Broadly Neutralizing Antibodies. *Trends Immunol.* **2016**, *37*, 221–232. [CrossRef] [PubMed]
16. Sattentau, Q.; Sattentau, J.Q. Envelope Glycoprotein Trimers as HIV-1 Vaccine Immunogens. *Vaccines* **2013**, *1*, 497–512. [CrossRef] [PubMed]
17. Escolano, A.; Steichen, J.M.; Dosenovic, P.; Kulp, D.W.; Golijanin, J.; Sok, D.; Freund, N.T.; Gitlin, A.D.; Oliveira, T.; Araki, T.; et al. Sequential Immunization Elicits Broadly Neutralizing Anti-HIV-1 Antibodies in Ig Knockin Mice. *Cell* **2016**, *166*, 1445–1458. [CrossRef] [PubMed]
18. Sharma, S.K.; deVal, N.; Bale, S.; Guenaga, J.; Tran, K.; Feng, Y.; Dubrovskaya, V.; Ward, A.B.; Wyatt, R.T. Cleavage-Independent HIV-1 Env Trimers Engineered as Soluble Native Spike Mimetics for Vaccine Design. *Cell Rep.* **2015**, *11*, 539–550. [CrossRef]
19. Wagner, R.; Graf, M.; Bieler, K.; Wolf, H.; Grunwald, T.; Foley, P.; Uberla, K. Rev-Independent Expression of Synthetic *gag-pol* Genes of Human Immunodeficiency Virus Type 1 and Simian Immunodeficiency Virus: Implications for the Safety of Lentiviral Vectors. *Hum. Gene Ther.* **2000**, *11*, 2403–2413. [CrossRef]
20. genannt Bonsmann, M.S.; Niezold, T.; Temchura, V.; Pissani, F.; Ehrhardt, K.; Brown, E.P.; Osei-Owusu, N.Y.; Hannaman, D.; Hengel, H.; Ackerman, M.E.; et al. Enhancing the Quality of Antibodies to HIV-1 Envelope by GagPol-Specific Th Cells. *J. Immunol.* **2015**, *195*, 4861–4872. [CrossRef]
21. Kozlova, D.; Chernousova, S.; Knuschke, T.; Buer, J.; Westendorf, A.M.; Epple, M. Cell targeting by antibody-functionalized calcium phosphatenanoparticles. *J. Mater. Chem.* **2012**, *22*, 396–404. [CrossRef]
22. Rojas-Sánchez, L.; Sokolova, V.; Riebe, S.; Voskuhl, J.; Epple, M. Covalent Surface Functionalization of Calcium Phosphate Nanoparticles with Fluorescent Dyes by Copper-Catalysed and by Strain-Promoted Azide-Alkyne Click Chemistry. *ChemNanoMat* **2019**, *5*, 436–446. [CrossRef]
23. Klesing, J.; Chernousova, S.; Epple, M. Freeze-dried cationic calcium phosphatenanorods as versatile carriers of nucleic acids (DNA, siRNA). *J. Mater. Chem.* **2012**, *22*, 199–204. [CrossRef]
24. Sanders, R.W.; Vesanen, M.; Schuelke, N.; Master, A.; Schiffner, L.; Kalyanaraman, R.; Paluch, M.; Berkhout, B.; Maddon, P.J.; Olson, W.C.; et al. Stabilization of the soluble, cleaved, trimeric form of the envelope glycoprotein complex of human immunodeficiency virus type 1. *J. Virol.* **2002**, *76*, 8875–8889. [CrossRef] [PubMed]
25. Finzi, A.; Pacheco, B.; Zeng, X.; Kwon, Y.D.; Kwong, P.D.; Sodroski, J. Conformational characterization of aberrant disulfide-linked HIV-1 gp120 dimers secreted from overexpressing cells. *J. Virol. Methods* **2010**, *168*, 155–161. [CrossRef] [PubMed]
26. Behzadi, S.; Serpooshan, V.; Tao, W.; Hamaly, M.A.; Alkawareek, M.Y.; Dreaden, E.C.; Brown, D.; Alkilany, A.M.; Farokhzad, O.C.; Mahmoudi, M. Cellular uptake of nanoparticles: Journey inside the cell. *Chem. Soc. Rev.* **2017**, *46*, 4218–4244. [CrossRef] [PubMed]

27. Klausen, M.S.; Jespersen, M.C.; Nielsen, H.; Jensen, K.K.; Jurtz, V.I.; Sønderby, C.K.; Sommer, M.O.A.; Winther, O.; Nielsen, M.; Petersen, B.; et al. NetSurfP-2.0, Improved prediction of protein structural features by integrated deep learning. *Proteins Struct. Funct. Bioinf.* **2019**, *87*, 520–527. [CrossRef]
28. Koniev, O.; Wagner, A. Developments and recent advancements in the field of endogenous amino acid selective bond forming reactions for bioconjugation. *Chem. Soc. Rev.* **2015**, *44*, 5495–5551. [CrossRef] [PubMed]
29. Lyumkis, D.; Julien, J.-P.; de Val, N.; Cupo, A.; Potter, C.S.; Klasse, P.-J.; Burton, D.R.; Sanders, R.W.; Moore, J.P.; Carragher, B.; et al. Cryo-EM Structure of a Fully Glycosylated Soluble Cleaved HIV-1 Envelope Trimer. *Science* **2013**, *342*, 1484–1490. [CrossRef]
30. Walker, L.M.; Huber, M.; Doores, K.J.; Falkowska, E.; Pejchal, R.; Julien, J.-P.; Wang, S.-K.; Ramos, A.; Chan-Hui, P.-Y.; Moyle, M.; et al. Broad neutralization coverage of HIV by multiple highly potent antibodies. *Nature* **2011**, *477*, 466–470. [CrossRef]
31. Gardt, O.; Grewe, B.; Tippler, B.G.; Überla, K.; Temchura, V.V. HIV-derived lentiviral particles promote T-cell independent activation and differentiation of naïve cognate conventional B2-cells in vitro. *Vaccine* **2013**, *31*, 5088–5098. [CrossRef] [PubMed]
32. Sveen, K.; Skaug, N. Comparative mitogenicity and polyclonal B cell activation capacity of eight oral or nonoral bacterial lipopolysaccharides in cultures of spleen cells from athymic (nu/nu-BALB/c) and thymic (BALB/c) mice. *Oral Microbiol. Immunol.* **1992**, *7*, 71–77. [CrossRef] [PubMed]
33. Ingale, J.; Stano, A.; Guenaga, J.; Sharma, S.K.; Nemazee, D.; Zwick, M.B.; Wyatt, R.T. High-Density Array of Well-Ordered HIV-1 Spikes on Synthetic Liposomal Nanoparticles Efficiently Activate B Cells. *Cell Rep.* **2016**, *15*, 1986–1999. [CrossRef] [PubMed]
34. Barnowski, C.; Kadzioch, N.; Damm, D.; Yan, H.; Temchura, V. Advantages and Limitations of Integrated Flagellin Adjuvants for HIV-Based Nanoparticle B-Cell Vaccines. *Pharmaceutics* **2019**, *11*, 204. [CrossRef] [PubMed]
35. Nabi, G.; Temchura, V.; Großmann, C.; Kuate, S.; Tenbusch, M.; Überla, K. T cell independent secondary antibody responses to the envelope protein of simian immunodeficiency virus. *Retrovirology* **2012**, *9*. [CrossRef] [PubMed]
36. Hu, J.K.; Crampton, J.C.; Cupo, A.; Ketas, T.; van Gils, M.J.; Sliepen, K.; de Taeye, S.W.; Sok, D.; Ozorowski, G.; Deresa, I.; et al. Murine Antibody Responses to Cleaved Soluble HIV-1 Envelope Trimers Are Highly Restricted in Specificity. *J. Virol.* **2015**, *89*, 10383–10398. [CrossRef] [PubMed]
37. Mosmann, T.R.; Coffman, R.L. TH1 and TH2 Cells: Different Patterns of Lymphokine Secretion Lead to Different Functional Properties. *Annu. Rev. Immunol.* **1989**, *7*, 145–173. [CrossRef]
38. Snapper, C.M.; Paul, W.E. Interferon-gamma and B cell stimulatory factor-1 reciprocally regulate Ig isotype production. *Science* **1987**, *236*, 944–947. [CrossRef]
39. Campbell, J.D. *Development of the CpG Adjuvant 1018, A Case Study*; Humana Press: New York, NY, USA, 2017; pp. 15–27. [CrossRef]
40. Tenbusch, M.; Ignatius, R.; Temchura, V.; Nabi, G.; Tippler, B.; Stewart-Jones, G.; Salazar, A.M.; Sauermann, U.; Stahl-Hennig, G.; Überla, K.; et al. Risk of Immunodeficiency Virus Infection May Increase with Vaccine-Induced Immune Response. *J. Virol.* **2012**, *86*, 10533–10539. [CrossRef]
41. Sokolova, V.; Knuschke, T.; Kovtun, A.; Buer, J.; Epple, M.; Westendorf, A.M. The use of calcium phosphate nanoparticles encapsulating Toll-like receptor ligands and the antigen hemagglutinin to induce dendritic cell maturation and T cell activation. *Biomaterials* **2010**, *31*, 5627–5633. [CrossRef]
42. Knuschke, T.; Sokolova, V.; Rotan, O.; Wadwa, M.; Tenbusch, M.; Hansen, W.; Staeheli, P.; Epple, M.; Buer, J.; Westendorf, A.M. Immunization with biodegradable nanoparticles efficiently induces cellular immunity and protects against influenza virus infection. *J. Immunol.* **2013**, *190*, 6221–6229. [CrossRef] [PubMed]
43. Turner, M.D.; Nedjai, B.; Hurst, T.; Pennington, D.J. Cytokines and chemokines: At the crossroads of cell signalling and inflammatory disease. *Biochim. Biophys. Acta* **2014**, *1843*, 2563–2582. [CrossRef] [PubMed]

7

Giant Viruses—Big Surprises

Nadav Brandes [1] **and Michal Linial** [2,*]

1 The Rachel and Selim Benin School of Computer Science and Engineering, The Hebrew University of Jerusalem, Jerusalem 91904, Israel; nadav.brandes@mail.huji.ac.il
2 Department of Biological Chemistry, The Alexander Silberman Institute of Life Sciences, The Hebrew University of Jerusalem, Jerusalem 91904, Israel
* Correspondence: michall@cc.huji.ac.il

Abstract: Viruses are the most prevalent infectious agents, populating almost every ecosystem on earth. Most viruses carry only a handful of genes supporting their replication and the production of capsids. It came as a great surprise in 2003 when the first giant virus was discovered and found to have a >1 Mbp genome encoding almost a thousand proteins. Following this first discovery, dozens of giant virus strains across several viral families have been reported. Here, we provide an updated quantitative and qualitative view on giant viruses and elaborate on their shared and variable features. We review the complexity of giant viral proteomes, which include functions traditionally associated only with cellular organisms. These unprecedented functions include components of the translation machinery, DNA maintenance, and metabolic enzymes. We discuss the possible underlying evolutionary processes and mechanisms that might have shaped the diversity of giant viruses and their genomes, highlighting their remarkable capacity to hijack genes and genomic sequences from their hosts and environments. This leads us to examine prominent theories regarding the origin of giant viruses. Finally, we present the emerging ecological view of giant viruses, found across widespread habitats and ecological systems, with respect to the environment and human health.

Keywords: Amebae viruses; viral evolution; protein domains; mimivirus; dsdna viruses; translation machinery; pandoravirus; NCLDV

---

1. Giant Viruses and the Viral World

Viruses are cell infecting agents present in almost every ecosystem. Questions the regarding viral origin and early evolution alongside all living organisms (bacteria, archaea and eukarya) are still wide open, and relevant theories remain speculative [1–5]. As viruses are exceptionally diverse and undergo rapid changes, it is impossible to construct an ancestral lineage tree for the viral world [6–10]. Instead, virus families are categorized according to the nature of their genetic material, mode of replication, pathogenicity, and structural properties [11].

At present, the viral world is represented by over 8,000 reference genomes [12]. The International Committee on Taxonomy of Viruses (ICTV) provides a universal virus taxonomical classification proposal that covers ~150 families and ~850 genera, with many viruses yet unclassified [13]. This collection provides a comprehensive, compact set of virus representatives.

Inspection of viral genomes reveals that most known viruses have genomes encoding only a few proteins. Actually, 69% of all known viruses have less than 10 proteins encoded in their genomes (Figure 1). It is a common assumption that viruses demonstrate near-optimal genome packing and information compression, presumably in order to maximize their replication rate, number of progenies, and other parameters that increase infectivity [14,15]. However, a debate is still ongoing over the generality of these phenomena [16], and there is a non-negligible percentage of larger viruses (Figure 1). On the far end of the distribution, there are viruses with hundreds of genes, most of them are considered giant viruses. While only 0.3% of the known viral proteomes contain 500 or more proteins, they encode as much as 7.5% of the total number of viral proteins (Figure 1B).

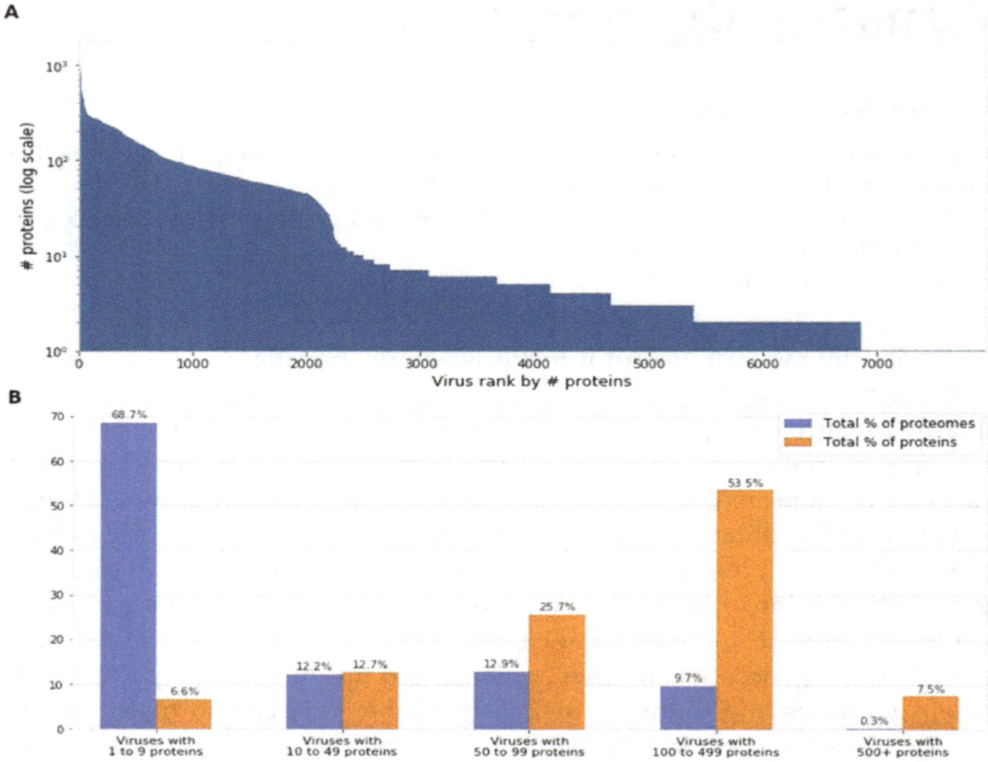

Figure 1. Number of proteins encoded by viruses. (**A**) The number of encoded proteins (y-axis) in all 7,959 viral representatives, ranked in descending order. (**B**) Partitioning of the 7959 viral proteomes by the number of encoded proteins. The 0.3% viral proteomes with the highest number of proteins (over 500) encode 7.5% of the total number of viral proteins.

2. The Discovery of Giant Viruses

The first giant virus, *Acanthamoeba polyphaga mimivirus* (APMV), was discovered in 2003 [17]. Its size was unprecedented, being on the scale of small bacteria or archaea cells [18]. Unlike any previously identified virus, APMV could be seen with a light microscope [19,20]. Initially it was mistaken for a bacterium and recognized as a virus only ten years after its isolation [21]. Up to this day, most of its proteins remain uncharacterized [22,23]. Notably, even more than a decade after the discovery of APMV, the identification of giant viruses still sometimes involves confusion, as illustrated in the discovery of the *Pandoravirus inopinatum* [24], which was initially described as an endoparasitic organism, and *Pithovirus sibericum* [25], which was also misinterpreted as an archaeal endosymbiont (see discussion in References [21,26]).

In the following years after the initial discovery of APMV, many additional giant viral species have been identified and their genomes fully sequenced. Most giant viral genomes have been obtained from large-scale metagenomic sequencing projects covering aquatic ecosystems (e.g., oceans, pools, lakes and cooling wastewater units) [27,28]; others have been sequenced from samples extracted from underexplored geographical and ecological niches (e.g., the Amazon River, deep seas and forest soils) [29–32]. Despite the accumulation of many more giant virus representatives, the fraction of uncharacterized proteins in their proteomes remains exceptionally high [33]. Many of these uncharacterized proteins were also considered orphan genes (ORFans), i.e., no significant match to any other sequence was identified. For example, 93% of the *Pandoravirus salinus* proteins, the first representative of this family [34] were reported as ORFans.

At present, there are over a hundred giant virus isolates, which reveal fascinating and unexpected characteristics. These extreme instances on the viral landscape challenge the current theories on genome size and compactness in viruses, and provide a new perspective on the very concept of a virus and viral origin [4,20,28,35–37].

3. Definition of Giant Viruses

Attempts to distinguish giant viruses from other large viruses remain somewhat fuzzy [38,39]. Any definition for giant viruses would necessarily involve some arbitrary threshold, as virus size, whether physical, genomic or proteomic, is clearly a continuum (Figure 2). Giant viruses were initially defined by their physical size as allowing visibility by a light microscope [33]. In this report, we prefer a proteomic definition, even if somewhat arbitrary. We consider giant viruses as eukaryote-infecting viruses with at least 500 protein-coding genes (Figure 2). Of the 7,959 curated viral genomes (extracted from NCBI Taxonomy complete genomes), 24 represented genomes meet this threshold. Of these, we consider the 19 eukaryote-infecting viruses to be the giant virus representatives (Table 1), excluding the five bacteria-infecting viruses.

Recall that reported proteome sizes are primarily based on automatic bioinformatics tools, which may differ from the experimental expression measurements (e.g., mimivirus APMV [40]). Moreover, physical dimensions are not in perfect correlation with the number of proteins or genome size. For example, *Pithovirus sibericum*, which was recovered from a 30,000-year-old permafrost sample [25], is one of the largest viruses by its physical dimensions (1.5 μm in length and 0.5 μm in diameter). However, it is excluded from this report, as its genome encodes only 467 proteins.

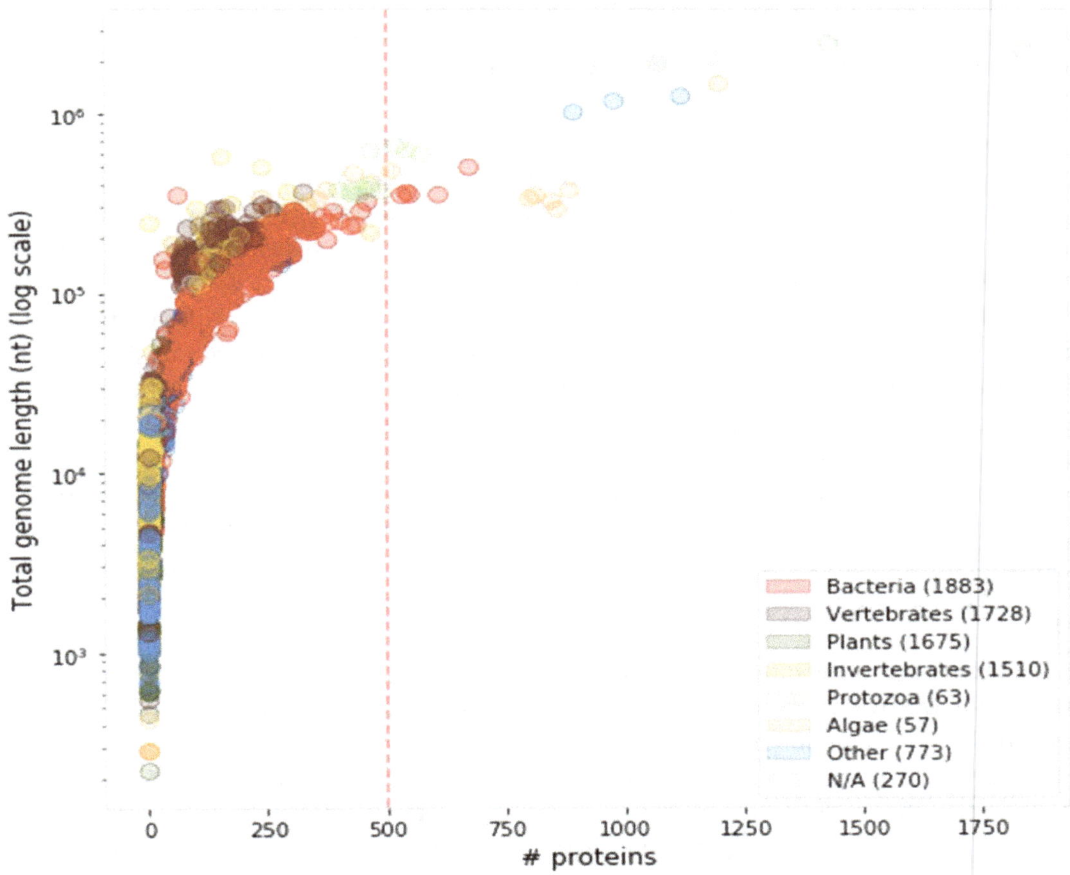

Figure 2. Distribution of viral proteome and genome sizes, colored by host taxonomy. There are 24 represented genomes that meet the threshold of ≥500 proteins (dashed red line), comprising five bacteria-infecting and 19 eukaryote-infecting viruses.

Table 1. Giant viruses.

Genome [a]	Accession	Genome Length (kb)	# of Proteins	Host [b]	Year [c]
Mi-Acanthamoeba polyphaga mimivirus	NC_014649	1181.5	979	Pz, Ver	2010
Mi-Acanthamoeba polyphaga moumouvirus	NC_020104	1021.3	894	Pz, Ver	2013
Ph-Acanthocystis turfacea chlorella virus 1	NC_008724	288.0	860	Algae	2006
Mi-Cafeteria roenbergensis virus BV-PW1	NC_014637	617.5	544	Pz	2010
Pi-Cedratvirus A11	NC_032108	589.1	574	Pz	2016
Ph-Chrysochromulina ericina virus	NC_028094	473.6	512	Algae	2015
Mi-Megavirus chiliensis	NC_016072	1259.2	1120	Pz, Ver	2011
UC-Mollivirus sibericum	NC_027867	651.5	523	Pz	2015
Ph-Orpheovirus IHUMI-LCC2	NC_036594	1473.6	1199	Algae	2017
Pa-Pandoravirus dulcis	NC_021858	1908.5	1070	Pz	2013
Pa-Pandoravirus inopinatum	NC_026440	2243.1	1839	Pz	2015
Pa-Pandoravirus macleodensis	NC_037665	1838.3	926	Pz	2018
Pa-Pandoravirus neocaledonia	NC_037666	2003.2	1081	Pz	2018
Pa-Pandoravirus quercus	NC_037667	2077.3	1185	Pz	2018
Pa-Pandoravirus salinus	NC_022098	2473.9	1430	Pz	2013
Ph-Paramecium bursaria Chlorella virus 1	NC_000852	330.6	802	Algae	1995
Ph-Paramecium bursaria Chlorella virus AR158	NC_009899	344.7	814	Algae	2007
Ph-Paramecium bursaria Chlorella virus FR483	NC_008603	321.2	849	Algae	2006
Ph-Paramecium bursaria Chlorella virus NY2A	NC_009898	368.7	886	Algae	2007

[a] Families: Mi, *Mimiviridae*; Ph, *Phycodnaviridae*; Pi, *Pithoviridae*; Pa, *Pandoraviridae*; UC, uncharacterized; [b] Pz, protozoa; Ver, vertebrates; [c] Year of genome submission to NCBI.

4. Classification of Giant Viruses and the Question of Origin

All giant viruses belong to the superfamily of nucleocytoplasmic large DNA viruses (NCLDV), which was substantially expanded following the discoveries of giant viruses [41,42]. The NCLDV superfamily had traditionally been comprised of the following families: *Phycodnaviridae*, *Iridoviridae*, *Poxviridae*, *Asfarviridae* and *Ascoviridae* [43,44], for which a common ancestor had been proposed [45,46]. Following the inclusion of additional giant virus taxonomy groups (*Mimiviradae*, *Pandoravirus* and *Marseillevirus*) into the NCLDV superfamily, there remained only a handful of genes shared by the entire superfamily. Additional disparities in virion shapes and replication modes among NCLDV has led to the conclusion that the superfamily is not necessarily a taxonomic group, and that NCLDV families are more likely to have evolved separately [47–49].

Two models have been proposed for the evolution of giant viruses. According to the reductive model, an ancestral cellular genome became reduced in size, leading to the dependence of the resulting genome on host cells. The presence of genes carrying cellular functions in almost any giant virus (e.g., translation components) [50] is consistent with this model. An alternative and more accepted theory argues for an expansion model. According to this model, current giant viruses originated from smaller ancestral viruses carrying only a few dozens of genes, and through gene duplications and horizontal gene transfer (HGT), have rapidly expanded and diversified [48,51–53]. This model agrees with metagenomic studies and the wave of giant virus discoveries in recent years, suggesting massive gene exchange between giant viruses and a variety of organisms sharing the same ecosystems (e.g., Reference [32]).

Of special interest is the degree of similarity between giant viruses and their hosts. The amebae host in particular is often described as a melting pot for DNA exchange [54] that leads to chimeric genomes. The majority of genes in giant viruses and specifically *Mimiviridae*, have originated from the cells they parasitize (mostly amoeba and bacteria). Based on phylogenetic trees, it is likely that extensive HGT events have led to their chimeric genomes. It was also suggested that the spectrum of viral hosts may be larger than anticipated, including yet unknown species [55]. Therefore, comparative genomics of giant viruses infecting the same host is unlikely to unambiguously resolve questions of gene origin, namely, whether shared genes have originated from a common viral ancestor. Thus, the degree of similarity among giant viruses infecting different hosts is of special interest. For example, the phyletic relationship between *Mimiviridae* (which infect *Acanthamoeba*) and *Phycodnaviridae* (which infect algae) was investigated, and it was found that the algae-infecting *Chrysochromulina ericina* virus (CeV, Table 1) showed moderate resemblance to the amebae-infecting mimivirus [56]. As a result, suggestions were made to reclassify CeV as a new clade of *Mimiviridae*, rather than *Phycodnaviridae*. However, a later discovery of another algae-infecting *Phycodnaviridae* virus (*Heterosigma akashiwo* virus, HaV53) has provided a coherent phyletic relationship among *Phycodnaviridae*, thereby questioning this reclassification [52].

In summary, the taxonomy of giant viruses, like all viruses, is still unstable, and rapidly updated with new discoveries [31,57]. The origin and ancestry of giant viruses have remained controversial with questions of origin also unresolved [39]. Many newly discovered giant viruses are not compatible with the notion of a single common ancestor, as some giant viruses remain taxonomically isolated [4].

5. Common Features

Despite the ongoing debate on their origin, giant viruses still share some important features. All giant viruses belong to the dsDNA group, as do all NCLDV families. The total genome size of all the giant viruses listed in Table 1 is at least 288 Kbp (Figure 2). These giant viruses are classified into several families: *Mimiviridae*, *Pithoviridae*, *Pandoraviridae*, *Phycodnaviridae* and the *Mollivirus* genus [21,25].

All amoebae-infecting giant viruses rely on the non-specific phagocytosis by the amoebae host [55]. Interestingly, a necessary condition for phagocytosis is a minimal particle size (~0.6 μm [58]), as amebae (and related protozoa) naturally feed on bacteria. It is likely that this minimal size for inducing phagocytosis has become an evolutionary driving force for giant viruses. This fact, together with the largely uncharacterized genomic content of giant viruses, may suggest that much of the content in the genomes of giant viruses serves only for volume filling to increase their physical size.

Giant viruses share not only the cell entry process. When they exit the host cells during lysis, as many as 1000 virions are released from each lysed host via membrane fusion and active exocytosis [59], which are relatively rare exit mechanisms in viruses.

Other than these genomic and cell-biology similarities, other features of giant viruses are mostly family-specific. For example, virion shapes and symmetries, nuclear involvement, duration of the infection cycle, and the stages of virion assembly—all substantially vary among giant viruses from different families [21,60,61].

6. Proteome Complexity and Functional Diversity

The majority of the giant virus proteomes remain with no known function (Figure 3). Actually, the fraction of uncharacterized proteins reaches 65–85% of all reported proteins in giant viral proteomes, many of them are ORFans. However, when proteomes of closely related species are considered, the fraction of ORFans obviously drops (by definition). For example, 93% of the proteins were reported as ORFans for the first representative of the *Pandoraviridae* family (*P. salinus*) [34]. But later, following the completion of five additional *Pandoravirus* proteomes (of the species *inopinatum*, *macleodensis*, *neocaledonia*, *dulcis*, and *quercus*), the number of *P. salinus* ORFans dropped to 29% (i.e., 71% of its genes now had a significant similarity to at least one other *Pandoravirus* protein sequence). Still, the vast majority of *Pandoravirus* proteins remain uncharacterized.

The most striking finding regarding the proteomes of giant viruses is the presence of protein functions that are among the hallmarks of cellular organisms, and are never detected in other viruses. To exemplify the complexity of proteome functions in giant viruses, we examined the proteome of the *Cafeteria roenbergensis* virus (CroV), which infects the marine plankton community in the Gulf of Mexico. *CroV* was sequenced in 2010 as the first algae-infecting virus in the *Mimiviridae* family. Unexpectedly, despite its affiliation with a recognized viral family, the majority of its proteins showed no significant similarity to any other known protein sequence. Of the remaining proteins that show significant basic local alignment search tool (BLAST) hits to other proteins from all domains of life, 45% are eukaryotic sequences, 22% are from bacteria, and the rest are mostly from other viruses, including other mimivirus strains. A similar partition of protein origin applies across other members of the *Mimiviridae* family [33].

The CroV proteome includes a rich set of genes involved in protein translation [62]. These genes include multiple translation factors, a dozen of ribosomal proteins, tRNA synthetases, and 22 sequences encoding five different tRNAs [62]. As a lack of translation potential is considered a hallmark of the virosphere, the presence of translation machinery components raised a debate on the very definition of viruses [63,64]. Similar findings were replicated in two other giant virus strains of the *Tupanvirus* genus in the same *Mimiviridae* family, which were recently isolated in Brazil [65]. The two strains have 20 open reading frames (ORFs) related to tRNA aminoacylation (aaRS), ~70 tRNA sequences decoding the majority of the codons, eight translation initiation factors, and elongation and release factors. The theory that translation optimization is an evolutionary driving force in viruses [66] may in part explain the curious presence of translation machinery in giant viruses.

In addition to translation, numerous CroV proteins are associated with the transcription machinery. Specifically, the CroV proteome contains several subunits of the DNA-dependent RNA polymerase II, initiation, elongation, and termination factors, the mRNA capping enzyme, and a poly(A) polymerase. Presumably, the virus can activate its own transcription in the viral factory foci in the cytoplasm of its host cell [47].

Another unexpected function detected in CroV is DNA repair, specifically of UV radiation damage and base-excision repair. Other DNA-maintenance functions found in CroV include helicase and topoisomerases (type I and II), suggesting regulation of DNA replication, recombination and chromatin remodeling.

Another rich set of functions related to protein maintenance include chaperons [67] and the ubiquitin-proteasome system [68]. Interestingly, some of these genes seem to be acquired from bacteria (e.g., a homolog of the Escherichia coli heat-shock chaperon). In addition, a rich collection of sugar-, lipid- and amino acid-related metabolic enzymes were also found [18,69], which occupy 13% of the CroV proteome (Figure 3).

It appears that the CroV proteome covers most functions traditionally attributed to cellular organisms, including: Protein translation, RNA maturation, DNA maintenance, proteostasis and metabolism. Although CroV exemplifies many widespread functions in giant viruses, each strain has its own unique functional composition. For example, the most abundant group of giant viruses in ocean metagenomes, the *Bodo saltans* virus (BsV), was recently identified and classified into the same microzooplankton-infecting *Mimiviridae* family [70]. Unlike the other family members, BsV does not have an elaborate translation apparatus or tRNA genes, but it carries proteins active in cell membrane trafficking and phagocytosis, yet more unprecedented functions discovered in viruses.

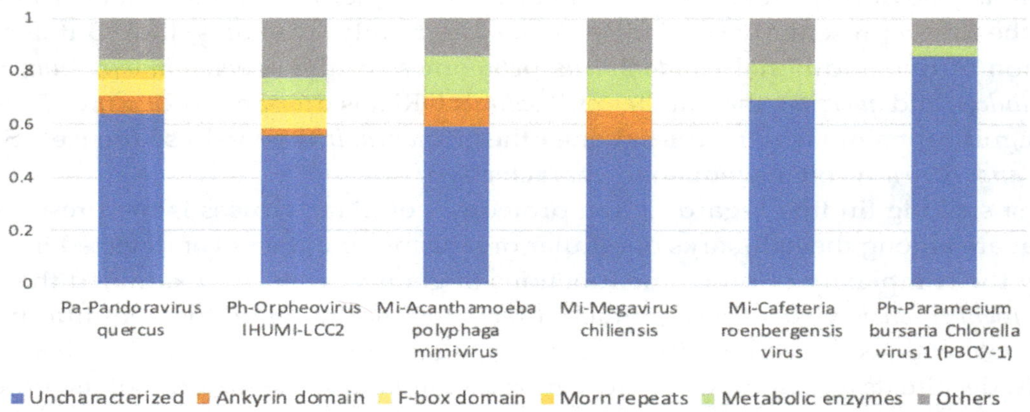

Figure 3. Protein function categories in six giant virus representatives from three families: *Mimiviridae* (Mi), *Pandoviridae* (Pa) and *Phycodnaviridae* (Ph). In all proteomes, the majority of proteins are uncharacterized. Short repeated domains such as ankyrin, F-box and MORM repeats are abundant in the proteomes of amebae-infecting giant viruses [71].

7. Virophages and Defense Mechanism in Giant Viruses

Additional important players of genome dynamics in giant viruses are the virophages [72]. These are small double-stranded DNA (dsDNA) viruses that hitchhike the replication system of giant viruses following coinfection of the host, and are considered parasites of the coinfecting giant viruses [73]. Virophages (e.g., Sputnik 1–3, Zamilion) are associated with *Mimiviridae* representatives and their specific viral strain infectivity [21,74]. Additionally, short mobile genetic agents, called transpovirions (combining features of a transposon and a virion) [73], together with other mobile elements display complex ecological interactions with their hosts. Indeed, similar to eukaryotic transposons and endogenous viruses, sequences of a virophage (*Mavirus*) of *Cafeteria roenbergensis virus* (CroV), that were integrated to the genome of the protozoan host, serves an antiviral defense mechanism, which is activated by giant virus infection [75].

An even more unexpected finding is the discovery of a nucleic acid-based immunity in mimiviruses, resembling the adaptive (clustered regularly interspaced short palindromic repeats) CRISPR-Cas system in bacteria and archaea. Despite the differences to the canonical CRISPR-Cas system, an operon-like cluster of sequences derived from the *Zamilion* virophage was identified in mimivirus and experimentally validated to govern virophage coinfection. This cluster, coined MIMIVIRE, acts as a mimivirus virophage resistance element system with an exonuclease, helicase and RNase III identified in its vicinity [76]. The homology between the MIMIVIRE-associated exonuclease to the bacterial Cas-4 exonuclease was revealed by 3D protein structure analysis [77]. This CRISPR-Cas related function in mimiviruses are assumed to degrade foreign DNA, thereby constituting an antiviral innate immune system. It is likely that the CRISPR-Cas immune system in mimiviruses contributes to its sequence diversification as well by removing unessential host sequences. Alternative mechanisms that govern viral-host infection specificity and immunity may be discovered as our knowledge on virophages and other mobile elements is expending [72,78].

Altogether, a rich network of mobile genetic elements contributes to the host-virus coevolution and interviral gene transfer [78]. Virophages and other mobile elements could facilitate gene transfer, thereby having the potential to shape the genomes of giant viruses and impact their diversity [21,79,80].

8. The Emerging Ecological View

Viruses are the most abundant entities in nature. In marine and fresh water habitats, there are millions of viruses in each milliliter of water [81]. However, the collection of virus isolates is often sporadic, especially for those without clinical or agricultural relevance. The accelerated pace in the discovery of giant viruses reflects the increasing number of sequencing projects of exotic environments, including metagenomic projects [32,82].

Giant viruses have been isolated from numerous environmental niches and distant geographic locations, revealing their global distribution and diversity. Current evidence suggests that the representation of giant viruses is underexplored, especially in soil ecosystems [31] and unique ecological niches [83,84]. In fact, ~60% of the giant viral genomes were completed after 2013 (Table 1). Many more virus–host systems, most of them reported in the last five years, still await isolation, characterization and classification [84,85].

The hosts of contemporary isolates include mainly protozoa, specifically amoeba (Table 1). However, the prevalence of amoeba as hosts may in part be attributed to sampling bias, specifically to the widespread use of amoebal coculture methods for testing ecological environments [28,85].

Despite their prevalence, the impact of giant viruses on human health deserves further investigation [22]. Some initial reports show that APMV giant virus is able to replicate in human peripheral blood cells and to induce the interferon system [86]. Sequences of numerous giant viruses were identified as part of large-scale human gut microbiome sequencing projects [87], but their abundance, compositions and ecological roles are yet to be determined [88]. Reports are accumulating on the presence of giant viral sequences in human blood, as well as antibodies against giant viral proteins. Some reports associate mimirus and marseillevirus with a broad collection of human diseases (e.g., rheumatoid arthritis, adenitis, unexplained pneumonia, lymphoma), yet causal relationship is mostly missing [89]. The presence of giant viruses in almost any environment, including extreme niches and manmade sites (e.g., sewage and wastewater plants), suggests that the ecological role of these fascinating entities and their impact on human health are yet to be fully explored.

Author Contributions: The authors contributed to writing and manuscript design.

Acknowledgments: The authors wish to thank the support to M.L. from Yad Hanadiv. We are grateful to Elixir infrastructure tools and resources.

References

1. Holmes, E.C. What does virus evolution tell us about virus origins? *J. Virol.* **2011**, *85*, 5247–5251. [CrossRef] [PubMed]
2. Forterre, P. The origin of viruses and their possible roles in major evolutionary transitions. *Virus Res.* **2006**, *117*, 5–16. [CrossRef]
3. Yutin, N.; Koonin, E.V. Evolution of DNA ligases of nucleo-cytoplasmic large DNA viruses of eukaryotes: A case of hidden complexity. *Biol. Direct.* **2009**, *4*, 51. [CrossRef] [PubMed]
4. Forterre, P. Viruses in the 21st Century: From the Curiosity-Driven Discovery of Giant Viruses to New Concepts and Definition of Life. *Clin. Infect. Dis.* **2017**, *65*, S74–S79. [CrossRef] [PubMed]
5. Moelling, K.; Broecker, F. Viruses and Evolution - Viruses First? A Personal Perspective. *Front. Microbiol.* **2019**, *10*, 523. [CrossRef] [PubMed]
6. Moreira, D.; Lopez-Garcia, P. Ten reasons to exclude viruses from the tree of life. *Nat. Rev. Microbiol.* **2009**, *7*, 306–311. [CrossRef] [PubMed]
7. Brüssow, H. The not so universal tree of life or the place of viruses in the living world. *Philos. Trans. R. Soc. B Biol. Sci.* **2009**, *364*, 2263–2274.
8. Bamford, D.H.; Grimes, J.M.; Stuart, D.I. What does structure tell us about virus evolution? *Curr. Opin. Struct. Biol.* **2005**, *15*, 655–663. [CrossRef] [PubMed]
9. Domingo, E.; Escarmis, C.; Sevilla, N.; Moya, A.; Elena, S.; Quer, J.; Novella, I.; Holland, J. Basic concepts in RNA virus evolution. *FASEB J.* **1996**, *10*, 859–864. [CrossRef]
10. Koonin, E.V.; Yutin, N. Origin and evolution of eukaryotic large nucleo-cytoplasmic DNA viruses. *Intervirology* **2010**, *53*, 284–292. [CrossRef]
11. Murphy, F.A.; Fauquet, C.M.; Bishop, D.H.; Ghabrial, S.A.; Jarvis, A.W.; Martelli, G.P.; Mayo, M.A.; Summers, M.D. *Virus taxonomy: Classification and nomenclature of viruses*; Springer Science & Business Media: Berlin, Germany, 2012; Volume 10.
12. Stano, M.; Beke, G.; Klucar, L. viruSITE-integrated database for viral genomics. *Database (Oxford)* **2016**, *2016*. [CrossRef]
13. Lefkowitz, E.J.; Dempsey, D.M.; Hendrickson, R.C.; Orton, R.J.; Siddell, S.G.; Smith, D.B. Virus taxonomy: The database of the International Committee on Taxonomy of Viruses (ICTV). *Nucleic Acids. Res.* **2018**, *46*, D708–D717. [CrossRef] [PubMed]
14. Pybus, O.G.; Rambaut, A. Evolutionary analysis of the dynamics of viral infectious disease. *Nat. Rev. Genet.* **2009**, *10*, 540–550. [CrossRef]
15. Pérez-Losada, M.; Arenas, M.; Galán, J.C.; Palero, F.; González-Candelas, F. Recombination in viruses: Mechanisms, methods of study, and evolutionary consequences. *Infect. Genet. Evol.* **2015**, *30*, 296–307. [CrossRef] [PubMed]
16. Brandes, N.; Linial, M. Gene overlapping and size constraints in the viral world. *Biol. Direct.* **2016**, *11*, 26. [CrossRef] [PubMed]
17. La Scola, B.; Audic, S.; Robert, C.; Jungang, L.; de Lamballerie, X.; Drancourt, M.; Birtles, R.; Claverie, J.M.; Raoult, D. A giant virus in amoebae. *Science* **2003**, *299*, 2033. [CrossRef]
18. Raoult, D.; Audic, S.; Robert, C.; Abergel, C.; Renesto, P.; Ogata, H.; La Scola, B.; Suzan, M.; Claverie, J.M. The 1.2-megabase genome sequence of Mimivirus. *Science* **2004**, *306*, 1344–1350. [CrossRef]
19. Xiao, C.; Chipman, P.R.; Battisti, A.J.; Bowman, V.D.; Renesto, P.; Raoult, D.; Rossmann, M.G. Cryo-electron microscopy of the giant Mimivirus. *J. Mol. Biol.* **2005**, *353*, 493–496. [CrossRef]
20. Van Etten, J.L.; Meints, R.H. Giant viruses infecting algae. *Annu. Rev. Microbiol.* **1999**, *53*, 447–494. [CrossRef]

21. Abergel, C.; Legendre, M.; Claverie, J.M. The rapidly expanding universe of giant viruses: Mimivirus, Pandoravirus, Pithovirus and Mollivirus. *FEMS Microbiol. Rev.* **2015**, *39*, 779–796. [CrossRef]
22. Abrahão, J.S.; Dornas, F.P.; Silva, L.C.; Almeida, G.M.; Boratto, P.V.; Colson, P.; La Scola, B.; Kroon, E.G. Acanthamoeba polyphaga mimivirus and other giant viruses: An open field to outstanding discoveries. *Virol. J.* **2014**, *11*, 120. [CrossRef] [PubMed]
23. Campos, R.K.; Boratto, P.V.; Assis, F.L.; Aguiar, E.R.; Silva, L.C.; Albarnaz, J.D.; Dornas, F.P.; Trindade, G.S.; Ferreira, P.P.; Marques, J.T. Samba virus: A novel mimivirus from a giant rain forest, the Brazilian Amazon. *Virol. J.* **2014**, *11*, 95. [CrossRef] [PubMed]
24. Antwerpen, M.H.; Georgi, E.; Zoeller, L.; Woelfel, R.; Stoecker, K.; Scheid, P. Whole-genome sequencing of a pandoravirus isolated from keratitis-inducing acanthamoeba. *Genome Announc.* **2015**, *3*. [CrossRef] [PubMed]
25. Legendre, M.; Bartoli, J.; Shmakova, L.; Jeudy, S.; Labadie, K.; Adrait, A.; Lescot, M.; Poirot, O.; Bertaux, L.; Bruley, C.; et al. Thirty-thousand-year-old distant relative of giant icosahedral DNA viruses with a pandoravirus morphology. *Proc. Natl. Acad. Sci. USA* **2014**, *111*, 4274–4279. [CrossRef] [PubMed]
26. Claverie, J.M.; Abergel, C. From extraordinary endocytobionts to pandoraviruses. Comment on Scheid et al.: Some secrets are revealed: Parasitic keratitis amoebae as vectors of the scarcely described pandoraviruses to humans. *Parasitol. Res.* **2015**, *114*, 1625–1627. [CrossRef] [PubMed]
27. Andreani, J.; Verneau, J.; Raoult, D.; Levasseur, A.; La Scola, B. Deciphering viral presences: Two novel partial giant viruses detected in marine metagenome and in a mine drainage metagenome. *Virol. J.* **2018**, *15*, 66. [CrossRef]
28. Aherfi, S.; Colson, P.; La Scola, B.; Raoult, D. Giant Viruses of Amoebas: An Update. *Front. Microbiol.* **2016**, *7*, 349. [CrossRef] [PubMed]
29. Andrade, A.; Arantes, T.S.; Rodrigues, R.A.L.; Machado, T.B.; Dornas, F.P.; Landell, M.F.; Furst, C.; Borges, L.G.A.; Dutra, L.A.L.; Almeida, G.; et al. Ubiquitous giants: A plethora of giant viruses found in Brazil and Antarctica. *Virol. J.* **2018**, *15*, 22. [CrossRef]
30. Dornas, F.P.; Assis, F.L.; Aherfi, S.; Arantes, T.; Abrahao, J.S.; Colson, P.; La Scola, B. A Brazilian Marseillevirus Is the Founding Member of a Lineage in Family Marseilleviridae. *Viruses* **2016**, *8*, 76. [CrossRef] [PubMed]
31. Schulz, F.; Alteio, L.; Goudeau, D.; Ryan, E.M.; Yu, F.B.; Malmstrom, R.R.; Blanchard, J.; Woyke, T. Hidden diversity of soil giant viruses. *Nat. Commun.* **2018**, *9*, 4881. [CrossRef]
32. Backstrom, D.; Yutin, N.; Jorgensen, S.L.; Dharamshi, J.; Homa, F.; Zaremba-Niedwiedzka, K.; Spang, A.; Wolf, Y.I.; Koonin, E.V.; Ettema, T.J.G. Virus Genomes from Deep Sea Sediments Expand the Ocean Megavirome and Support Independent Origins of Viral Gigantism. *MBio* **2019**, *10*. [CrossRef] [PubMed]
33. Colson, P.; La Scola, B.; Levasseur, A.; Caetano-Anolles, G.; Raoult, D. Mimivirus: Leading the way in the discovery of giant viruses of amoebae. *Nat. Rev. Microbiol.* **2017**, *15*, 243–254. [CrossRef] [PubMed]
34. Philippe, N.; Legendre, M.; Doutre, G.; Coute, Y.; Poirot, O.; Lescot, M.; Arslan, D.; Seltzer, V.; Bertaux, L.; Bruley, C.; et al. Pandoraviruses: Amoeba viruses with genomes up to 2.5 Mb reaching that of parasitic eukaryotes. *Science* **2013**, *341*, 281–286. [CrossRef] [PubMed]
35. Sharma, V.; Colson, P.; Pontarotti, P.; Raoult, D. Mimivirus inaugurated in the 21st century the beginning of a reclassification of viruses. *Curr. Opin. Microbiol.* **2016**, *31*, 16–24. [CrossRef] [PubMed]
36. Rolland, C.; Andreani, J.; Louazani, A.C.; Aherfi, S.; Francis, R.; Rodrigues, R.; Silva, L.S.; Sahmi, D.; Mougari, S.; Chelkha, N.; et al. Discovery and Further Studies on Giant Viruses at the IHU Mediterranee Infection That Modified the Perception of the Virosphere. *Viruses* **2019**, *11*. [CrossRef] [PubMed]
37. Colson, P.; La Scola, B.; Raoult, D. Giant Viruses of Amoebae: A Journey Through Innovative Research and Paradigm Changes. *Annu. Rev. Virol.* **2017**, *4*, 61–85. [CrossRef]
38. Claverie, J.M.; Abergel, C. Mimiviridae: An Expanding Family of Highly Diverse Large dsDNA Viruses Infecting a Wide Phylogenetic Range of Aquatic Eukaryotes. *Viruses* **2018**, *10*. [CrossRef]
39. Colson, P.; Levasseur, A.; La Scola, B.; Sharma, V.; Nasir, A.; Pontarotti, P.; Caetano-Anolles, G.; Raoult, D. Ancestrality and Mosaicism of Giant Viruses Supporting the Definition of the Fourth TRUC of Microbes. *Front. Microbiol.* **2018**, *9*, 2668. [CrossRef] [PubMed]
40. Legendre, M.; Santini, S.; Rico, A.; Abergel, C.; Claverie, J.M. Breaking the 1000-gene barrier for Mimivirus using ultra-deep genome and transcriptome sequencing. *Virol. J.* **2011**, *8*, 99. [CrossRef] [PubMed]
41. Yutin, N.; Wolf, Y.I.; Raoult, D.; Koonin, E.V. Eukaryotic large nucleo-cytoplasmic DNA viruses: Clusters of orthologous genes and reconstruction of viral genome evolution. *Virol. J.* **2009**, *6*, 223. [CrossRef] [PubMed]

42. Koonin, E.V.; Yutin, N. Evolution of the Large Nucleocytoplasmic DNA Viruses of Eukaryotes and Convergent Origins of Viral Gigantism. *Adv. Virus Res.* **2019**, *103*, 167–202. [CrossRef] [PubMed]
43. Andreani, J.; Aherfi, S.; Bou Khalil, J.Y.; Di Pinto, F.; Bitam, I.; Raoult, D.; Colson, P.; La Scola, B. Cedratvirus, a Double-Cork Structured Giant Virus, is a Distant Relative of Pithoviruses. *Viruses* **2016**, *8*. [CrossRef]
44. Wilhelm, S.W.; Coy, S.R.; Gann, E.R.; Moniruzzaman, M.; Stough, J.M. Standing on the Shoulders of Giant Viruses: Five Lessons Learned about Large Viruses Infecting Small Eukaryotes and the Opportunities They Create. *PLoS Pathog.* **2016**, *12*, e1005752. [CrossRef] [PubMed]
45. Iyer, L.M.; Aravind, L.; Koonin, E.V. Common origin of four diverse families of large eukaryotic DNA viruses. *J. Virol.* **2001**, *75*, 11720–11734. [CrossRef] [PubMed]
46. Nasir, A.; Kim, K.M.; Caetano-Anolles, G. Giant viruses coexisted with the cellular ancestors and represent a distinct supergroup along with superkingdoms Archaea, Bacteria and Eukarya. *BMC Evol. Biol.* **2012**, *12*, 156. [CrossRef] [PubMed]
47. Fischer, M.G.; Allen, M.J.; Wilson, W.H.; Suttle, C.A. Giant virus with a remarkable complement of genes infects marine zooplankton. *Proc. Natl. Acad. Sci. USA* **2010**, *107*, 19508–19513. [CrossRef]
48. Yutin, N.; Wolf, Y.I.; Koonin, E.V. Origin of giant viruses from smaller DNA viruses not from a fourth domain of cellular life. *Virology* **2014**, *466–467*, 38–52. [CrossRef]
49. Koonin, E.V.; Yutin, N. Multiple evolutionary origins of giant viruses. *F1000Research* **2018**, *7*. [CrossRef]
50. Arslan, D.; Legendre, M.; Seltzer, V.; Abergel, C.; Claverie, J.M. Distant Mimivirus relative with a larger genome highlights the fundamental features of Megaviridae. *Proc. Natl. Acad. Sci. USA* **2011**, *108*, 17486–17491. [CrossRef]
51. Iyer, L.M.; Balaji, S.; Koonin, E.V.; Aravind, L. Evolutionary genomics of nucleo-cytoplasmic large DNA viruses. *Virus Res.* **2006**, *117*, 156–184. [CrossRef] [PubMed]
52. Maruyama, F.; Ueki, S. Evolution and Phylogeny of Large DNA Viruses, Mimiviridae and Phycodnaviridae Including Newly Characterized Heterosigma akashiwo Virus. *Front. Microbiol.* **2016**, *7*, 1942. [CrossRef]
53. Filee, J.; Chandler, M. Gene exchange and the origin of giant viruses. *Intervirology* **2010**, *53*, 354–361. [CrossRef]
54. Boyer, M.; Yutin, N.; Pagnier, I.; Barrassi, L.; Fournous, G.; Espinosa, L.; Robert, C.; Azza, S.; Sun, S.; Rossmann, M.G.; et al. Giant Marseillevirus highlights the role of amoebae as a melting pot in emergence of chimeric microorganisms. *Proc. Natl. Acad. Sci. USA* **2009**, *106*, 21848–21853. [CrossRef]
55. Moreira, D.; Brochier-Armanet, C. Giant viruses, giant chimeras: The multiple evolutionary histories of Mimivirus genes. *BMC Evol. Biol.* **2008**, *8*, 12. [CrossRef]
56. Gallot-Lavallee, L.; Blanc, G.; Claverie, J.M. Comparative Genomics of Chrysochromulina Ericina Virus and Other Microalga-Infecting Large DNA Viruses Highlights Their Intricate Evolutionary Relationship with the Established Mimiviridae Family. *J. Virol.* **2017**, *91*. [CrossRef]
57. Yoosuf, N.; Pagnier, I.; Fournous, G.; Robert, C.; La Scola, B.; Raoult, D.; Colson, P. Complete genome sequence of Courdo11 virus, a member of the family Mimiviridae. *Virus Genes.* **2014**, *48*, 218–223. [CrossRef]
58. Rodrigues, R.A.L.; Abrahao, J.S.; Drumond, B.P.; Kroon, E.G. Giants among larges: How gigantism impacts giant virus entry into amoebae. *Curr. Opin. Microbiol.* **2016**, *31*, 88–93. [CrossRef]
59. Silva, L.; Andrade, A.; Dornas, F.P.; Rodrigues, R.A.L.; Arantes, T.; Kroon, E.G.; Bonjardim, C.A.; Abrahao, J.S. Cedratvirus getuliensis replication cycle: An in-depth morphological analysis. *Sci. Rep.* **2018**, *8*, 4000. [CrossRef]
60. Suzan-Monti, M.; La Scola, B.; Barrassi, L.; Espinosa, L.; Raoult, D. Ultrastructural characterization of the giant volcano-like virus factory of Acanthamoeba polyphaga Mimivirus. *PLoS ONE* **2007**, *2*, e328. [CrossRef]
61. Diesend, J.; Kruse, J.; Hagedorn, M.; Hammann, C. Amoebae, Giant Viruses, and Virophages Make Up a Complex, Multilayered Threesome. *Front. Cell. Infect. Microbiol.* **2017**, *7*, 527. [CrossRef]
62. Schulz, F.; Yutin, N.; Ivanova, N.N.; Ortega, D.R.; Lee, T.K.; Vierheilig, J.; Daims, H.; Horn, M.; Wagner, M.; Jensen, G.J.; et al. Giant viruses with an expanded complement of translation system components. *Science* **2017**, *356*, 82–85. [CrossRef] [PubMed]
63. Abrahao, J.S.; Araujo, R.; Colson, P.; La Scola, B. The analysis of translation-related gene set boosts debates around origin and evolution of mimiviruses. *PLoS Genet.* **2017**, *13*, e1006532. [CrossRef] [PubMed]
64. Raoult, D.; Forterre, P. Redefining viruses: Lessons from Mimivirus. *Nat. Rev. Microbiol.* **2008**, *6*, 315–319. [CrossRef] [PubMed]

65. Abrahao, J.; Silva, L.; Silva, L.S.; Khalil, J.Y.B.; Rodrigues, R.; Arantes, T.; Assis, F.; Boratto, P.; Andrade, M.; Kroon, E.G.; et al. Tailed giant Tupanvirus possesses the most complete translational apparatus of the known virosphere. *Nat. Commun.* **2018**, *9*, 749. [CrossRef] [PubMed]
66. Bahir, I.; Fromer, M.; Prat, Y.; Linial, M. Viral adaptation to host: A proteome-based analysis of codon usage and amino acid preferences. *Mol. Syst. Biol.* **2009**, *5*, 311. [CrossRef] [PubMed]
67. Barik, S. A Family of Novel Cyclophilins, Conserved in the Mimivirus Genus of the Giant DNA Viruses. *Comput. Struct. Biotechnol. J.* **2018**, *16*, 231–236. [CrossRef]
68. Yutin, N.; Colson, P.; Raoult, D.; Koonin, E.V. Mimiviridae: Clusters of orthologous genes, reconstruction of gene repertoire evolution and proposed expansion of the giant virus family. *Virol. J.* **2013**, *10*, 106. [CrossRef]
69. Van Etten, J.L. Unusual life style of giant chlorella viruses. *Annu. Rev. Genet.* **2003**, *37*, 153–195. [CrossRef] [PubMed]
70. Deeg, C.M.; Chow, C.T.; Suttle, C.A. The kinetoplastid-infecting Bodo saltans virus (BsV), a window into the most abundant giant viruses in the sea. *Elife* **2018**, *7*. [CrossRef]
71. Shukla, A.; Chatterjee, A.; Kondabagil, K. The number of genes encoding repeat domain-containing proteins positively correlates with genome size in amoebal giant viruses. *Virus Evol.* **2018**, *4*, vex039. [CrossRef]
72. Bekliz, M.; Colson, P.; La Scola, B. The Expanding Family of Virophages. *Viruses* **2016**, *8*. [CrossRef]
73. Desnues, C.; La Scola, B.; Yutin, N.; Fournous, G.; Robert, C.; Azza, S.; Jardot, P.; Monteil, S.; Campocasso, A.; Koonin, E.V.; et al. Provirophages and transpovirons as the diverse mobilome of giant viruses. *Proc. Natl. Acad. Sci. USA* **2012**, *109*, 18078–18083. [CrossRef]
74. Fischer, M.G. Sputnik and Mavirus: More than just satellite viruses. *Nat. Rev. Microbiol.* **2012**, *10*, 78. [CrossRef]
75. Fischer, M.G.; Hackl, T. Host genome integration and giant virus-induced reactivation of the virophage mavirus. *Nature* **2016**, *540*, 288–291. [CrossRef] [PubMed]
76. Levasseur, A.; Bekliz, M.; Chabriere, E.; Pontarotti, P.; La Scola, B.; Raoult, D. MIMIVIRE is a defence system in mimivirus that confers resistance to virophage. *Nature* **2016**, *531*, 249–252. [CrossRef] [PubMed]
77. Dou, C.; Yu, M.; Gu, Y.; Wang, J.; Yin, K.; Nie, C.; Zhu, X.; Qi, S.; Wei, Y.; Cheng, W. Structural and Mechanistic Analyses Reveal a Unique Cas4-like Protein in the Mimivirus Virophage Resistance Element System. *iScience* **2018**, *3*, 1–10. [CrossRef] [PubMed]
78. Koonin, E.V.; Krupovic, M. Polintons, virophages and transpovirons: A tangled web linking viruses, transposons and immunity. *Curr. Opin. Virol.* **2017**, *25*, 7–15. [CrossRef]
79. Campbell, S.; Aswad, A.; Katzourakis, A. Disentangling the origins of virophages and polintons. *Curr. Opin. Virol.* **2017**, *25*, 59–65. [CrossRef] [PubMed]
80. La Scola, B.; Desnues, C.; Pagnier, I.; Robert, C.; Barrassi, L.; Fournous, G.; Merchat, M.; Suzan-Monti, M.; Forterre, P.; Koonin, E.; et al. The virophage as a unique parasite of the giant mimivirus. *Nature* **2008**, *455*, 100–104. [CrossRef] [PubMed]
81. Suttle, C.A. Marine viruses–major players in the global ecosystem. *Nat. Rev. Microbiol.* **2007**, *5*, 801–812. [CrossRef]
82. Chatterjee, A.; Sicheritz-Ponten, T.; Yadav, R.; Kondabagil, K. Genomic and metagenomic signatures of giant viruses are ubiquitous in water samples from sewage, inland lake, waste water treatment plant, and municipal water supply in Mumbai, India. *Sci. Rep.* **2019**, *9*, 3690. [CrossRef]
83. Yoshikawa, G.; Blanc-Mathieu, R.; Song, C.; Kayama, Y.; Mochizuki, T.; Murata, K.; Ogata, H.; Takemura, M. Medusavirus, a novel large DNA virus discovered from hot spring water. *J. Virol.* **2019**. [CrossRef]
84. Wilhelm, S.W.; Bird, J.T.; Bonifer, K.S.; Calfee, B.C.; Chen, T.; Coy, S.R.; Gainer, P.J.; Gann, E.R.; Heatherly, H.T.; Lee, J.; et al. A Student's Guide to Giant Viruses Infecting Small Eukaryotes: From Acanthamoeba to Zooxanthellae. *Viruses* **2017**, *9*. [CrossRef]
85. Watanabe, T.; Yamazaki, S.; Maita, C.; Matushita, M.; Matsuo, J.; Okubo, T.; Yamaguchi, H. Lateral Gene Transfer Between Protozoa-Related Giant Viruses of Family Mimiviridae and Chlamydiae. *Evol. Bioinform. Online* **2018**, *14*. [CrossRef]
86. Silva, L.C.; Almeida, G.M.; Oliveira, D.B.; Dornas, F.P.; Campos, R.K.; La Scola, B.; Ferreira, P.C.; Kroon, E.G.; Abrahao, J.S. A resourceful giant: APMV is able to interfere with the human type I interferon system. *Microbes Infect.* **2014**, *16*, 187–195. [CrossRef]
87. Scarpellini, E.; Ianiro, G.; Attili, F.; Bassanelli, C.; De Santis, A.; Gasbarrini, A. The human gut microbiota and virome: Potential therapeutic implications. *Dig. Liver Dis.* **2015**, *47*, 1007–1012. [CrossRef]

8

A Needle in a Haystack: Tracing Bivalve-Associated Viruses in High-Throughput Transcriptomic Data

Umberto Rosani [1], Maxwell Shapiro [2], Paola Venier [1] and Bassem Allam [3,*]

1 Department of Biology, University of Padua, 35121 Padua, Italy; umberto.rosani@unipd.it (U.R.); paola.venier@unipd.it (P.V.)
2 Department of Applied Mathematics and Statistics, Stony Brook University, Stony Brook, NY 11794-5000, USA; maxwell.shapiro@stonybrook.edu
3 School of Marine and Atmospheric Sciences, Stony Brook University, Stony Brook, NY 11794-5000, USA
* Correspondence: bassem.allam@stonybrook.edu

Abstract: Bivalve mollusks thrive in environments rich in microorganisms, such as estuarine and coastal waters, and they tend to accumulate various particles, including viruses. However, the current knowledge on mollusk viruses is mainly centered on few pathogenic viruses, whereas a general view of bivalve-associated viromes is lacking. This study was designed to explore the viral abundance and diversity in bivalve mollusks using transcriptomic datasets. From analyzing RNA-seq data of 58 bivalve species, we have reconstructed 26 nearly complete and over 413 partial RNA virus genomes. Although 96.4% of the predicted viral proteins refer to new viruses, some sequences belong to viruses associated with bivalve species or other marine invertebrates. We considered short non-coding RNAs (sncRNA) and post-transcriptional modifications occurring specifically on viral RNAs as tools for virus host-assignment. We could not identify virus-derived small RNAs in sncRNA reads obtained from the oyster sample richest in viral reads. Single Nucleotide Polymorphism (SNP) analysis revealed 938 A-to-G substitutions occurring on the 26 identified RNA viruses, preferentially impacting the AA di-nucleotide motif. Under-representation analysis revealed that the AA motif is under-represented in these bivalve-associated viruses. These findings improve our understanding of bivalve viromes, and set the stage for targeted investigations on the specificity and dynamics of identified viruses.

Keywords: bivalve; virome; RNA-seq; RNA viruses; sncRNA; ADAR; RNAi

1. Introduction

Viruses are the most abundant biological entity on the Earth, likely outnumbering bacteria and eukaryotic cells [1], with the oceans being the most likely richest reservoir of virus biodiversity [2]. The only constraint that viruses have is the need for a host for their replication, either to take advantage of the host replication machinery, or to hijack the genome to freely replicate as selfish genetic elements [3]. The evolutionary success of viruses is supported by highly dynamic genomes, which can undergo punctual changes or integration events that enable the circumvention of host immune defenses, the capture of new genes, and even host switching, among other events [4,5]. The frequent exchange of genetic material is evident in the highly variable sizes of viral genomes. While RNA viruses seem to have a ~32 kb size constraint [6], the genomes of DNA viruses can be uncommonly large, with the giant *Mimiviruses* genomes being in the order of megabases and far exceeding the few kilobases of circular single stranded DNA genomes of *cress viruses* [7,8]. The presence of an antiviral system in every living organism further supports the global distribution of viruses [9], although their biological roles go beyond pathogenicity [10]. In fact, viruses are responsible for selective pressures causing evolutionary transitions [11] as they drive the dynamics of host populations and interfere with biological invasions [12]. The gene flux from viruses to eukaryotic organisms is suggested to drive the long-term evolution of host genomes [13]. Conversely, the evolutionary pressure of host antiviral defenses shapes viral genomes in a never-ending arms race [14–17]. According to the sequence data currently available, the viruses identified so far represent a numerically insignificant portion of viral biodiversity, possibly no more than 1% of the extant viruses [1,18]. Thanks to an unprecedented level of sensitivity and accuracy, high-throughput sequencing (HTS) has become the gold standard for viral discovery and for advancements in the characterization of viral metagenomes [19,20], although most of the so-generated viral sequences remain "unclassified" due to uncertainty about authentic virus hosts. As viruses are mostly hidden in the host nucleic acids, and an unusually high sequencing coverage would be necessary to allow their detection, the current representations of the extant virosphere are fragmentary, and they crucially depend on sample preparation strategy, sequencing technology, and sequencing depth [21]. Nevertheless, DNA and RNA sequencing datasets often contain viral sequences, and committed analyses can provide snapshots of the viromes associated with a given organism [5,22].

Although it is steadily expanding, virus discovery and the study of antiviral immunity in invertebrates is biased towards a few model organisms and arthropods of economic and medical importance [23]. In this respect, highly speciose and ecologically important groups like mollusks, and the Lophotrochozoa more widely, remain largely unstudied, leaving huge taxonomic gaps in our knowledge. Since their initial diversification in the early Cambrian (Paleozoic Era), bivalves successfully colonized a variety of aquatic environments, from cold-water seas, to freshwater basins and deep anoxic vents, with some species showing an invasive behavior [24–27]. A number of bivalve species have been investigated for their peculiar adaptation strategies [28], innate immune systems [29], and bio-inspired applications [30] as well as for their use as models for human health [31]. Today, few bivalve genome drafts are available, whereas more than 2,100 transcriptomic datasets have been deposited in public databases (NCBI SRA archive, accessed in November 2018). So far, very few viruses of bivalve mollusks have been described, mainly those that have major negative economic impacts on farmed species. In particular, a herpesvirus (*Ostreid herpesvirus-1*, OsHV-1) associated with temperature-related oyster mortalities [32] has become a pressing issue for the production sector, and today, OsHV-1 and its variants are described in numerous studies [33–38]. Viruses belonging to the *Papovaviridae* and *Iridoviridae* families have been associated with bivalve diseases, whereas a few members of the *Togaviridae*, *Reoviridae*, *Birnaviridae*, and *Picornaviridae* virus families have been reported without evidence of associated disease [39,40]. Until the advent of HTS technologies, the identification of these viruses was mainly based on electron microscopy, and seldom validated by molecular studies [41].

Virome discovery through RNA HTS is challenging when applied to bivalve samples. According to the ability to detect minute quantities of viral nucleic acids, HTS also catches sequences that possibly derive from tissue surface contamination, or from the simple transit of another virus host in bivalve tissues [42,43]. The identification of giant viruses and human viruses in bivalve samples mainly accumulated in the gills and gut by filter-feeding [44–46], and the presence of an algal virus (Heterosigma akashiwo RNA virus-1) in the gills of both *Crassostrea gigas* and *Mytilus galloprovincialis* growing up in association [47], exemplify the importance of developing new approaches for assigning a virus to its authentic host. Ecological role and economic importance, peculiar genome features, differential susceptibility to pathogens, as well as their tendency to accumulate microbes highlights filter-feeding bivalves as fascinating models for virus–host interaction studies. The objective of this work was to explore the diversity and distribution of bivalve RNA viruses through the analysis of available RNA-seq samples. To do so, we performed an extensive analysis of the HTS transcriptome data of bivalves, we recovered bivalve-associated RNA viruses, and we traced their distribution over many bivalve RNA-seq samples. Moreover, we investigated how different RNA selection methods applied during library preparation can affect the performance of viral-oriented HTS analysis.

Finally, we tested two different in silico approaches for assigning bivalve-associated viruses to their genuine hosts.

2. Materials and Methods

The overall analysis pipeline is summarized in Figure 1, and it is detailed through the following paragraphs. The analyses are based either on the available RNA-seq datasets, or on newly produced data that we submitted to the NCBI SRA archive (the corresponding IDs are cited in the text and tables).

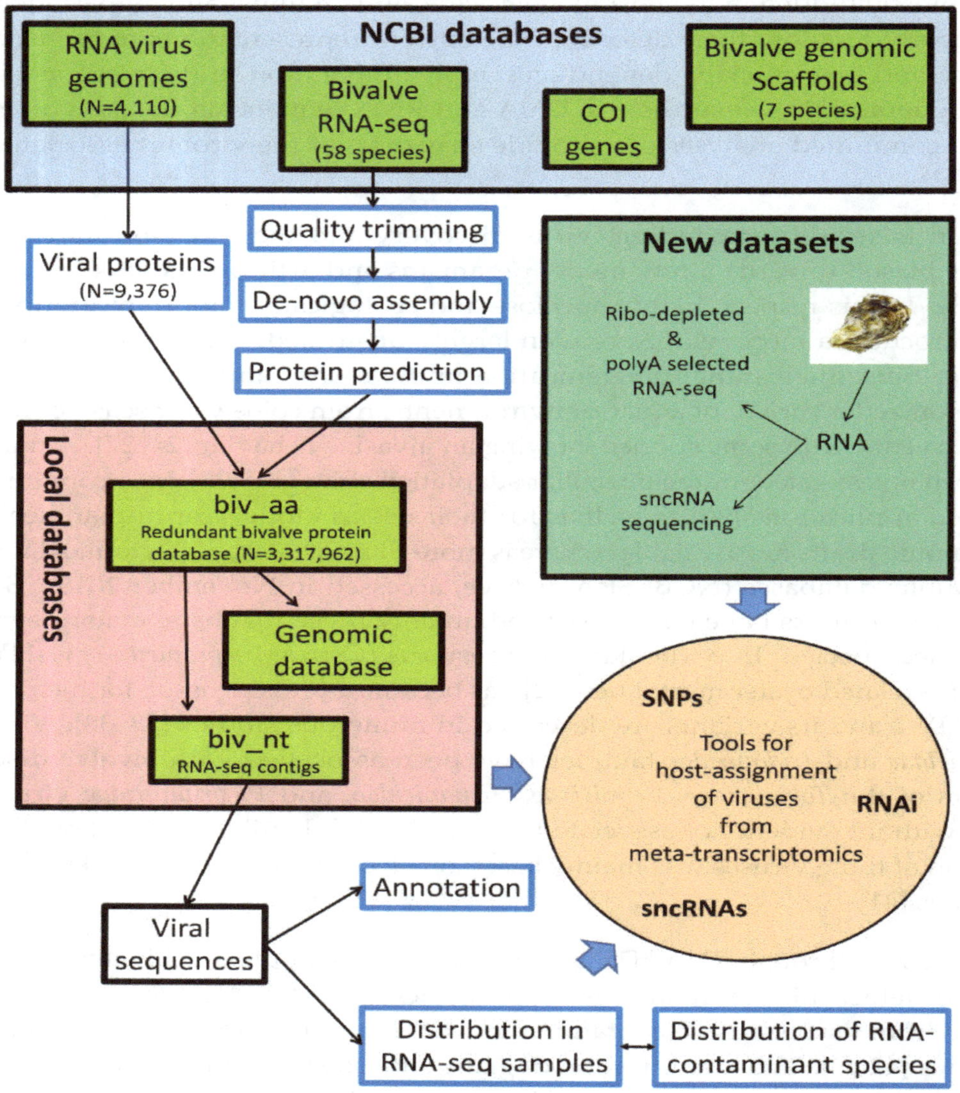

Figure 1. Summary of the analysis pipeline. The graph summarizes all of the steps that were used to extract viral sequences from bivalve RNA-seq datasets. Additional details are reported in corresponding sections of Materials and Methods.

2.1. Data Retrieval

Public sequence datasets were retrieved from NCBI databases in April 2017. A total of 7125 viral genomes, including 3008 RNA viruses, were downloaded from the NCBI Genome database. Additionally 1102 invertebrate-associated RNA virus genomes were downloaded from the NCBI nucleotide database [43,47] for a total of 4110 genomes of RNA viruses. RNA-seq samples referring to 58 bivalve species and four pooled bivalve or gastropod meta-transcriptomic samples were obtained from NCBI SRA archives. Genome drafts of five bivalve species (*Bathymodiolus platifrons*, *C. virginica*, *Mizuhopecten yessoensi*, *Modiolus philippinarum*, and *M. galloprovincialis*) were downloaded from the NCBI WGS database, while *C. gigas* and *Pinctada fucata* genomes were obtained from EnsembleMetazoa release35 and from [48], respectively. Supplementary Table S1 summarizes the sequence datasets used in this work. *Cytochrome C Oxidase Subunit I* (COI) sequences were downloaded from the NCBI nucleotide archive, and their redundancies were reduced using *cd-hit-est* [49], applying a cut-off of 95% similarity. In order to compare two RNA selection methods for their aptitudes of viral read recovery, we retrieved two RNA-seq datasets obtained from a single *C. gigas* sample (described elsewhere, SRR8237210 and SRR7636587 for polyA and ribo-depleted data, respectively).

2.2. Transcriptome de novo Assembly, ORF Prediction, and Protein Domain Mapping

RNA-seq reads were trimmed for the presence of adaptor sequences, and for quality, using TrimGalore! [50], allowing for a maximum of two ambiguous bases and a quality threshold of PHRED20. Trimmed reads were de novo assembled using CLC Genomic Workbench v.10 (Qiagen, Hilden, Germany), setting automatic word and bubble sizes, and a minimal contig length of 200 bp. The resulting contigs were subjected to open reading frame (ORF) prediction, using the transdecoder tool included in the Trinity suite [51], applying a minimal ORF length of 100 codons. HMMer v.3.1 [52] was used to identify the presence of conserved protein domains (Pfam-A models, v.29 [53], applying a cut-off *E*-value of 10^{-5}.

2.3. Identification of Viral Sequences

A redundant BLAST database [54] was built, using the predicted proteins obtained from all of the transcriptomic assemblies of bivalve species. All the annotated ORFs encoded by 4110 RNA virus genomes were translated into amino acids, and the resulting 9376 protein sequences were used as blast queries against the bivalve database (blastp, cut-off *E*-value 10^{-50}). Moreover, all the bivalve-derived protein sequences encoding a viral RNA-dependent RNA-polymerase (vRdRp) domain were selected. A vRdRp was identified by using six different PFAM Hidden Markov Models, corresponding to IDs: PF00680, PF00978, PF00998, PF02123, PF07925, and PF04197. Bivalve genome scaffolds were used to compose a genomic database to discriminate between host-encoded and viral sequences. Sequences identified from the redundant protein database, and from the search of the vRdRp domains, and showing no matches against bivalve genomes, were further processed to reduce the redundancy, applying a cut-off of 90% of similarity (*cd-hit*). The resulting protein sequences were used to recover the corresponding nucleotidic contigs from the initial transcriptome assemblies and they were considered as complete or partial genomes of RNA viruses. For the purposes of this paper, a viral genome was considered to be "nearly complete" if it was composed of a unique contig that was longer than 5 kb and encoding at least one complete ORF.

2.4. Distribution of Viruses among RNA-Seq Samples, Expression Analysis, and SNP Calling

The amount of reads mapping to the "nearly complete" viral genomes in selected RNA-seq samples was determined by stringent mapping of the trimmed reads on the viral genome sequences (0.9 both for length and similarity fraction, CLC mapper tool). For a selection of informative RNA-seq samples, the total read counts were used to calculate the percentage of reads mapping to each virus over the total numbers of reads of the sample, thus providing a comparison of the amount of viral RNA between RNA-seq samples that was not biased by different sequencing depths or read layouts. To obtain the expression profiles of selected oyster RNAi-related genes, 183 RNA-seq datasets (Supplementary Table S1) were mapped onto *C. gigas* gene models [55] and used to compute expression values such as transcripts per million (TPM) [56]. RNA-seq data were also used to call single-nucleotide variations (single-nucleotide polymorphisms, SNPs) across viral genomes. Specifically, to detect genuine SNPs, the trimmed reads were mapped onto the "nearly complete" viral genomes, setting 0.5 and 0.8 for the length and similarity fractions, respectively. A SNP was called if it was present in at least 1% of the locally aligned reads after using the following parameters: minimum average quality of the five surrounding bases, PHRED30; minimum required coverage, 50×; minimum required count, 5. The SNPs were annotated according to the neighbor base.

2.5. Estimation of the Contamination Levels of RNA-Seq Samples

To provide an estimation of the fraction of reads that were not related to the declared biological sample (as indicated in the SRA details), we mapped the RNA-seq reads onto a collection of 205,357 non-redundant COI sequences. Reads were mapped applying a similarity fraction of 0.8, over 0.8 of the read length (CLC mapper tool) and the TPM values were computed. Similarly, to estimate the presence of known viruses among the RNA-seq datasets, the amount of reads mapping to the 7125 virus genomes obtained from NCBI was also computed (in this case, by applying 0.9 and 0.9 for the length and similarity fractions, respectively).

2.6. Small RNA Sequencing and Reads Analysis

The fraction of small RNAs (<200 bp) of the *C. gigas* sample that was used to prepare the paired polyA and ribo-depleted libraries were extracted using the Mirvana kit (Thermofisher, Waltham, MA, USA). RNA was quantified by using a Qubit fluorimeter instrument, and the RNA size profile was determined with an Agilent small RNA chip (Agilent, Santa Clara, CA, USA). Library preparation and sequencing (PE150) was outsourced and carried out on an HiSeq Illumina platform (Admera Health, New York, NY, USA), and submitted to the NCBI SRA archive, under the accession ID SRR8587800. The paired reads were trimmed for quality, and for the presence of adaptors, as described for mRNA reads, and the correctly paired reads were joined into fragments. The resulting clean fragments, in a length range of 15–50 nt, were used for the detection of viral-derived small RNAs (vsRNAs) by direct mapping on the identified viral contigs or by using the VirusDetect pipeline [57]. To discriminate between genuine vsRNAs versus RNA degradation products, we correlated the number of mapped sRNA reads with the viral expression levels.

2.7. Analysis of Viral Genome Editing

The genomes of the RNA viruses retrieved from NCBI were analyzed for the distribution of the frequency of di-nucleotides as hallmarks of the virus genome fitness (i.e., adaptive genome changes). To look for these adaptive genome changes, we used the cytidine deaminase under-representation reporter (CDUR) [58]. Given the user-defined motifs and an input coding sequence, CDUR effectively utilizes a permutation test to determine whether the given motif is significantly depleted in the input sequence than one would expect by chance (details below). The two main metrics that are analyzed are: 1. the "below" metric, which determines whether the number of occurrences of a motif is significantly fewer than expected, and 2. the "repTrFrac" metric, which determines the ratio of motifs that would incur a non-synonymous transition mutation, against the total number of those motifs in the sequence, which is significantly fewer than expected. Both of these metrics are determined by shuffling the coding sequence at the third position of each codons, so that the underlying amino acid sequence is unchanged. We chose this method of shuffling as it also preserves the GC content of the input sequence, as changing the GC content has been reported to yield biased results [14]. This shuffling is repeated 1000 times; in each shuffled sequence, we counted the number of user-defined motifs ("below" metric), as well as the ratio of nonsynonymous transition mutations that occur at those motifs, compared to the number of motifs ("repTrFrac" metric). In both cases, we determined the percentage of shuffled sequences with fewer motif counts and repTrFrac counts than that of the input, to yield a statistical p-value. A sequence with a p-value of <0.05 is said to be under-represented in that motif metric, whereas a p-value of >0.95 is said to be over-represented in that motif metric (see Figure 2 in [58]).

A particularly interesting case is when, for a given motif, a sequence is under-represented in the "below" metric, and is over-represented in the "repTrFrac" metric for that motif. This suggests that this sequence has maximally tried to reduce the number of occurrences of that motif, as any further reduction would result in amino acid changes, which may negatively impact that coding sequence. Recent studies have shown that certain gammaherpesvirus may be under such pressures [59]. In this case, the sequence is considered to have attained maximal under-representation. We performed CDUR analysis on 3872 RNA viral genomes with a known host obtained from NCBI (Supplementary Table S1), as well as on the newly recovered "nearly complete viral genomes" presented in this paper.

2.8. Phylogenetic Analysis

Protein sequences referring to vRdRP domains were aligned using MUSCLE [60], and trees were generated with MEGA 6 [61], using neighbor-joining clustering methods with 100 bootstrap replicates. The phylogenetic tree was uploaded to the iTOL server for easier visualization [62].

3. Results

We produced transcriptomic assemblies of 58 bivalve species, and we used all of the predicted proteins to produce a redundant BLAST database, including more than 3 million entries (hereinafter called biv_aa). To identify the putative virus sequences associated with the analyzed RNA-seq samples from different bivalve species, we queried biv_aa with 9376 protein sequences belonging to 4110 known RNA virus genomes (Supplementary Table S1). We extracted additional viral sequences from the same database by searching all six available PFAM domains of viral *RNA-dependent RNA polymerase* (vRdRp). In the absence of a conserved viral gene, we used these domains to identify RNA viruses, since vRdRp is needed for the transcription of the viral genome during productive virus replication [63,64]. For the BLAST searches, we applied a conservative *E*-value of 10^{-50}, and to further limit false positive results, we discharged the BLAST matches with less than five hits. Moreover, we screened all of the positive hits against a database composed of available genomic scaffolds of bivalve species, to remove genomically encoded sequences (Supplementary Table S1). As a result, 708 biv_aa entries showed a genuine similarity to viral sequences, and the conserved domains included in these proteins further sustained their viral origin, since we found 253 vRdRp, 80 *CRPV capsid protein like*, 73 RNA *helicase*, 78 *Dicistroviridae minor capsid protein*, and 69 *Picornavirus capsid protein* domains (Supplementary Table S2). The removal of similar sequences (>90% of similarity) resulted in 413 unique sequences. Most of the redundant sequences were found in transcriptomes of the same bivalve species, either in RNA-seq samples originating from the same geographical location, as in the case of *M. galloprovincialis* or *C. gigas* samples from Goro (Italy), or obtained from geographically unrelated samples of the same species, although few exceptions are present, and they are discussed below. Despite most sequences being retrieved by BLAST searches using viral sequences as queries, they showed a limited similarity to known viruses (Supplementary Table S2). We could confidently assign only 15 sequences (3.6% of the total) to 11 known viruses (BLASTn with an *E*-value lower than 10^{-100} and identity >95%), namely, six bivalve-associated RNA viruses from the lagoon of Goro (Italy), two viruses associated with marine invertebrates from China, plus three other RNA viruses, the algal *Heterosigma akashiwo RNA virus*, the plant virus *Zygocactus virus X*, and the *Sacbrood virus* (Supplementary Table S2).

3.1. Effect of RNA-Seq Library Preparation Protocols on the Detection of Viral Sequences

In order to evaluate the effect of the RNA selection method applied during library preparation on the recovery of viral reads, we analyzed two different datasets, each of them derived from a single biological sample by using alternative RNA selection approaches: polyA RNA selection or ribosomal RNA depletion. The first dataset was prepared specifically for viral meta-transcriptomic analysis, starting from two biological samples (named "mix of bivalves", sample ID: SAMN04625952 and "mix of gastropods", sample ID: SAMN04625958 [43]). We analyzed a second dataset obtained from a single *C. gigas* specimen naturally infected with OsHV-1, using the same two RNA selection methods (sample ID: SAMN09760011). The analysis of the four meta-transcriptomic samples showed that, despite a considerable variability in the numbers of raw reads, the assembled contigs, as well as the number of predicted proteins yielded somewhat comparable values, except for SRR3401755, for which only few proteins could be predicted. For these datasets, polyA-selection allowed for the recovery of a higher ratio of viral to total proteins, particularly for the bivalve samples (Table 1). On the contrary, the analysis of the oyster dataset clearly showed the opposite trend in terms of viral read recovery, since we identified 46 contigs encoding a vRdRP domain in the ribo-depleted sample, compared to 10 in the polyA selected one. Although five out of the 10 polyA viral contigs were also found in the ribo-depleted dataset, the longer contig was always generated from the ribo-depleted dataset.

Table 1. Assembly statistics of four viral-metagenomic samples. SRA and sample IDs, RNA selection methods, number of reads in million, and the number of assembled contigs and predicted proteins are reported. The number of viral RNA-dependent-RNA-polymerase (vRdRp) domains, the ratio of viral protein to total proteins, as well as the number of complete viral genomes, are also indicated.

SRA ID	Sample ID	RNA Selection Method	No. of Reads (M)	No. Of Assembled Contig	No. of Predicted Protein	No. of vRdRp	Ratio of Viral Proteins	No. of Complete Genomes *
SRR3401648	SAMN04625952	Ribo-depletion	99.7	96,102	45,343	30	0.00007	2
SRR3401653		polyA-selection	58.3	120,399	38,498	129	0.00034	9
SRR3401753	SAMN04625958	Ribo-depletion	47.9	180,272	48,687	43	0.00009	6
SRR3401755		polyA-selection	60.3	105,611	14,661	54	0.00037	4
SRR7637587	SAMN09760011	Ribo-depletion	54.1	156,166	41,785	46	0.00011	5
SRR8237210		polyA-selection	52.0	93,172	40,301	10	0.00002	0

* additional details on the complete viral genomes are reported in Table 2.

3.2. *Identification of "Nearly Complete" Viral Genomes*

As mentioned above, we putatively identified 413 viral protein sequences in 364 nucleotidic contigs, indicating that some contigs included more than one viral ORF. Theoretically, each contig can be considered as a viral genome, but if we compare their average length (1.39 kb) with the median lengths of the known RNA virus genomes (4.8 kb), a realistic assumption is that most of these represent incomplete genomes. For the purpose of this paper, we considered "nearly full-length viral genomes" as only being contigs that are longer than 5 kb and encoding at least one complete ORF. Therefore, we identified 26 contigs ranging in length between 5.4 and 9.7 kb as being "nearly complete viral genomes", with 12 contigs encoding two ORFs corresponding to one replicative and one structural protein, while the other contigs (14) encoded a single ORF. These "nearly complete" viruses were named according to the species from which they were assembled (for instance, viruses identified in RNA-seq samples rich in viral sequences and referring to *C. gigas*, *M. galloprovincialis*, *Ruditapes philippinarum*, and *M. edulis*), whereas the unique viral contig found in the *Elliptio complanata* transcriptome (Table 2) was a complete viral genome (Elicom_virus1, 7106 nt). Notably, five nearly complete viral genomes sequences were identified in our ribo-depleted *C. gigas* RNA-seq sample. A total of 10 nearly complete viral genomes sequences could be assigned to a known virus, while other eight other sequences displayed an intermediate/low similarity to known viral sequences, and eight other different sequences referred to completely unknown viruses (the latter being associated with RNA-seq samples of *C. gigas* (4), *E. complanata* (1), *M. galloprovincialis* (1), *Mizuhopecten yessoensis* (1) and *R. philippinarum* (1)). In three viral genomes (*Bivalve RNA virus G1*, *Rudphi virus 4*, and *Heterosigma akashiwo RNA virus*-1) we could identify a polyA tail at the 3′ end of the sequence (Supplementary Figure S1). New sequences or sequences not fully matching the known viral genomes have been deposited in the NCBI database, and the accession IDs are reported in Table 2.

Table 2. Summary of the 26 nearly complete viral genomes. Bivalve species, total number of viral sequences, non-redundant (nr) sequences, and number of "nearly complete viral genomes" are reported. Sequenced tissue and geographical origins of the RNA samples, library type, virus name, and virus distribution in transcriptomes of other bivalves and virus coverage are also reported (as is the total number of viral reads, and as a percentage over the total reads). The BLAST similarities are reported with the *E*-values, NCBI ID, and the description and percentage of identity.

Species	Viral Sequences			Tissue	Geographic Origin	Library Type	Virus Name	Virus Distribution among Bivalve RNA-seq	Total Viral Reads	% of Viral Reads *	NCBI ID	*Blastp*		
	Total	nr	Nearly Complete									*E*-Value	Description	Identity %
Atrina pectinata	17	15	1	mixed	China	PA	Atrpec_virus1	\	2104	0.00202	MG210792	0	Wenzhou picorna-like virus 26	99.57
Crassostrea gigas	148	109	13	gills	Italy	PA	Cragig_virus3	mytgal	898	0.00135	MG210795	\	\	\
						PA	Bivalve hepelivirus G	mytgal	4058	0.00611	KX158876	0	Bivalve hepelivirus G	99.95
						PA	Bivalve RNA virus G5	\	1382	0.00173	KX158874	0	Bivalve RNA virus G5	100
						PA	Bivalve RNA virus G3	mytgal	1286	0.00161	KX158873	0	Bivalve RNA virus G3	100
						PA	Cragig_virus1	\	890	0.00134	MG210793	\	\	\
						PA	Cragig_virus2	mytgal	468	0.00058	MG210794	0	Wenzhou picorna-like virus 24	93.6
						PA	Bivalve RNA virus G1	mytgal	37902	0.04732	KX158871	0	Bivalve RNA virus G1	100
						PA	AY337486	mytgal	5156	0.00644	AY337486	0	Heterosigma akashiwo RNA virus	100
						RD	Cragig_virus6	\	30047	0.05554	MK561968	1^-108	Beihai picorna-like virus 21	66
						RD	Cragig_virus7	\	5493	0.01015	MK561969	3^-40	Wenzhou picorna-like virus 41	70
						RD	Cragig_virus8	\	955	0.00177	MK561970	0	Rhizosolenia setigera RNA virus	69
						RD	Cragig_virus9	\	9200	0.01701	MK561971	\	\	\
						RD	Cragig_virus10	\	568	0.00105	MK561972	\	\	\
Elliptio complanata	2	1	1	mixed	USA	PA	Elicom_virus1	\	2268	0.00552	MG210796	\	\	\

Table 2. *Cont.*

Species	Viral Sequences			Tissue	Geographic Origin	Library Type	Virus Name	Virus Distribution among Bivalve RNA-seq	Total Viral Reads	% of Viral Reads *	NCBI ID	*E*-Value	*Blastp*	
	Total	nr	Nearly Complete										Description	Identity %
Mizuhopecten yessoensis	25	25	1	mixed	China	PA	Mizyes_virus1	\	1974	0.00522	MG210800	\	\	\
Mytilus coruscus			1	mixed	China	PA	Mytcor_virus1	\	4460	0.01028	MG210801	0	Pitaya virus X isolate P37	98.24
Mytilus edulis	37	33	1	mixed	France	PA	Mytedu_virus1	\	1936	0.00694	MG210802	0	Barns Ness breadcrumb sponge aquatic picorna-like virus 2	99
Mytilus galloprovincialis	115	52	3	gills	Italy	PA	Bivalve RNA virus G4	cragig, atrpec	4818	0.00713	KX158875	0	Bivalve RNA virus G4	99.77
						PA	Mytgal_virus1	cragig	\	\	MG210803	\	\	\
						PA	Mytgal_virus2	cragig	1432	0.00212	MG210804	0	Wenzhou picorna-like virus 51	78.7
Ruditapes philippinarum	121	49	5	gills	China	PA	Rudphi_virus1	\	9842	0.02869	MG210805	0	Wenzhou picorna-like virus 38	72.9
				gills	China	PA	Rudphi_virus2	\	1031	0.00301	MG210806	\	\	\
				gills	China	PA	Rudphi_virus3	\	181992	0.29884	MG210807	0	Wenzhou gastropodes virus 2	97.3
				gills	China	PA	Rudphi_virus4	ruddec, ostste, ostlur, cracor, cragig, mytedu	5388365	3.06157	MG210808	0	Wenzhou gastropodes virus 1	92
				larvae	USA	PA	Rudphi_virus5	\	19965	0.02936	MG210809	0	Marine RNA virus BC-4	70

Abbreviations: ruddec, *R. decussatus*; rudphi, *R. philippinarum*; ostste, *O. stentina*; ostlur, *O. lurida*; cragig, *C. gigas*; atrpec, *A. pectinata*; mytgal, *M. galloprovincialis*; mytedu, *M. edulis*; PA, polyadenylated RNA library; RD, ribo-depleted RNA library; * percentage of reads from the RNA-seq sample with the highest number of mapped reads, mapping to the given virus (counting only correctly paired reads).

Subsequently, we evaluated the distribution of these 26 viruses in 226 RNA-seq samples, referring to their putative host species. Since the initial removal of redundant viral proteins suggested that some of these viruses are distributed over RNA-seq samples of multiple species, or they originated from samples that were possibly contaminated by pathogen-associated (e.g., *Perkinsus spp.*) RNAs, we included additional 24 RNA-seq samples in the distribution analysis, for a total of 250 datasets (Supplementary Table S3).

More than eight million reads were mapped onto the 26 viral genomes, with 862,949 and 7,092,869 reads that matched the *Rudphi_virus3* and *Rudphi_virus4* genomes, respectively. Twenty-two viral genomes were covered by at least 1000 reads, and 82 out of 250 RNA-seq samples included more than 1000 viral reads, and for this reason, they were selected for further consideration (Supplementary Table S3). In these 82 samples, the fraction of viral reads over the total ones per single virus usually did not exceed 1‰, except for *Rudphi_virus4*, which was covered by 30‰ of total reads for a larval *R. philippinarum* RNA-seq sample, and *Rudphi_virus3*, which reached 6‰ in one gastropod meta-transcriptomic sample (Table 2, and Supplementary Table S3). Few viruses showed a distribution over samples of different bivalves, e.g., *Rudphi_virus4* (present in *R. decussatus*, *R. philippinarum*, *C. cortenzinesis*, *C. gigas*, and *M. edulis* samples) and *Bivalve RNA virus G4* (present in *C. gigas*, *M. galloprovincialis*, and *Atrina pectinata*, Figure 2). Moreover, the occurrences of *Rudphi_virus4* and *Rudphi_virus3* go beyond bivalve species, since we traced them both in metagenomic gastropod samples. Since *Rudphi_virus3* originated from a *Perkinsus*-infected sample of *R. philippinarum*, and it was traced in 12 clam datasets, we further investigated the presence of this virus in the publicly available *Perkinsus* transcriptome data (11 RNA-seq samples, Supplementary Table S1). As a result, some reads (3.9‰) of a sample of *Perkinsus olseni* trophozoites exposed to clam plasma (SRR2094558) were mapped to this virus, whereas, only 22 viral reads (<0.00001‰) were detected in the paired control (Supplementary Table S3). In contrast, other viruses were associated to the unique RNA sample, for instance, *Elicom_virus1*, *Rudphi_virus5*, and *Mytedu_virus1* (Figure 2 and Supplementary Table S3).

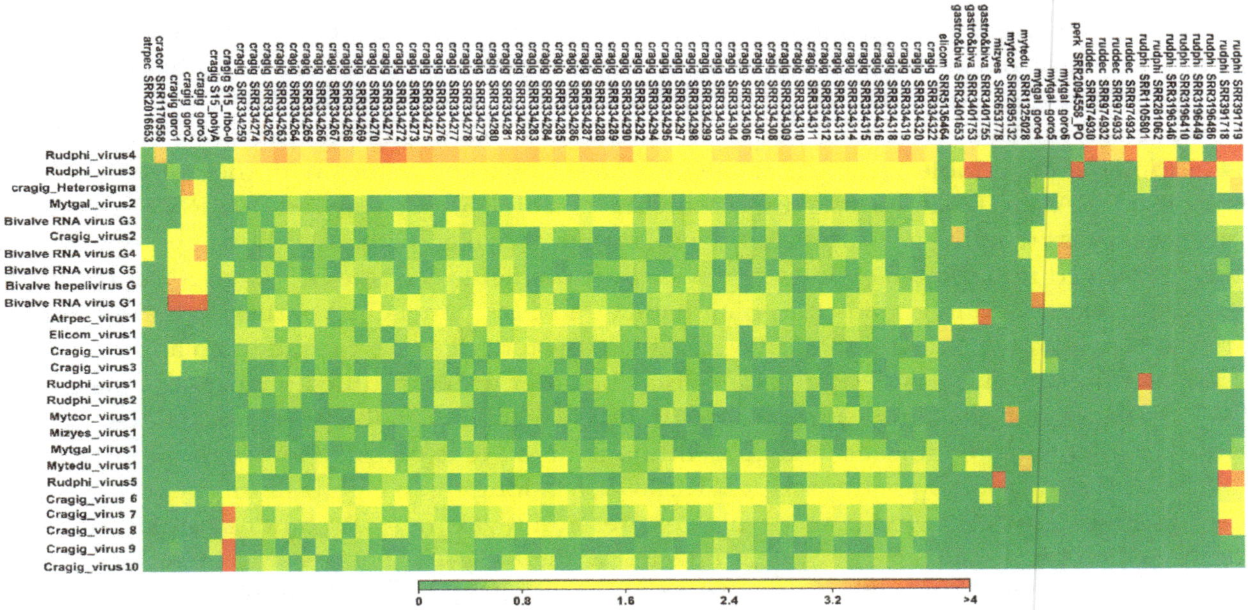

Figure 2. Coverage heat map for the 26 bivalve-associated viruses over 82 RNA-seq samples with at least 1000 viral reads. Data are reported as $\log_{10}$, as depicted in the colored scale. Raw data are reported in Supplementary Table S3.

3.3. Evaluation of Contaminant RNAs in RNA-Seq Samples

We mapped the reads of 16 selected RNA-seq samples to a collection of COI gene sequences, namely, *C. gigas* S15 (Ribo-0 and polyA), a "mix of gastropods" (Ribo-0 and polyA), one *P. olseni*, four *R. decussatus*, and seven *R. philippinarum* samples, since they included reads of multi-host viruses (Figure 2). We calculated the fraction of reads mapping to each COI entry over the total reads that mapped onto the whole COI dataset, and we used it as a tool to evaluate the contribution of biological contaminants in each RNA-seq dataset. As result, 159 COI entries showed at least 0.01% of mapped reads (Supplementary Table S4). Obviously, the first COI entry of each sample corresponded to the sequenced biological sample, thus confirming that the gastropod mix samples were composed of multiple species. For some samples, we observed additional COI entries with lower percentages, as in the two *R. philippinarum* samples with the highest numbers of viral reads (SRR391718-19), where we detected several contaminant species (56 and 57 COI entries with at least 1% of mapped reads, respectively). We noted the presence of a known bivalve-associated tunicate (*Diplosoma listerianum*) in 10 of the tested RNA-seq samples. Although the *D. listerianum* COI value is equal to 100% in the *P. olseni* sample, due to the absence of the Alveolata entries in the COI dataset, our analysis confirmed the absence of clam RNAs in the *Perkinsus* samples, including a high level of *Rudphi_virus3*. Intriguingly, both the Ribo-0 and polyA S15 datasets showed a low contamination of *Lacconectus peguensis* (Coleoptera).

3.4. Tools for the Host-Assignment of Bivalve-Associated Viruses

Our analysis further demonstrated that most of the transcriptome-derived viruses could be only tentatively assigned to a specific host, due to their occurrence in samples of even phylogenetically distant species. Under this context, the application of coverage cut-offs appeared to be an unreliable approach for host-assignment. Therefore, we investigated the feasibility of two alternative approaches for the host-assignment of bivalve-associated viruses obtained from transcriptomic data, as follows.

The first approach investigates the presence of virus-derived RNAi products (vsRNAs), and it is used to reconstruct full-length genomes of viruses infecting arthropods [65–67]. Since the antiviral role of the RNAi system of bivalves has never been demonstrated, we firstly investigated the expression patterns of selected RNAi-related genes (*DICER*, *DROSHA*, *ARGONAUTE*, *PIWI*, and *RNA-dependent RNA polymerase*) in 184 *C. gigas* RNA-seq samples, including some samples that were very rich in viral reads (Supplementary Table S1), to correlate the gene expression values with the presence of actively transcribing RNA and DNA viruses (Supplementary Table S5). We showed that RNAi-related genes are mostly expressed in the early developmental stages of oyster, when two PIWI and one Argonaute transcript showed remarkable expression levels (Supplementary Table S5, panel A), and PIWI1 was preferentially expressed in gonads (Supplementary Table S5, panel C). Apart from these samples, we reported a considerable expression of PIWI1 in three oyster gill samples, and in an additional sample referring to adductor muscles (SRR334286). While the latter result is difficult to explain, the expression of PIWI1 in the oyster gill samples from Goro (Italy) correlated with the presence of RNA viruses (see Figure 2, Supplementary Tables S3 and S5). Although at lower expression levels we reported that one *RNA-dependent RNA polymerase* transcript (EKC38952), belonging to a gene family typically expressed in the digestive gland, showed considerable expression levels in a few other samples, namely two out of three biological replicates of oysters infected with OsHV-1 (12 hours after infection, gills) and a spat sample highly infected by the same virus (Supplementary Table S5, panel B; sample G1). Taken together, these results provide limited evidence for an active role of some components of the RNAi pathway during viral infections in oyster. To further investigate the functionality of RNAi as antiviral system, we sequenced the fraction of small RNAs of the *C. gigas* sample used for library comparison, and found a high number of viral reads belonging both to DNA and RNA viruses (see Figure 2). Small non-coding RNA (sncRNA) sequencing yielded 10.1 million clean fragments in a length range of 15–50 nt. A total of 22,587 sncRNA reads matched the viral contigs identified in this sample, plus the OsHV-1 genome. However, we observed a positive correlation between the expressions of viral genes (using both Ribo-0 and polyA datasets) and the number of sncRNA reads that matched these ORFs (r^2 of 0.994 and 0.996, respectively), suggesting that the reads mostly originated from RNA degradation products, instead of being genuine vsRNAs. We further analyzed the coverage of the sncRNA reads along the five "nearly complete viral genomes" originating from the S15 oyster ribo-depleted data. To do this, we mapped to the viral genomes the sncRNA reads that did not match to the oyster genome (1.436 M reads), and we calculated the size profiles of each of the mapped subsets (Figure 3). Notably, comparing the size profile of the whole sncRNA library with the profile of the sncRNA reads that did not match to the oyster genome, we showed that *C. gigas* sncRNA reads peaked at 21 nt (microRNAs), whereas the unmapped sncRNA reads peaked at 30 nt, indicative of their Piwi-interacting RNA (piRNA) nature. However, only 199 sncRNA reads mapped to one of the five viral genomes reconstructed by using the paired Ribo-depleted RNA-seq reads, and the size profiles showed a low enrichment of 29–30 nt reads with a distribution over the whole viral genome (Figure 3).

Moreover, we subjected the sncRNA reads to VirusDetect, a bioinformatics pipeline that is designed for the identification and reconstruction of viral genomes starting from short reads [57]. Although 169,970 sncRNA reads could be aligned to the viral reference database, and the tool could assemble 40 contigs, the 20–22 nt enrichment fraction was always low, and it did not support their vsRNA nature. According to the presence of numerous OsHV-1 reads in the paired RNA-seq data (the polyA and Ribo-0 datasets), VirusDetect identified several matches to the OsHV-1 genomes, but again, with a low 20–22 nt enrichment fraction.

The second approach that we tested leveraged on the identification of single-nucleotide modifications (SNPs) occurring specifically on viral transcripts produced by the action of host enzymes acting as antiviral defenses. Therefore, we attempted to select and count the subset of total SNPs generated by the host double-stranded RNA (dsRNA) editor enzyme *adenosine deaminase acting on dsRNA* (ADAR), which is assumed to specifically modify viral dsRNAs through A-to-I editing [68]. For each of the 26 viruses, we selected the RNA-seq sample with the higher number of reads, and we called these low-frequency SNPs; among the identified SNPs, we selected the ADAR-compatible ones (A-to-G). We identified 7569 SNPs located on viral coding sequences, and we classified 938 of them as being ADAR-compatible. Considering the 5′ position, we showed that 31% of the selected SNPs had an adenine at the flanking position, while 42% had a thymine (Figure 4a). Also, we searched for the evolutionary footprint of the action of ADAR on viral genomes in parallel. To do this, we used the CDUR tool [58] (see Materials and Methods) to determine under- or over-representation of a motif in a given sequence. Firstly, we used a training set of 3872 genomes of RNA viruses with a known host (Figure 4b and Supplementary Table S1). By analyzing the WA (W = A/T), AA, CA, GA, and TA motifs, the CDUR analysis showed that the TA motif is under-represented in 62.7% of the analyzed ORFs, while the AA, GA, and CA motifs are under-represented in 32.9, 8.1, and 1.5% of ORFs, respectively. Intriguingly, 4% of TA-under-represented ORFs maximized this under-representation, since additional variations will cause non-synonymous SNPs. Although we have to take into consideration that the viral representatives of each of the host classes are variable (Figure 4b), by linking the under-representation values with the viral host, we showed that most (>70%) of the algae, invertebrate, and vertebrate viruses reduced the TA motifs in their coding regions, while we observed moderate percentages (50–60%) for fungal and plant viruses, and lower percentages for bacterial and protozoa viruses (Figure 4c). Accordingly, the sequences with a maximization of the TA reduction were only a small fraction of the ones for fungal, plant, invertebrate, protozoa and vertebrate viruses (Figure 4c). Subsequently, we used the CDUR package to investigate the under-representation of the motif in the ORFs of the nearly complete RNA virus genomes described in this paper (Figure 4d). Consistent with the previous results, only the TA and AA motifs were statistically significantly under-represented. However, we did not observe ORFs with maximized TA reductions, while eight out of 11 ORFs showing AA being under-represented, significantly maximized the AA motif reduction (Figure 4d).

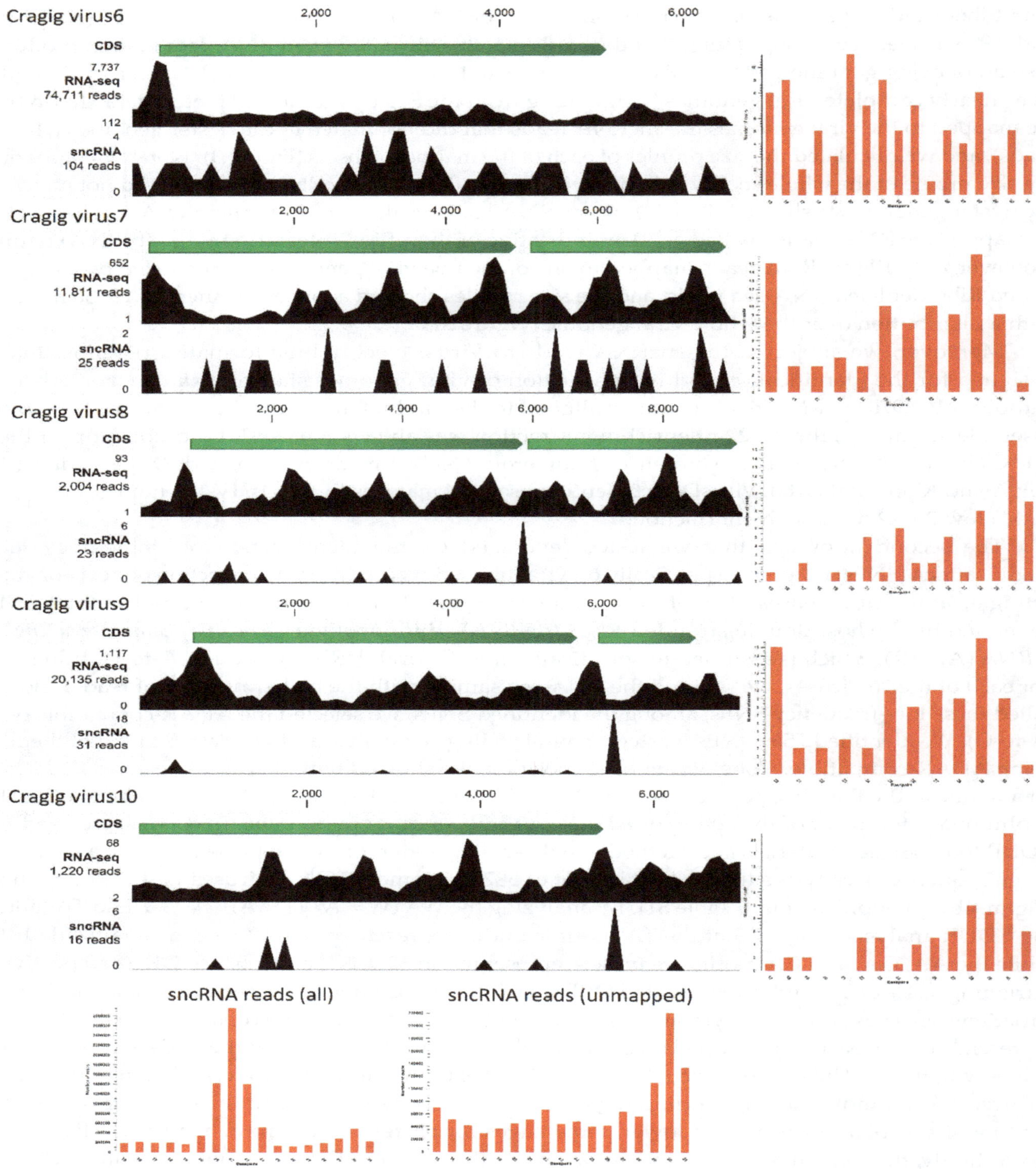

Figure 3. Small non-coding RNA (sncRNA) reads analysis. The RNA-seq and sncRNA read distribution for each of the five "nearly complete viral genomes" reconstructed from the ribo-depleted oyster RNA-seq data are shown. The open reading frames (ORFs) for each virus are shown in green, while the number of mapped reads are reported on the left. The histograms on the right represent the size distributions of the mapped sncRNA reads (in the range of 15–31 nt). The bottom histograms show the size distribution for the whole library (left), and for the reads that did not match the oyster genome (right).

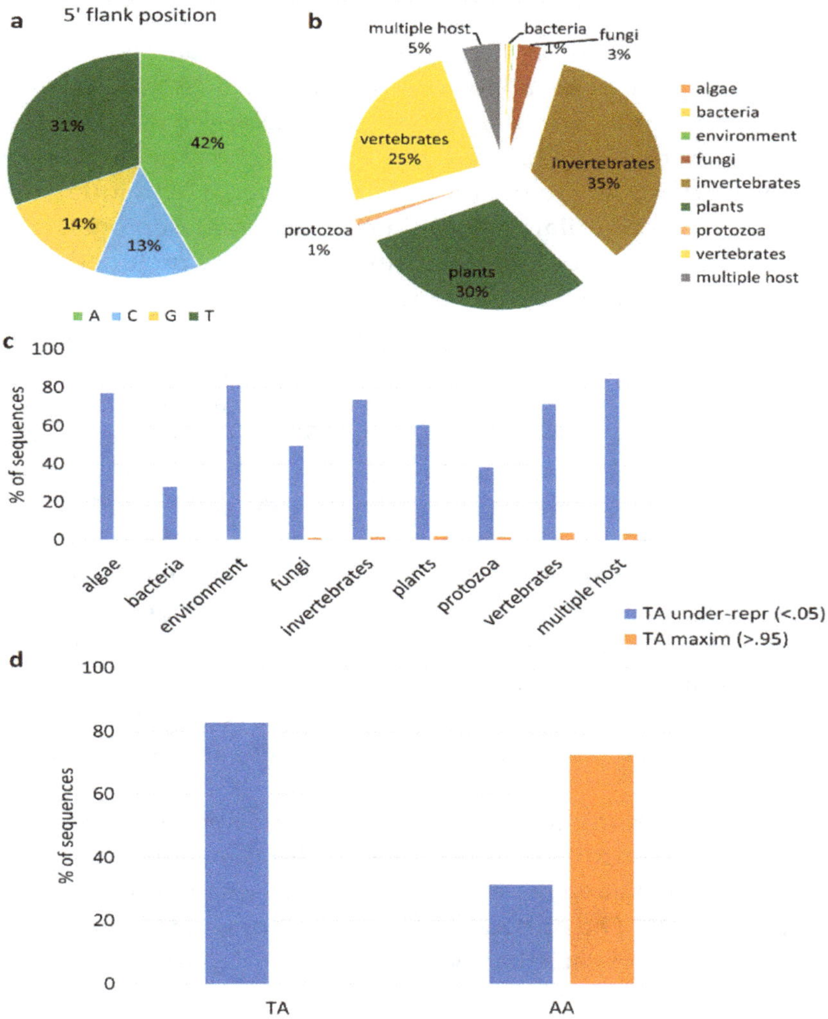

Figure 4. Analysis of virus editing. (**a**). Distribution of the flanking base of ADAR-compatible SNPs. (**b**). Host distribution of the 3872 viral genomes used as a training set. (**c**). Percentage of viral sequences (N = 3872) showing statistically significant TA under-representation (<0.05) and maximization (>0.95), divided per host class. (**d**). Percentage of TA and AA under-representation (<0.05) and maximization (>0.95), measured on the coding sequences of the nearly complete viral genomes reported in this paper.

To better contextualize our results, and to assign the 26 "nearly complete viral genomes" to a viral group, we attempted a phylogenetic analysis based on the regions corresponding to the vRdRP domains (the phylogenetic tree can be visualized at [69], or as Supplementary Data S1). The phylogenetic tree obtained by the comparison of 2019 sequences of viral origins showed poor bootstrap support for most of the nodes, due to the high heterogenicity of the vRdDP sequences. Several of the sequences of the 26 viruses reported herein clustered with picoRNA-like viruses obtained from meta-transcriptomic surveys of mollusk species [43]. These viruses included *Mytedu virus1*, *Myzyes virus1*, *Cragig virus1*, *Cragig virus2*, *Cragig virus3*, *Cragig virus6*, *Rudphi virus4*, *Rudphi virus5*, and *Bivalve RNA virus G3*. *Cragig virus 10* showed similarities with *Bivalve hepelivirus G* (herpes-like viral family, as defined by [43]). Although *Cragig virus7* also clustered in a group of picoRNA-like viruses, it appeared to be separated from the other marine picoRNA-like sequences. Similarly, *Cragig virus8* and *Cragig virus9* formed a cluster including picoRNA-like viruses and one diatom virus (*Chaetoceros socialis f. radians RNA virus1*). None of our viruses grouped in clusters were characterized by the presence of abundant vertebrate viruses, while *Mytcor virus1* was grouped with plant viruses, supporting its BLASY similarity to *Pitaya virus X*.

4. Discussion

Viruses can infect almost every living organism, and viral nucleic acids, either DNA or RNA, are often found when the host sequences are analyzed, making host RNA-seq samples suitable targets for viral discovery [21,70]. In this study, we analyzed bivalve RNA-seq data, and recovered both partial and complete RNA virus genomes, exploiting them to investigate the limits of meta-transcriptomics approaches oriented to viral discovery in bivalves. We identified 413 unique sequences of viral origin, most of them showing a limited similarity with known viruses, demonstrating that bivalve RNA sequencing allows for the identification of viral sequences. These included 26 nearly complete viral genomes. Although the factors that mostly influence the number of viral reads in an RNA-seq sample seem to be the strategies used for sample collection (e.g., the inclusion of water present in the shell cavity), we showed that the RNA selection method used during library preparation also contributed to the recovery of viral reads. For the samples prepared specifically for viral meta-transcriptomics, it was demonstrated that it is possible to recover multiple RNA virus genomes from single samples [43], while analyzing RNA-seq data (polyA-selected) originally designed for host expression analysis, we could identify at most, one complete viral genome per sample, over multiple (partial) viral genomes. Analyzing available meta-transcriptomic data, we found that polyA-enrichment is somewhat more effective than ribosomal-depletion in term of viral read recovery. However, we demonstrated that ribo-depletion is capable of higher performance, since we could reconstruct five nearly complete viral genomes from a RNA sample prepared for oyster expression survey. Arguably, polyA-selection would bias the virus sequence identification in the case of polyadenylated viral genomes (e.g., *Picornavirales*), although we could find evidence of the presence of polyA-tails only in three out of 26 viruses. Overall, our result strongly enforced the use of ribo-depletion for the preparation of RNA-seq libraries targeting viral discovery.

In agreement with a recent study reporting a wide host distribution of invertebrate viruses [43], we traced six out of 26 viruses in RNA-seq samples of different bivalve species, and we reported three bivalve-associated viruses that were very similar to viruses identified from gastropod or sponge meta-transcriptomics data [70,71]. The presence of identical viruses in different bivalve species, or even in phylogenetically distant invertebrates has two possible explanations: either these viruses infect a broad-range of animals, or the species hosting these viruses is shared by different (marine) animals. In support of the first hypothesis, even if invertebrates (arthropods in particular) are rich in viruses [72], strong evidence for host–virus co-evolution was rarely reported [17], and host jumping seems to be common for invertebrate viruses [5]. These attributes are in agreement with the new concepts of RNA virus phylogenesis that are inferred by viral metagenomics, suggesting extensive horizontal virus transfer events and a broad host range for protostome viruses [73]. The second hypothesis, i.e., that these viruses are hosted by an organism that is common in the marine environment, may be the easiest explanation for the presence of identical viruses in samples of different species, and can be further supported by the filter-feeding activity of bivalves. In fact, given the functions that are exerted by the gut and gills (the latter tissue is commonly used for RNA-seq experiments), contamination by RNA originating from waterborne bacteria, fungi, microalgae, or even microeukaryotes, is common in bivalve RNA-seq samples. This situation is well-depicted by one of the complete genomes that we recovered, the algal virus *Heterosigma akashiwo RNA virus*, which we traced in RNA-seq samples of co-cultured *C. gigas* and *M. galloprovincialis*, and even in an unrelated *R. philippinarum* sample. In this study, we exploited the COI reads to identify possible co-occurring organisms of the RNA-seq samples rich in viral reads. Although COI is not a universal gene marker, such an analysis can provide an immediate view of the purity of the samples [71]. In our study, the COI analysis did not identify a contaminant organism that completely matched the distributions of multi-species viruses. The contamination with *D. listerianum* RNA present in several bivalve RNA-seq samples confirmed the wide distribution of this fouling tunicate, but it could represent only a partial explanation for the multi-species distribution of *Rudphi_virus3*. The *R. philippinarum* sample, including 3% of *Rudphi_virus4* reads was shown to be heavily contaminated by this tunicate, and by other non-bivalve species, but none of these species correlated with the distribution of this virus.

The determination of the host of meta-transcriptomics-derived viruses is likely one of the main challenges of viromics based on high-throughput data [23]. At the tissue level, both Transmission Electron Microscope (TEM) imaging of viral particles and in situ hybridization techniques are suitable to confirm host assignment only if infection intensity is sufficiently high, but they are unfeasible in the case of a very large number of samples, or for already-sequenced RNA-seq samples, such as the ones that we analyzed. Therefore, we investigated the feasibility of two alternative approaches that, exploiting the same RNA-seq samples used for viral discovery, can provide evidence that is useful for host-assigning bivalve-associated viruses. RNAi is used as an antiviral defense in plants, insects and nematodes, where efficient RNAi processing of viral genomes into virus-derived small RNAs (vsRNAs) perfectly matching to the original genome activates the RNA-induced silencing complex (RISC), which in turn catalytically marks viral sequences for degradation [74]. Differently from immuno-recognition mechanisms based on antibodies, RNAi is not impacted by viral mutations, and in *C. elegans*, it functions at nano-molar concentrations thanks to the amplification of vsRNA signals using RNA-dependent RNA polymerase [75]. Recently, an RdRP-independent mode of RNAi amplification has been reported in *Drosophila*, although the underpinning genetic mechanism is still unknown [76]. At least in worms, RNAi-based antiviral immunity can be generationally transmitted, and provide a kind of epigenetic immune-memory [77] that recalls the prokaryotic CRISPR-Cas-based adaptive immunity [78]. Recently, the identification of vsRNAs in a molluscan gastropod (*Nucella lapillus*) opens intriguing questions about the phylogenetic distribution of the antiviral defense system, and about the mechanism itself [71]. In insects, the analysis of the fraction of vsRNAs among sncRNA datasets allowed for an unbiased reconstruction of pathogenic viruses [65–67] but, although bivalve antiviral immunity partially resembles that of arthropods [79], the antiviral role of RNAi has never been directly demonstrated [80]. Our analysis suggested that RNAi exerted limited importance in antiviral defense in bivalves, since even if we showed that *C. gigas* PIWI1 is induced in RNA-seq samples containing RNA viruses and a more limited induction of one oyster RDR gene was consistent with the active transcription of OsHV-1, we were not able to clearly identify vsRNAs among sncRNA reads obtained from the same oyster sample, including abundant reads of RNA viruses and of OsHV-1 (dsDNA virus). Considering all of the viral contigs obtained from this RNA-seq dataset, our analyses strongly suggest that the sncRNA reads that mapped on these viral contigs are due to RNA degradation. Differently, looking only at the five complete viral genomes, we could not exclude that weak RNAi activity generated few vsRNAs. Improving the power of our analysis by increasing the coverage or by using chemical treatments to specifically enrich the sncRNA fraction [71], it would be possible to detect genuine vsRNAs, even in bivalves. In particular, the size profiles that we reported for the putative vsRNAs seemed to be biased through a non-Dicer production mechanism [65,71].

In vertebrates, the antiviral role of RNAi is superseded by the interferon pathway, which through the activation of interferon-stimulated genes (ISGs), promotes the recognition of viral-derived products and inhibits viral propagation. Among ISGs, powerful sequence editors like ADAR and *apolipoprotein B mRNA editing enzyme, catalytic polypeptide-like* (APOBEC), enzymatically mutate viral transcripts and genomes [81,82] by acting on target sites that are highly conserved throughout the metazoan evolution [14,83–85]. To counteract these host-mediated editing mechanisms, some viral genomes have evolved to reduce the frequency of sites that are more vulnerable to targeting by the host immune system [58,86,87]. Surprisingly, we demonstrated a diffuse under-representation of the "TA" motif in most of the known RNA viruses, although only few of them maximized this reduction, and we could not link this result to a specific class of hosts. A similar trend of TA under-representation was also present in the 26 viruses described herein, but for these viruses, we showed that there was a tendency to maximize the reduction of the AA motif. Similar to the 26 viruses reported herein, a similar trend characterizes, among others, the *Antarctic picorna-like virus 1* and 3, *Acute bee paralysis virus* and *Aphid lethal paralysis virus*, which represented *Picornavirales* with a phylogenetic vicinity with some bivalve-derived viruses [47]. SNP analysis did not highlight a predominant fraction of ADAR-compatible variations over the total SNPs in the 26 viruses. Arguably, CDUR analysis only determined under-/over-representation, and the possible role of ADAR as source of these shifts should be confirmed by dedicated experiments.

Overall, This study underlines the heterogeneity and variability of RNA viruses that are associated with marine mollusks, and the limited data that is available on environmental RNA viruses. While the simultaneous analysis of viral products, antiviral host defense processes, and products in the RNA-seq samples could support host assignment, this alone is not enough when dealing with suspension-feeders that are able to accumulate environmental microbes, and their viral symbionts. Given the growing body of knowledge on the role of viruses in host fitness, targeted investigations aimed at unraveling the diversity of "genuine" bivalve viruses are needed for a better understanding of factors affecting the health and well-being of these ecologically- and economically-important species.

Supplementary Materials:
Supplementary Table S1. Summary of the sequence data used for this work. 1. Genome data of bivalves. Species name, order, genome status and source are reported. 2. Transcriptome data. Class, Organism abbreviation, project ID (SRA archive), species name and order, sample origin, tissue, read layout, and number of predicted CDS sequences are reported. 3. *C. gigas* transcriptome data used for RNA-seq expression analysis. Project and SRR IDs, description and millions of reads are reported. 4. Viral genomes. NCBI ID, genome size, virus name, and taxonomical classification are reported. Supplementary Table S2. Annotation of the 413 viral sequences. The E-value, accession, description, and identity of the first BLAST match are reported, as well as the PFAM domain identified on these sequences (region, E-value, PFAM accession, and name). Supplementary Table S3. Virus distribution over 250 RNA-seq samples. The number of reads mapping to the 26 viral genomes are reported for 250 RNA-seq samples. The 82 samples, including more than 1000 viral reads are highlighted in red. Supplementary Table S4. COI mapping results. Mapped reads (reported as per millions) for 16 RNA-seq samples on COI entries (only the entries with at least 0.01 % of mapped reads are reported). Supplementary Table S5. Expression data of 10 RNAi-associated genes in oyster RNA-seq samples. Expression data are reported as TPM. Gene ID, description, and RNA-seq sample IDs, grouped per experimental treatment, are reported. (Panel A-C) Amount of OsHV-1 reads or RNA virus reads are depicted as blue or red dots, respectively, and referred to a secondary $\log_{10}$ bar scale (on the right of each graph). Expression patterns of RNAi-associated genes in oyster RNA-seq samples. Supplementary Figure S1. Genome construction of the 26 "nearly complete viral genomes" described in this work. ORFs are depicted as green arrows, and polyadenylation tails as black boxes. Supplementary Data S1. Phylogenetic tree of vRdRP regions in nexus format.

Author Contributions: Conceptualization, P.V. and B.A.; Data curation, U.R.; Formal analysis, U.R. and M.S.; Funding acquisition, P.V.; Methodology, U.R. and M.S.; Resources, P.V.; Writing—original draft, U.R. and B.A.; Writing—review & editing, U.R., P.V. and B.A.

References

1. Simmonds, P.; Adams, M.J.; Benkő, M.; Breitbart, M.; Brister, J.R.; Carstens, E.B.; Davison, A.J.; Delwart, E.; Gorbalenya, A.E.; Harrach, B.; et al. Consensus statement: Virus taxonomy in the age of metagenomics. *Nat. Rev. Microbiol.* **2017**, *15*, 161–168. [CrossRef] [PubMed]
2. Suttle, C.A. Marine viruses — major players in the global ecosystem. *Nat. Rev. Microbiol.* **2007**, *5*, 801–812. [CrossRef] [PubMed]
3. Koonin, E.V.; Dolja, V.V. Virus world as an evolutionary network of viruses and capsidless selfish elements. *Microbiol. Mol. Biol. Rev. MMBR* **2014**, *78*, 278–303. [CrossRef] [PubMed]
4. Schulz, F.; Yutin, N.; Ivanova, N.N.; Ortega, D.R.; Lee, T.K.; Vierheilig, J.; Daims, H.; Horn, M.; Wagner, M.; Jensen, G.J.; et al. Giant viruses with an expanded complement of translation system components. *Science* **2017**, *356*, 82–85. [CrossRef] [PubMed]
5. Zhang, Y.-Z.; Shi, M.; Holmes, E.C. Using Metagenomics to Characterize an Expanding Virosphere. *Cell* **2018**, *172*, 1168–1172. [CrossRef] [PubMed]
6. Mahy, B.W.J. The Evolution and Emergence of RNA Viruses. *Emerg. Infect. Dis.* **2010**, *16*, 899. [CrossRef]
7. Colson, P.; La Scola, B.; Levasseur, A.; Caetano-Anollés, G.; Raoult, D. Mimivirus: Leading the way in the discovery of giant viruses of amoebae. *Nat. Rev. Microbiol.* **2017**, *15*, 243–254. [CrossRef] [PubMed]
8. Rosario, K.; Duffy, S.; Breitbart, M. A field guide to eukaryotic circular single-stranded DNA viruses: Insights gained from metagenomics. *Arch. Virol.* **2012**, *157*, 1851–1871. [CrossRef] [PubMed]
9. Iranzo, J.; Puigbò, P.; Lobkovsky, A.E.; Wolf, Y.I.; Koonin, E.V. Inevitability of Genetic Parasites. *Genome Biol. Evol.* **2016**, *8*, 2856–2869. [CrossRef] [PubMed]

10. TenOever, B.R. The Evolution of Antiviral Defense Systems. *Cell Host Microbe* **2016**, *19*, 142–149. [CrossRef] [PubMed]
11. Koonin, E.V. Viruses and mobile elements as drivers of evolutionary transitions. *Philos. Trans. R. Soc. Lond. B. Biol. Sci.* **2016**, *371*. [CrossRef] [PubMed]
12. Faillace, C.A.; Lorusso, N.S.; Duffy, S. Overlooking the smallest matter: Viruses impact biological invasions. *Ecol. Lett.* **2017**, *20*, 524–538. [CrossRef] [PubMed]
13. Forterre, P.; Prangishvili, D. The major role of viruses in cellular evolution: Facts and hypotheses. *Curr. Opin. Virol.* **2013**, *3*, 558–565. [CrossRef] [PubMed]
14. Chen, J.; MacCarthy, T. The preferred nucleotide contexts of the AID/APOBEC cytidine deaminases have differential effects when mutating retrotransposon and virus sequences compared to host genes. *PLoS Comput. Biol.* **2017**, *13*. [CrossRef] [PubMed]
15. Coffin, J.M. Virions at the gates: Receptors and the host-virus arms race. *PLoS Biol.* **2013**, *11*, e1001574. [CrossRef] [PubMed]
16. Nakano, Y.; Aso, H.; Soper, A.; Yamada, E.; Moriwaki, M.; Juarez-Fernandez, G.; Koyanagi, Y.; Sato, K. A conflict of interest: The evolutionary arms race between mammalian APOBEC3 and lentiviral Vif. *Retrovirology* **2017**, *14*, 31. [CrossRef] [PubMed]
17. Obbard, D.J.; Dudas, G. The genetics of host–virus coevolution in invertebrates. *Curr. Opin. Virol.* **2014**, *8*, 73–78. [CrossRef] [PubMed]
18. Mokili, J.L.; Rohwer, F.; Dutilh, B.E. Metagenomics and future perspectives in virus discovery. *Curr. Opin. Virol.* **2012**, *2*, 63–77. [CrossRef] [PubMed]
19. Van Aerle, R.; Santos, E.M. Advances in the application of high-throughput sequencing in invertebrate virology. *J. Invertebr. Pathol.* **2017**. [CrossRef] [PubMed]
20. Penaud-Budloo, M.; Lecomte, E.; Guy-Duché, A.; Saleun, S.; Roulet, A.; Lopez-Roques, C.; Tournaire, B.; Cogné, B.; Léger, A.; Blouin, V.; et al. Accurate identification and quantification of DNA species by next-generation sequencing in adeno-associated viral vectors produced in insect cells. *Hum. Gene Ther. Methods* **2017**. [CrossRef] [PubMed]
21. Greninger, A.L. A decade of RNA virus metagenomics is (not) enough. *Virus Res.* **2018**, *244*, 218–229. [CrossRef] [PubMed]
22. Moniruzzaman, M.; Wurch, L.L.; Alexander, H.; Dyhrman, S.T.; Gobler, C.J.; Wilhelm, S.W. Virus-host relationships of marine single-celled eukaryotes resolved from metatranscriptomics. *Nat. Commun.* **2017**, *8*, 16054. [CrossRef] [PubMed]
23. Obbard, D.J. Expansion of the metazoan virosphere: Progress, pitfalls, and prospects. *Curr. Opin. Virol.* **2018**, *31*, 17–23. [CrossRef] [PubMed]
24. Bielen, A.; Bošnjak, I.; Sepčić, K.; Jaklič, M.; Cvitanić, M.; Lušić, J.; Lajtner, J.; Simčič, T.; Hudina, S. Differences in tolerance to anthropogenic stress between invasive and native bivalves. *Sci. Total Environ.* **2016**, *543*, 449–459. [CrossRef] [PubMed]
25. Darrigran, G.; Damborenea, C. Ecosystem engineering impact of Limnoperna fortunei in South America. *Zoolog. Sci.* **2011**, *28*, 1–7. [CrossRef] [PubMed]
26. Karatayev, A.Y.; Burlakova, L.E.; Mastitsky, S.E.; Padilla, D.K. Predicting the spread of aquatic invaders: Insight from 200 years of invasion by zebra mussels. *Ecol. Appl. Publ. Ecol. Soc. Am.* **2015**, *25*, 430–440. [CrossRef]
27. Plazzi, F.; Ceregato, A.; Taviani, M.; Passamonti, M. A molecular phylogeny of bivalve mollusks: Ancient radiations and divergences as revealed by mitochondrial genes. *PLoS ONE* **2011**, *6*, e27147. [CrossRef] [PubMed]
28. Faure, B.; Schaeffer, S.W.; Fisher, C.R. Species distribution and population connectivity of deep-sea mussels at hydrocarbon seeps in the Gulf of Mexico. *PloS ONE* **2015**, *10*, e0118460. [CrossRef] [PubMed]
29. Buchmann, K. Evolution of Innate Immunity: Clues from Invertebrates via Fish to Mammals. *Front. Immunol.* **2014**, *5*. [CrossRef] [PubMed]
30. Petrone, L.; Kumar, A.; Sutanto, C.N.; Patil, N.J.; Kannan, S.; Palaniappan, A.; Amini, S.; Zappone, B.; Verma, C.; Miserez, A. Mussel adhesion is dictated by time-regulated secretion and molecular conformation of mussel adhesive proteins. *Nat. Commun.* **2015**, *6*, 8737. [CrossRef] [PubMed]
31. Fernández Robledo, J.A.; Yadavalli, R.; Allam, B.; Pales Espinosa, E.; Gerdol, M.; Greco, S.; Stevick, R.J.; Gómez-Chiarri, M.; Zhang, Y.; Heil, C.A.; et al. From the raw bar to the bench: Bivalves as models for human health. *Dev. Comp. Immunol.* **2019**, *92*, 260–282. [CrossRef] [PubMed]

32. Farley, C.A.; Banfield, W.G.; Kasnic, G.; Foster, W.S. Oyster herpes-type virus. *Science* **1972**, *178*, 759–760. [CrossRef] [PubMed]
33. Abbadi, M.; Zamperin, G.; Gastaldelli, M.; Pascoli, F.; Rosani, U.; Milani, A.; Schivo, A.; Rossetti, E.; Turolla, E.; Gennari, L.; et al. Identification of a newly described OsHV-1 µvar from the North Adriatic Sea (Italy). *J. Gen. Virol.* **2018**. [CrossRef] [PubMed]
34. Bai, C.; Wang, C.; Xia, J.; Sun, H.; Zhang, S.; Huang, J. Emerging and endemic types of Ostreid herpesvirus 1 were detected in bivalves in China. *J. Invertebr. Pathol.* **2015**, *124*, 98–106. [CrossRef] [PubMed]
35. Burioli, E. a. V.; Prearo, M.; Riina, M.V.; Bona, M.C.; Fioravanti, M.L.; Arcangeli, G.; Houssin, M. Ostreid herpesvirus type 1 genomic diversity in wild populations of Pacific oyster Crassostrea gigas from Italian coasts. *J. Invertebr. Pathol.* **2016**, *137*, 71–83. [CrossRef] [PubMed]
36. Chang, P.H.; Kuo, S.T.; Lai, S.H.; Yang, H.S.; Ting, Y.Y.; Hsu, C.L.; Chen, H.C. Herpes-like virus infection causing mortality of cultured abalone Haliotis diversicolor supertexta in Taiwan. *Dis. Aquat. Organ.* **2005**, *65*, 23–27. [CrossRef] [PubMed]
37. Davison, A.J.; Trus, B.L.; Cheng, N.; Steven, A.C.; Watson, M.S.; Cunningham, C.; Le Deuff, R.-M.; Renault, T. A novel class of herpesvirus with bivalve hosts. *J. Gen. Virol.* **2005**, *86*, 41–53. [CrossRef] [PubMed]
38. Martenot, C.; Lethuillier, O.; Fourour, S.; Oden, E.; Trancart, S.; Travaillé, E.; Houssin, M. Detection of undescribed ostreid herpesvirus 1 (OsHV-1) specimens from Pacific oyster, Crassostrea gigas. *J. Invertebr. Pathol.* **2015**, *132*, 182–189. [CrossRef] [PubMed]
39. Arzul, I.; Corbeil, S.; Morga, B.; Renault, T. Viruses infecting marine molluscs. *J. Invertebr. Pathol.* **2017**. [CrossRef] [PubMed]
40. Zannella, C.; Mosca, F.; Mariani, F.; Franci, G.; Folliero, V.; Galdiero, M.; Tiscar, P.G.; Galdiero, M. Microbial Diseases of Bivalve Mollusks: Infections, Immunology and Antimicrobial Defense. *Mar. Drugs* **2017**, *15*, 182. [CrossRef] [PubMed]
41. Renault, T.; Novoa, B. Viruses infecting bivalve molluscs. *Aquat. Living Resour.* **2004**, *17*, 397–409. [CrossRef]
42. Brum, J.R.; Sullivan, M.B. Rising to the challenge: Accelerated pace of discovery transforms marine virology. *Nat. Rev. Microbiol.* **2015**, *13*, 147–159. [CrossRef] [PubMed]
43. Shi, M.; Lin, X.-D.; Tian, J.-H.; Chen, L.-J.; Chen, X.; Li, C.-X.; Qin, X.-C.; Li, J.; Cao, J.-P.; Eden, J.-S.; et al. Redefining the invertebrate RNA virosphere. *Nature* **2016**, *540*, 539–543. [CrossRef] [PubMed]
44. Andrade, K.R.; Boratto, P.P.V.M.; Rodrigues, F.P.; Silva, L.C.F.; Dornas, F.P.; Pilotto, M.R.; La Scola, B.; Almeida, G.M.F.; Kroon, E.G.; Abrahão, J.S. Oysters as hot spots for mimivirus isolation. *Arch. Virol.* **2015**, *160*, 477–482. [CrossRef] [PubMed]
45. Iaconelli, M.; Purpari, G.; Della Libera, S.; Petricca, S.; Guercio, A.; Ciccaglione, A.R.; Bruni, R.; Taffon, S.; Equestre, M.; Fratini, M.; et al. Hepatitis A and E Viruses in Wastewaters, in River Waters, and in Bivalve Molluscs in Italy. *Food Environ. Virol.* **2015**, *7*, 316–324. [CrossRef] [PubMed]
46. Suffredini, E.; Proroga, Y.T.R.; Di Pasquale, S.; di Maro, O.; Losardo, M.; Cozzi, L.; Capuano, F.; de Medici, D. Occurrence and Trend of Hepatitis A Virus in Bivalve Molluscs Production Areas Following a Contamination Event. *Food Environ. Virol.* **2017**, *9*, 423–433. [CrossRef] [PubMed]
47. Rosani, U.; Gerdol, M. A bioinformatics approach reveals seven nearly-complete RNA-virus genomes in bivalve RNA-seq data. *Virus Res.* **2016**. [CrossRef] [PubMed]
48. Okinawa Institute of Science and Technology (OIST) Marine Genomics Unit Genome Projects. Available online: http://marinegenomics.oist.jp/ (accessed on 1 January 2019).
49. Fu, L.; Niu, B.; Zhu, Z.; Wu, S.; Li, W. CD-HIT: Accelerated for clustering the next-generation sequencing data. *Bioinforma. Oxf. Engl.* **2012**, *28*, 3150–3152. [CrossRef] [PubMed]
50. TrimGalore! Available online: https://www.bioinformatics.babraham.ac.uk/projects/trim_galore/ (accessed on 1 January 2019).
51. Haas, B.J.; Papanicolaou, A.; Yassour, M.; Grabherr, M.; Blood, P.D.; Bowden, J.; Couger, M.B.; Eccles, D.; Li, B.; Lieber, M.; et al. De novo transcript sequence reconstruction from RNA-seq using the Trinity platform for reference generation and analysis. *Nat. Protoc.* **2013**, *8*, 1494–1512. [CrossRef] [PubMed]
52. Eddy, S.R. Accelerated Profile HMM Searches. *PLoS Comput. Biol.* **2011**, *7*, e1002195. [CrossRef] [PubMed]
53. Finn, R.D.; Bateman, A.; Clements, J.; Coggill, P.; Eberhardt, R.Y.; Eddy, S.R.; Heger, A.; Hetherington, K.; Holm, L.; Mistry, J.; et al. Pfam: The protein families database. *Nucleic Acids Res.* **2014**, *42*, D222–D230. [CrossRef] [PubMed]
54. Camacho, C.; Coulouris, G.; Avagyan, V.; Ma, N.; Papadopoulos, J.; Bealer, K.; Madden, T.L. BLAST+: Architecture and applications. *BMC Bioinf.* **2009**, *10*, 421. [CrossRef] [PubMed]

55. Zhang, G.; Fang, X.; Guo, X.; Li, L.; Luo, R.; Xu, F.; Yang, P.; Zhang, L.; Wang, X.; Qi, H.; et al. The oyster genome reveals stress adaptation and complexity of shell formation. *Nature* **2012**, *490*, 49–54. [CrossRef] [PubMed]
56. Wagner, G.P.; Kin, K.; Lynch, V.J. A model based criterion for gene expression calls using RNA-seq data. *Theory Biosci. Theor. Den Biowiss.* **2013**, *132*, 159–164. [CrossRef] [PubMed]
57. Zheng, Y.; Gao, S.; Padmanabhan, C.; Li, R.; Galvez, M.; Gutierrez, D.; Fuentes, S.; Ling, K.-S.; Kreuze, J.; Fei, Z. VirusDetect: An automated pipeline for efficient virus discovery using deep sequencing of small RNAs. *Virology* **2017**, *500*, 130–138. [CrossRef] [PubMed]
58. Shapiro, M.; Meier, S.; MacCarthy, T. The cytidine deaminase under-representation reporter (CDUR) as a tool to study evolution of sequences under deaminase mutational pressure. *BMC Bioinf.* **2018**, *19*, 163.
59. Martinez, T.; Shapiro, M.; Bhaduri-McIntosh, S.; MacCarthy, T. Evolutionary effects of the AID/APOBEC family of mutagenic enzymes on human gamma-herpesviruses. *Virus Evol.* **2019**, *5*. [CrossRef] [PubMed]
60. Edgar, R.C. MUSCLE: A multiple sequence alignment method with reduced time and space complexity. *BMC Bioinf.* **2004**, *5*, 113. [CrossRef] [PubMed]
61. Tamura, K.; Stecher, G.; Peterson, D.; Filipski, A.; Kumar, S. MEGA6: Molecular Evolutionary Genetics Analysis version 6.0. *Mol. Biol. Evol.* **2013**, *30*, 2725–2729. [CrossRef] [PubMed]
62. Interactive Tree Of Life (iTOL). Available online: https://itol.embl.de/ (accessed on 28 February 2019).
63. Venkataraman, S.; Prasad, B.V.L.S.; Selvarajan, R. RNA Dependent RNA Polymerases: Insights from Structure, Function and Evolution. *Viruses* **2018**, *10*, 76. [CrossRef] [PubMed]
64. Dolan, P.T.; Whitfield, Z.J.; Andino, R. Mechanisms and Concepts in RNA Virus Population Dynamics and Evolution. *Annu. Rev. Virol.* **2018**, *5*, 69–92. [CrossRef] [PubMed]
65. Miesen, P.; Joosten, J.; van Rij, R.P. PIWIs Go Viral: Arbovirus-Derived piRNAs in Vector Mosquitoes. *PLoS Pathog.* **2016**, *12*, e1006017. [CrossRef] [PubMed]
66. Vijayendran, D.; Airs, P.M.; Dolezal, K.; Bonning, B.C. Arthropod viruses and small RNAs. *J. Invertebr. Pathol.* **2013**, *114*, 186–195. [CrossRef] [PubMed]
67. Vodovar, N.; Goic, B.; Blanc, H.; Saleh, M.-C. In silico reconstruction of viral genomes from small RNAs improves virus-derived small interfering RNA profiling. *J. Virol.* **2011**, *85*, 11016–11021. [CrossRef] [PubMed]
68. Samuel, C.E. Adenosine deaminases acting on RNA (ADARs) are both antiviral and proviral. *Virology* **2011**, *411*, 180–193. [CrossRef] [PubMed]
69. Interactive Tree Of Life (iTOL). Available online: https://itol.embl.de/shared/umbertorosani (accessed on 28 February 2019).
70. Shi, M.; Zhang, Y.-Z.; Holmes, E.C. Meta-transcriptomics and the evolutionary biology of RNA viruses. *Virus Res.* **2018**, *243*, 83–90. [CrossRef] [PubMed]
71. Waldron, F.M.; Stone, G.N.; Obbard, D.J. Metagenomic sequencing suggests a diversity of RNA interference-like responses to viruses across multicellular eukaryotes. *PLoS Genet.* **2018**, *14*, e1007533. [CrossRef] [PubMed]
72. Li, C.-X.; Shi, M.; Tian, J.-H.; Lin, X.-D.; Kang, Y.-J.; Chen, L.-J.; Qin, X.-C.; Xu, J.; Holmes, E.C.; Zhang, Y.-Z. Unprecedented genomic diversity of RNA viruses in arthropods reveals the ancestry of negative-sense RNA viruses. *eLife* **2015**, *4*, 5378. [CrossRef] [PubMed]
73. Dolja, V.V.; Koonin, E.V. Metagenomics reshapes the concepts of RNA virus evolution by revealing extensive horizontal virus transfer. *Virus Res.* **2017**. [CrossRef] [PubMed]
74. Chin, W.-X.; Ang, S.K.; Chu, J.J.H. Recent advances in therapeutic recruitment of mammalian RNAi and bacterial CRISPR-Cas DNA interference pathways as emerging antiviral strategies. *Drug Discov. Today* **2017**, *22*, 17–30. [CrossRef] [PubMed]
75. Tsai, H.-Y.; Chen, C.-C.G.; Conte, D.; Moresco, J.J.; Chaves, D.A.; Mitani, S.; Yates, J.R.; Tsai, M.-D.; Mello, C.C. A ribonuclease coordinates siRNA amplification and mRNA cleavage during RNAi. *Cell* **2015**, *160*, 407–419. [CrossRef] [PubMed]
76. Petit, M.; Mongelli, V.; Frangeul, L.; Blanc, H.; Jiggins, F.; Saleh, M.-C. piRNA pathway is not required for antiviral defense in Drosophila melanogaster. *Proc. Natl. Acad. Sci. USA* **2016**, *113*, E4218–E4227. [CrossRef] [PubMed]
77. Gammon, D.B.; Ishidate, T.; Li, L.; Gu, W.; Silverman, N.; Mello, C.C. The Antiviral RNA Interference Response Provides Resistance to Lethal Arbovirus Infection and Vertical Transmission in Caenorhabditis elegans. *Curr. Biol. CB* **2017**, *27*, 795–806. [CrossRef] [PubMed]

78. Koonin, E.V. Evolution of RNA- and DNA-guided antivirus defense systems in prokaryotes and eukaryotes: Common ancestry vs convergence. *Biol. Direct* **2017**, *12*, 5. [CrossRef] [PubMed]
79. Green, T.J.; Rolland, J.-L.; Vergnes, A.; Raftos, D.; Montagnani, C. OsHV-1 countermeasures to the Pacific oyster's anti-viral response. *Fish Shellfish Immunol.* **2015**, *47*, 435–443. [CrossRef] [PubMed]
80. Green, T.J.; Speck, P. Antiviral Defense and Innate Immune Memory in the Oyster. *Viruses* **2018**, *10*, 133. [CrossRef] [PubMed]
81. Bass, B.L. RNA editing by adenosine deaminases that act on RNA. *Annu. Rev. Biochem.* **2002**, *71*, 817–846. [CrossRef] [PubMed]
82. Liu, M.-C.; Liao, W.-Y.; Buckley, K.M.; Yang, S.Y.; Rast, J.P.; Fugmann, S.D. AID/APOBEC-like cytidine deaminases are ancient innate immune mediators in invertebrates. *Nat. Commun.* **2018**, *9*, 1948. [CrossRef] [PubMed]
83. Carpenter, J.A.; Keegan, L.P.; Wilfert, L.; O'Connell, M.A.; Jiggins, F.M. Evidence for ADAR-induced hypermutation of the Drosophila sigma virus (Rhabdoviridae). *BMC Genet.* **2009**, *10*, 75. [CrossRef] [PubMed]
84. Porath, H.T.; Carmi, S.; Levanon, E.Y. A genome-wide map of hyper-edited RNA reveals numerous new sites. *Nat. Commun.* **2014**, *5*, 4726. [CrossRef] [PubMed]
85. Porath, H.T.; Schaffer, A.A.; Kaniewska, P.; Alon, S.; Eisenberg, E.; Rosenthal, J.; Levanon, E.Y.; Levy, O. A-to-I RNA Editing in the Earliest-Diverging Eumetazoan Phyla. *Mol. Biol. Evol.* **2017**, *34*, 1890–1901. [CrossRef] [PubMed]
86. Piontkivska, H.; Matos, L.F.; Paul, S.; Scharfenberg, B.; Farmerie, W.G.; Miyamoto, M.M.; Wayne, M.L. Role of Host-Driven Mutagenesis in Determining Genome Evolution of Sigma Virus (DMelSV.; Rhabdoviridae) in Drosophila melanogaster. *Genome Biol. Evol.* **2016**, *8*, 2952–2963. [CrossRef] [PubMed]
87. Piontkivska, H.; Frederick, M.; Miyamoto, M.M.; Wayne, M.L. RNA editing by the host ADAR system affects the molecular evolution of the Zika virus. *Ecol. Evol.* **2017**, *7*, 4475–4485. [CrossRef] [PubMed]

9

Purple: A Computational Workflow for Strategic Selection of Peptides for Viral Diagnostics using MS-Based Targeted Proteomics

Johanna Lechner [1,†], Felix Hartkopf [1,†], Pauline Hiort [1], Andreas Nitsche [2], Marica Grossegesse [2], Joerg Doellinger [3], Bernhard Y. Renard [1,*] and Thilo Muth [1]

1 Bioinformatics Unit (MF 1), Department for Methods Development and Research Infrastructure, Robert Koch Institute, 13353 Berlin, Germany; LechnerJ@rki.de (J.L.); HartkopfF@rki.de (F.H.); p.hiort@web.de (P.H.); MuthT@rki.de (T.M.)

2 Centre for Biological Threats and Special Pathogens, Highly Pathogenic Viruses (ZBS1), Robert Koch Institute, 13353 Berlin, Germany; NitscheA@rki.de (A.N.); GrossegesseM@rki.de (M.G.)

3 Centre for Biological Threats and Special Pathogens, Proteomics and Spectroscopy (ZBS 6), Robert Koch Institute, 13353 Berlin, Germany; DoellingerJ@rki.de (J.D.)

* Correspondence: RenardB@rki.de

† These authors contributed equally to this work.

Abstract: Emerging virus diseases present a global threat to public health. To detect viral pathogens in time-critical scenarios, accurate and fast diagnostic assays are required. Such assays can now be established using mass spectrometry-based targeted proteomics, by which viral proteins can be rapidly detected from complex samples down to the strain-level with high sensitivity and reproducibility. Developing such targeted assays involves tedious steps of peptide candidate selection, peptide synthesis, and assay optimization. Peptide selection requires extensive preprocessing by comparing candidate peptides against a large search space of background proteins. Here we present Purple (Picking unique relevant peptides for viral experiments), a software tool for selecting target-specific peptide candidates directly from given proteome sequence data. It comes with an intuitive graphical user interface, various parameter options and a threshold-based filtering strategy for homologous sequences. Purple enables peptide candidate selection across various taxonomic levels and filtering against backgrounds of varying complexity. Its functionality is demonstrated using data from different virus species and strains. Our software enables to build taxon-specific targeted assays and paves the way to time-efficient and robust viral diagnostics using targeted proteomics.

Keywords: virus proteomics; mass spectrometry; virus diagnostics; data analysis; targeted proteomics; peptide selection; parallel reaction monitoring

---

1. Introduction

Virus infections present serious health threats to millions of individuals worldwide. For public health, the accurate detection of pathogenic viruses is time-critical because reducing the time for diagnosis and treatment lowers the risk of disease transmission and patient mortality. Fast and robust diagnostic assays are therefore required to rapidly detect re-emerging and newly emerging viruses (e.g., Influenza, Ebola, Zika, or Hepatitis C virus). These diagnostic methods need to cover a broad spectrum of potentially disease-causing viral agents.

Classical diagnostic strategies for detecting viral infection can be divided into two different categories: on the one hand, virus detection can be established by targeted methods, such as agent-specific polymerase chain reaction (PCR) or immunological techniques. On the other hand, detection approaches exist that provide an open view, such as electron microscopy or next-generation sequencing (NGS). Besides their unbiased view, the latter methods have the advantage of identifying multiple pathogens in a single experimental run. Due to its specificity (hybridization and sequencing) and sensitivity (qPCR), the detection of nucleic acids is the gold standard in diagnostics. Conversely, the detection of viral proteins is used less frequently in diagnostic settings and is usually based on interaction with affine binding molecules such as antibodies or aptamers. However, producing these binding molecules is generally time-consuming and laborious, as is the validation of their specificity.

While in clinical microbiology the analysis of subproteomes (<12 kDa) using matrix assisted laser desorption/ionization-time of flight (MALDI-TOF) mass spectrometry (MS) has become a standard method for the identification of bacteria and fungi, no comparable proteomic approach exists in virology for technical reasons [1]. In recent years, MS-based targeted proteomics has evolved into a technique for detecting proteins in complex samples with high sensitivity, quantitative accuracy, and reproducibility [2,3]. Targeted proteomics is commonly used to test hypotheses on a subset of proteins of interest, in contrast to discovery shotgun proteomics. The latter provides global proteome profiling of thousands of proteins in a sample, however, at the expense of sensitivity and reproducibility. Unlike discovery methods, targeted methods of selected/multiple reaction monitoring (SRM/MRM) [4] and parallel reaction monitoring (PRM) [5] nowadays allow for detecting and analyzing preselected proteins and peptides in sensitive, specific, and time-efficient manner. Furthermore, the development of targeted proteomics assays has become easier in the past few years, owing to the advances of analytical methods, instrumental capabilities, and computational workflows [6].

Targeted MS-based proteomics assay development typically involves (i) peptide candidate selection, (ii) peptide synthesis, and (iii) assay optimization. This procedure now enables the transfer of a process highly similar to the design of multiplex PCRs to the proteome level for detecting pathogens. While MS-based targeted assays have not been used for detecting viruses in any diagnostic setting yet, promising findings could already be achieved for identifying and quantifying pathogenic bacterial species. For example, targeted proteomics methods were successfully used in previous studies on *Streptococcus pyogenes* [7] and *Mycobacterium tuberculosis* [8].

Although targeted proteomics has gained much popularity with many use cases in experimental research by now, relatively few research-oriented algorithms and software tools have been developed that support the user-defined selection of peptides for designing targeted SRM or PRM assays. In this context, Skyline [9] is a powerful and widely used software for designing targeted proteomics assays. Besides its wide applicability to different targeted methods and its intuitive use, it also has some internal limitations: first, Skyline is dependent on the operating system Windows, and can therefore not be used under a Linux cluster server environment, and second, it does perform only exact string matching during the peptide selection process without considering any homologies between related organisms. PeptidePicker [10] is a web-based workflow to select peptides by providing, amongst further options, the protein accession number and was designed for human and mouse proteomes. PeptideManager [11] is a tool developed to select peptide candidates as protein surrogates from a defined proteome. It was optimized for the use case of xenografts, i.e., human tumors orthotopically implanted into a different species. While this software allows for constructing a peptide database from any species-specific proteome, sequence homologies, and multiple taxonomic levels are disregarded. Picky [12]—a web-based method designer for targeted assays—only provides support for human and mouse sequences, while it relies on synthetic peptide data from the human-focused ProteomeTools project [13,14]. In the context of targeted metaproteomics, the Unique Peptide Finder of the UniPept web application [15] was developed to select unique peptides for user-defined taxa. Furthermore, various computational tools have been developed to predict proteotypic peptides for targeted proteomics experiments [16–18]. These methods often make use of machine learning training setups that incorporate the probability of observing a peptide in a standard proteomics analysis, referred to as peptide detectability [19] or observability [20], and commonly involve physicochemical properties of the proteins to select high-responding peptides [21]. To our best knowledge, however, no software tool is currently available to select taxon-specific peptides for targeted proteomics assays that also accounts for sequence homologies between different species or strain proteomes. Effectively considering homologies is crucial for accurate taxon-specific diagnostics, because proteins measured in virus samples frequently have a high sequence similarity either in closely related strains or due to highly conserved functional domains.

Here we present Purple (Picking unique relevant peptides for viral experiments), a platform-independent software that returns a set of taxon-specific peptides, after the user has specified the viral target (i.e., a particular virus species or genus), as candidates for targeted proteomics experiments. Equipped with a user-friendly graphical user interface and a threshold-based filtering strategy for homologous sequences, it simplifies the design of MS-based targeted proteomics assays for the end user. Purple enables peptide candidate selection and considers background sequence information, i.e., proteins that are not related to a specific virus target, at various taxonomic levels. Thus, all peptide candidates are validated against a user-defined database of virus proteomes. While the design of MS-based targeted assays requires further steps, our software greatly facilitates the cumbersome, yet important task of peptide selection and thereby paves the way to time-efficient and robust pathogen screening and viral diagnostics. Purple is open source software available at https://gitlab.com/rki_bioinformatics/Purple.

2. Materials and Methods

2.1. *Purple Workflow*

Purple is implemented in Python (version 3.6) and makes use of additional Python libraries such as tqdm (https://github.com/tqdm/tqdm) for process bar calculation and Biopython [22] to calculate the molecular weight of peptides. Purple is available as portable standalone version that already includes all required libraries or Purple can be installed using pip or conda, which are managing dependencies. The workflow of Purple is depicted in an overview diagram (Figure 1). Purple requires the input of protein sequence databases and a configuration file. The databases are automatically rearranged into a target and a background database. The "exact matching" step is used to remove exact sequence matches with the background from the target peptide set. The remaining target peptides are used to detect and remove homologous peptides. A result file containing the final unique peptides is created together with various intermediate result files. These are outputs of all Purple processing tasks, namely (i) digested peptides, (ii) exact matching peptides, (iii) non-homologous matching peptides and (iv) background shared peptides.

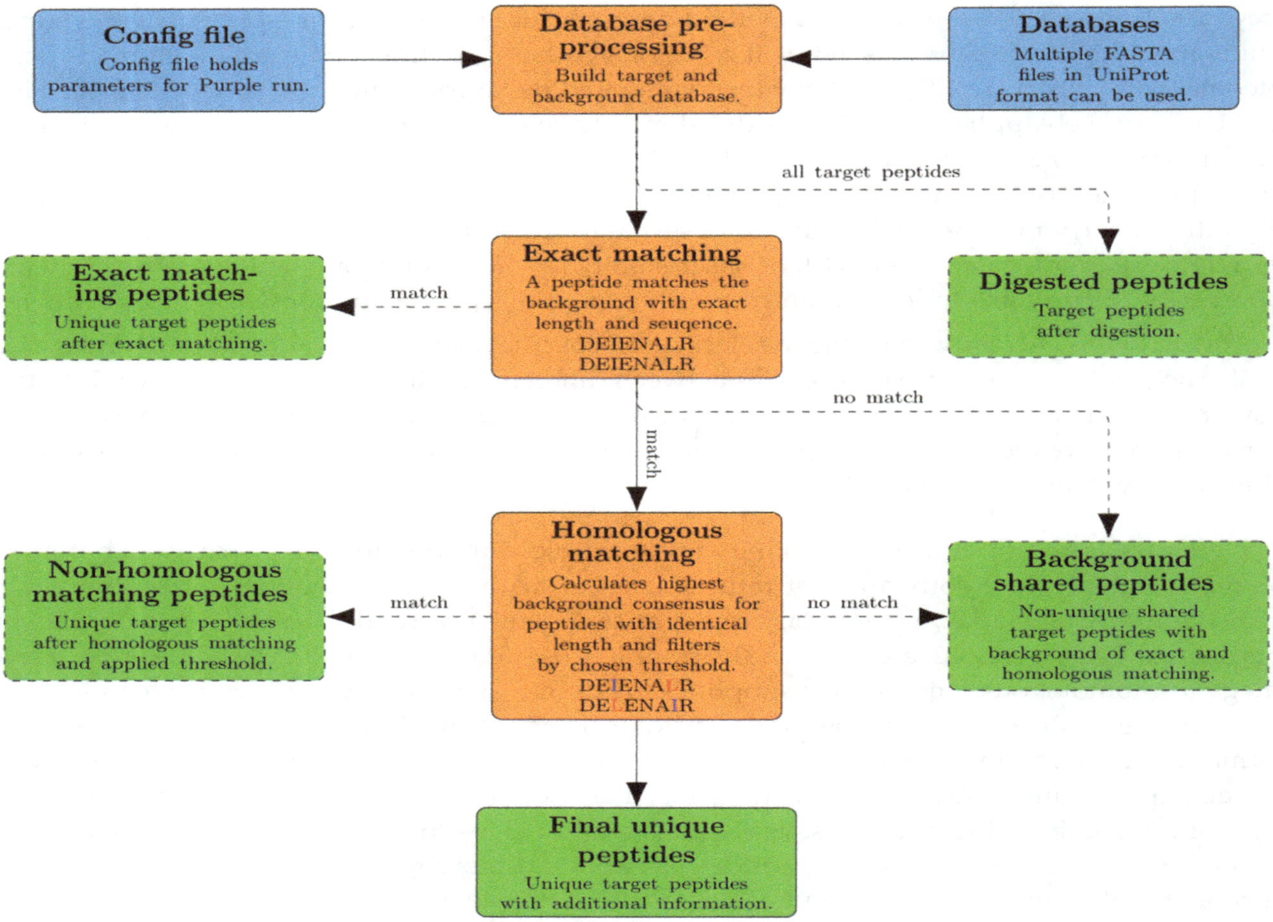

Figure 1. Overview of the Purple workflow. A configuration file and a directory path to the location of FASTA databases serve as input (blue). In the database preprocessing step, the databases are separated into target and background (orange). Any target peptides exactly matching to the background database are removed. In the homologous matching step, any target peptides that have similar sequences are filtered out (orange). All intermediate and final results are exported automatically to a user-defined output folder (green).

2.1.1. Preprocessing (Target Selection)

The selection of a target virus proteome is handled by input and preprocessing routines in Purple. For target selection, protein sequence databases in FASTA format serve as main input and are required to be provided in UniProt format. To select the database input, a directory needs to be specified by the user and multiple FASTA files can be considered for the processing. Two options of database specification are available in Purple: the first option is to explicitly define target species names as a list, which leads to the merging of all provided input databases. Each protein entry that contains one of the defined target species names in the protein header is considered as a target protein. The protein sequences not matching the defined target species are used as background database. The second option is to select a specific FASTA file in the database directory as target database. All remaining databases in the directory are then automatically merged to a single background database. As the background database may still consist of proteins originating from one of the target species, each protein in the background database is checked once more: if a protein header matches any species in the specified target database file, the protein entry is removed from further processing accordingly.

Both options result in two types of databases, namely a target and a background database. In the following, each protein sequence in these databases is *in silico*-digested using the enzymatic rule of trypsin with optional proline digestion. The in silico digest step results in multiple peptides for each protein entry, and peptide sequences beyond the user-defined length boundaries are filtered out. In addition, preprocessing includes the option of removing protein fragments and also allows replacing each isoleucine by leucine: this option was implemented because these amino acids share identical molecular masses and are therefore commonly not distinguishable in mass spectrometry. When the preprocessing is completed, both a target and a background database are provided for further analysis, which in this stage consist of peptides instead of proteins.

2.1.2. Exact Matching

Exact matching presents the first actual processing step in Purple: here, each of the previously *in silico*-digested target peptides is compared against the provided background database (see previous paragraph). In this procedure, target and background peptides of identical length are compared and only those target peptides that are not contained in the background are considered further; thus, peptide sequences with one or more exact sequence matches in the background database are filtered out at this stage, because they are not unique to the user-defined taxa of the target space. This procedure is performed iteratively until all *in silico*-digested peptides have been evaluated. The remaining peptides that have not been filtered out are stored as unique peptide candidates for further processing and are exported as intermediate result of the exact matching step.

2.1.3. Homologous Matching

Homologous matching is performed subsequently to the exact matching step. The goal is to evaluate each of the unique peptide candidates concerning its potential sequence consensus to homologous proteins in the background. The rationale behind this approach is that the more similar a target peptide is to the background, the less appropriate it is as candidate for a taxon-specific targeted assay. To assess the similarity of each peptide to the background proteomes, a sequence background consensus metric is introduced (see next paragraph). The target peptides that are discarded either during the exact or the homologous matching step are exported as so-called "shared" peptides. Shared peptides have either an exact sequence match with the background or have background consensus value above a user-defined threshold. To keep track of all processed data, target peptides with a background consensus below the threshold are exported as well.

2.1.4. Background Consensus Metric and Threshold Generation

Owing to mutational effects on conserved viral proteins, peptides can often be shared within a virus genus or family with minor sequence variations between them. This is problematic for targeted assays because such peptide candidates are not specific for species- or strain-level identification. To remove such taxon-unspecific peptides from the final sequence set, the background consensus metric $f(A, B)$ is used in Purple as the essential part of the homologous matching. Basically, the background consensus presents the Hamming distance of a target peptide A and background peptide B of the same length in relation to the length of the peptide n (Equation (2)). An amino acid is shared if the same amino acid ($d(x, y)$) is at the same position in A and B (Equation (1)).

$$d(x,y) = \begin{cases} 1, & if\ x = y \\ 0, & if\ x \neq y \end{cases} \tag{1}$$

In other words, the background consensus is the sum of shared amino acids at a specific position i divided by the number of amino acids in both (target and background) peptides. Even though the Hamming distance is a simple metric, it provides a proof-of-concept and validation of Purple, as adding more sophisticated methods should only slightly improve the homologous matching while increasing the computational effort and complexity.

$$\begin{aligned} For\ A = \{a_1, a_2, \ldots, a_n\}\ and\ B = \ A = \{b_1, b_2, \ldots, b_n\}\ and\ n = |A| = |B| : \\ f(A,B) = \frac{\sum_{i=1}^{n} d(a_i, b_i)}{n}, for\ a_i \in A\ and\ b_i \in B \end{aligned} \quad (2)$$

This metric is applied to each of the target peptides that are compared to all background peptides of the same length. For each target peptide, the maximum consensus is stored when being below a user-defined background consensus threshold. A target peptide with a high background consensus is likely to originate from a homologous protein or common protein domain. Therefore, the consensus metric evaluates the conservation of peptides in the target and background database. A low background consensus marks target peptides that are unique in sequence in the target species. All peptides with a high background consensus below the previously chosen threshold are exported into the final results file and the remaining shared peptides are exported as part of the intermediate output. The results are supplemented with the peptide weight, the number of occurrences in the target database, as well as species and proteins names. This enables the user to conduct further analysis with the previously retrieved unique peptides. The Purple documentation is available for a complete description of all output files and more details about the data interpretation.

2.2. Graphical User Interface

A graphical user interface (GUI) was developed for using Purple (Figure 2). This interface allows researchers with less expertise in handling bioinformatics methods on the command line to use Purple in a efficient and user-friendly manner. The Purple GUI makes software configuration and execution straightforward and complex tasks can be rapidly accomplished. Any parameter can be adjusted in the GUI, and the background consensus threshold can be set by the user. Furthermore, the processing status can be inspected in a logging panel and a file menu provides options for saving and loading configuration files. Note that configuration files are optional in Purple and a default configuration is provided; thus, only system-specific parameters must be set in the GUI. Using configuration files makes each task reproducible and the GUI-integrated configuration file choice allows for switching between multiple settings easily. Figure 3 shows the final output in the tab separated values (TSV) format that can be further processed and visualized using common spreadsheet software.

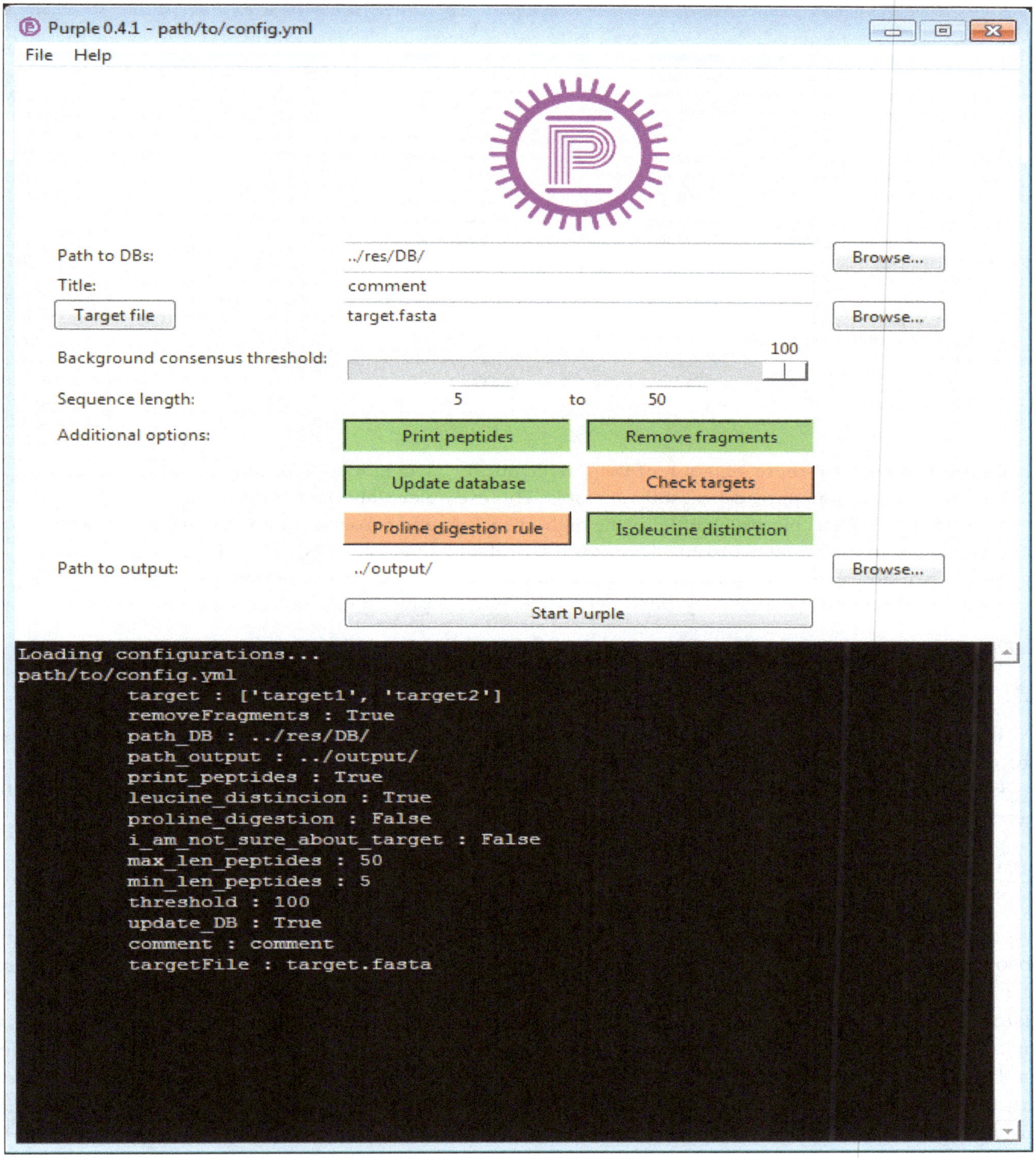

Figure 2. The graphical user interface of Purple. In the top file menu, configurations files can be loaded and saved. The top menu also includes a link to the documentation and manual. The listed GitLab page provides direct user support from the developers via an issue tracking system. The upper panel shows default parameters and allows modifying the configuration settings and processing start. The lower panel displays the current processing status with logging information on the current run, configuration, and progress of the analysis.

	A	B	C	D	E	F	G	H	I
1	peptide	peptide weight	highest background consensus	occurrences	species	proteins	protein names	fasta entries	descriptions
2	PSFLA	533.6171	80.00%	1	['Cowpox viru	1	['RP132_CWPXI	1	['>sp|P17474
3	LYVGGGISNDQTTT	2149.2249	70.00%	1	['Cowpox viru	1	['KBTB1_CWPX	1	['>sp|Q8QM
4	NISNLLDDDILCDVI	2158.4717	36.84%	1	['Cowpox viru	1	['KBTB2_CWPX	1	['>sp|Q8QM
5	QLCLVCHDTK	1159.3794	50.00%	1	['Cowpox viru	1	['KBTB2_CWPX	1	['>sp|Q8QM
6	YNVCNPCILVYNIN	6135.9066	14.55%	1	['Cowpox viru	1	['KBTB2_CWPX	1	['>sp|Q8QM
7	LLPDMPIALSSYGM	5877.5567	16.98%	1	['Cowpox viru	1	['KBTB2_CWPX	1	['>sp|Q8QM
8	YDTVNNIWETLPNF	2412.6495	35.00%	1	['Cowpox viru	1	['KBTB2_CWPX	1	['>sp|Q8QM
9	PGVVSHEDDIYVVC	1888.1022	41.18%	1	['Cowpox viru	1	['KBTB2_CWPX	1	['>sp|Q8QM
10	YIENK	665.735	80.00%	1	['Cowpox viru	1	['SPI1_CWPXB	1	['>sp|P42927
11	NIVTSVDMMVSTK	1424.6826	53.85%	1	['Cowpox viru	1	['SPI1_CWPXB	1	['>sp|P42927
12	NDLQYVHINELFGG	4999.556	25.00%	1	['Cowpox viru	1	['SPI1_CWPXB	1	['>sp|P42927
13	ESFGNFSIIELPYVGI	4776.24	23.26%	1	['Cowpox viru	1	['SPI2_CWPXB	1	['>sp|P07385
14	HQESEPASVPTSSR	1511.5499	42.86%	1	['Cowpox viru	1	['A18_CWPXB	1	['>sp|Q8QM
15	IIPIDNGSNMLILNP	3034.5026	28.57%	1	['Cowpox viru	1	['IL1BP_CWPXB	1	['>sp|Q0452
16	NETYCDMMSLNLT	3896.3331	23.53%	1	['Cowpox viru	1	['IL1BP_CWPXB	1	['>sp|Q0452
17	NDAGYYTCVLK	1246.3886	54.55%	1	['Cowpox viru	1	['IL1BP_CWPXB	1	['>sp|Q0452
18	YTYGDK	745.7766	66.67%	1	['Cowpox viru	1	['IL1BP_CWPXB	1	['>sp|Q0452
19	INPVK	569.694	80.00%	1	['Cowpox viru	1	['IL1BP_CWPXB	1	['>sp|Q0452

Figure 3. Graphical representation of the Purple output. The tabular TSV output of Purple can be imported into various spreadsheet software tools. This exemplary table shows the peptide sequence, the calculated theoretical mass weight (Da), the highest background consensus, and the number of peptide occurrences in the target proteome. The species, protein name and full description of the associated protein are stored in a list for further analysis. In addition, the number of proteins and FASTA entries are listed separately, because they can diverge, e.g., when a protein has multiple sequence variants.

2.3. Data

2.3.1. Target Virus Databases

To evaluate the performance of Purple, selected target virus species from sequence databases were used. This section provides an overview on the virus species used with respect to database composition and further background information on the virus type. The virus species were selected based on their relevance for current or upcoming diagnostic settings.

Arenaviruses

Arenaviruses are enveloped RNA viruses with an average diameter of 120 nanometers that have a bisegmented negative-strand RNA genome. The Latin term "arena" refers to the grainy ribosomal particles acquired from the virus-host cells that can be viewed in cross-section with electron microscopy imaging. Arenaviridae is a virus family whose members are generally associated with causing chronic infections in rodents and zoonotically acquired severe diseases, such as lymphocytic choriomeningitis or hemorrhagic fever, in humans. In this work, nine disease-causing Old and New World arenavirus species are taken as targets for evaluating the performance of Purple (Table 1). Besides Lymphocytic choriomeningitis virus, strain members of which cause aseptic meningitis, encephalitis, or meningoencephalitis, all listed arenaviruses are causative agents for viral hemorrhagic fever (VHF).

Table 1. Alphabetically ordered list of arenavirus species used for the performance benchmarking. The reader is referred to [23] for further details on these arenaviruses.

Virus Species	Abbreviation	NW/OW [2]	NW - Clade [3]	No. Proteins	No. Peptides [1]
Chapare mammarenavirus	CHAV	NW	B	4	252
Guanarito mammarenavirus	GTOV	NW	B	4	244
Junin mammarenavirus	JUNV	NW	B	4	246
Lassa virus	LASV	OW	-	4	242
Lujo mammarenavirus	LUJV	OW [4]	-	4	250
Lymphocytic choriomeningitis virus	LCMV	OW	-	4	245
Machupo virus	MACV	NW	B	4	237
Sabia mammarenavirus	SABV	NW	B	4	248
Whitewater Arroyo mammarenavirus	WWAV	NW	A/rec	4	240

[1] Number of in silico-digested peptide sequences, [2] New World (NW)/ Old World (OW), [3] New World clade [4] Based on genome sequence clustering, Lujo mammaarenavirus shows its own cluster [23].

Cowpox virus

Cowpox virus (CPXV) is a large double-stranded DNA virus with a proteome of over 200 proteins [24] that belongs to the genus Orthopoxvirus (OPV) of the Poxviridae family. CPXV has been described as the source of the first vaccine used by Edward Jenner, who was the first to scientifically describe the vaccination process against the smallpox-causing variola virus. Recent findings based on a conducted analysis on the smallpox vaccine gave evidence of the suspected role of horsepox (instead of cowpox) in the origin of the vaccine [25,26]. Since the pathogenicity and zoonotic potential of CPXV are investigated at the Robert Koch Institute, detailed data acquired from MS measurements were available (see Section 2.3.3). For performance evaluations, CPXV is further beneficial because this virus species has several close relatives. In addition to the cowpox strains Brighton Red and Grishak-90, four very close relatives with high sequence similarity are given: a genome comparison performed with BLAST [27] showed that variola virus, monkeypox virus, horsepox virus, and vaccinia virus share sequence identities of up to 98% (Supplementary Table S1).

Vaccinia virus (VACV Copenhagen and VACV Western Reserve)

Vaccinia virus is a member of the Orthopoxvirus (OPV) genus [28] and has been used for vaccination against smallpox since the 19th century. Due to the high sequence similarity of members of the OPV genus, it is possible to provide cross-protection vaccination by one member of the OPV genus. Hence, the classification can be an issue, because it can be challenging to find peptides to reliably classify a species or a strain. In this work, we investigate whether it is possible to distinguish between the two strains VACV Copenhagen and VACV Western Reserve by finding strain-specific peptides using Purple. Similar to CPXV, experimental data was publically available (see Section 2.3.3).

2.3.2. Background Virus Databases

The target databases mentioned above are species-specific and therefore cannot represent all available virus proteomes. From the target databases, Purple only yields to species-specific unique peptides. To extend this space to all virus proteomes and subsequently be able to find unique peptides in that relation, we added a database that consists of all reviewed virus proteins available on UniProt/Swiss-Prot [29]. In contrast to the target databases, this database is used exclusively as a background database. At the time of writing, UniProt/Swiss-Prot contains 16,846 reviewed viral proteins, which results in 301,387 in silico-digested tryptic peptides. In this work, we evaluate Purple with and without the use of the larger background database.

2.3.3. Background Human Databases

To account for samples mixed with human proteins we added a human database to the background. This database originates from UniProt/Swiss-Prot [29] and enables Purple to discard human peptides. Subsequently, this reduces false positives in experiments using virus-infected human samples. The database consists of 20,428 proteins and was used exclusively for the CPXV analysis in this work.

2.3.4. Experimental Data

The MS/MS datasets used for the benchmarking of Purple originate from a previous study published by Doellinger et al. in 2015 [24] (PRIDE project accession: PXD003013). In this work, a subset of the data available was used including three CPXV Brighton Red, three VACV Copenhagen, and three VACV Western Reserve MS/MS raw files. These raw files were acquired by an LTQ Orbitrap in data-dependent manner. Further experimental details are listed and described in the above-mentioned publication. Subsequently, three CPXV Brighton Red raw files were converted into MGF files using the MSConvert function of ProteoWizard [30] with the peak picking parameter of MS-level two and with zero sampling removal activated. Table 2 shows the number of MS/MS spectra for each virus strain (CPXV Brighton Red, VACV Copenhagen and VACV Western Reserve). For peptide and protein identification, these spectra were searched against proteome databases using the MS-GF+ [31] (version v20181015) database search engine. The database search was performed with eight threads, an activated decoy search, a chosen precursor with mass tolerance of five ppm, optimized for Orbitrap instruments, and trypsin was selected as digestion enzyme. The sequence databases used for protein identification are described in detail in Section 2.3.1. The database searches produced mzid output files that were converted into TSV files using the build-in MS-GF+ conversion tool. Afterwards, the results were filtered by applying a 1% false discovery rate (FDR) threshold at the PSM-level.

Table 2. This table shows the number of spectra from each sample replicate for CPXV Brighton Red, VACV Copenhagen, and VACV Western Reserve virus species/strains.

Species/Strain	No. Spectra in Replicate 1	No. Spectra in Replicate 2	No. Spectra in Replicate 3	No. Total Spectra
CPXV Brighton Red	19,396	19,352	18,920	57,668
VACV Copenhagen	19,740	19,265	19,170	58,175
VACV Western Reserve	19,421	19,453	19,076	57,950

3. Results

We here present three different use cases to illustrate the possibilities of targeted proteomics using Purple in viral diagnostic settings. The first analysis focuses on the species-level resolution for arenaviruses, the second evaluates the taxonomic classification using cowpox data from shotgun proteomics measurements, and the third tests the capabilities of strain-level differentiation using experimental data from two closely related vaccinia virus strains.

3.1. Analysis of Species-Level Resolution using Nine Arenavirus Species

In the first analysis, we aimed to evaluate the species-level resolution of our diagnostic approach using sequence data from the Arenaviridae family. For this purpose, we investigated the resolution of Purple by evaluating different viral species as target organisms against a proteome background of similar species and viruses in general. We used nine arenavirus species (MACV, JUNV, SABV, CHAV, GTOV, LASV, LCMV, WWAV, and LUJV; see Table 1) with proteomes containing four proteins, namely (1) RNA-directed RNA polymerase L, (2) nucleoprotein N, (3) pre-glycoprotein polyprotein GP complex and (4) RING finger protein Z. As background proteomes, we added all reviewed virus proteins available on UniProt/Swiss-Prot to remove frequently occurring peptides (e.g., from conserved sequences of functional domains). The removal of target peptides from similar virus proteomes intends to eliminate false positive detections (i.e., to increase the specificity). Since the protein sequences differ strongly between the arenavirus species, we expected to retrieve sufficient unique peptides for each species that serve as candidates for designing a targeted assay. For a benchmarking, we examined the relative amount of taxon-specific target peptides for each of the arenavirus species using both exact and homologous matching mode (Tables 3 and 4). The homologous matching was performed to evaluate the impact of sequence homologies for the arenaviruses and between these and all other virus species.

Table 3. This table shows the number of taxon-specific peptides from nine arenavirus species after (i) *in silico* digest, (ii) exact matching, and (iii) homologous matching (80% background consensus threshold). Each target species was compared against the background of eight remaining arenavirus species proteomes. The second column provides the number of nonspecific peptides, i.e., the ones being shared with the background.

Species	No. Digested Peptides	No. Background Shared	No. Exact Matching	No. Homologous Matching
MACV	237	119	178	**118**
SABV	248	127	191	**121**
LUJV	250	24	241	**226**
CHAV	252	121	197	**131**
GTOV	244	75	205	**169**
JUNV	246	123	187	**123**
LASV	242	35	227	**207**
LCMV	245	31	232	**214**
WWAV	240	31	226	**209**

Table 4. This table shows the number of taxon-specific peptides from nine arenavirus species after (i) *in silico* digest, (ii) exact matching, and (iii) homologous matching (80% background consensus threshold). Each target species was compared against the background of eight remaining arenavirus species proteomes and additionally against all reviewed virus proteomes (from UniProt/Swiss-Prot). The second column provides the number of nonspecific peptides, i.e., the ones being shared with the background.

Species	No. Digested Peptides	No. Background Shared	No. Exact Matching	No. Homologous Matching
MACV	237	143	162	**94**
SABV	248	144	183	**104**
LUJV	250	52	229	**198**
CHAV	252	137	190	**115**
GTOV	244	118	189	**126**
JUNV	246	139	171	**107**
LASV	242	126	171	**116**
LCMV	245	110	187	**135**
WWAV	240	130	181	**110**

First, we investigated the ratios of taxon-specific unique peptides and in silico-digested peptides with a background database consisting of the four arenavirus proteins, as mentioned above. The exact matching yielded to taxon-specific peptide ratios between 75.1% (MACV) and 96.4% (LUJV) (Figure 4). This can be explained by the high sequence diversity between the nine arenavirus species: when generating multiple sequence alignments (MSA) of these species for their four proteins, overall, a low consensus of the sequences was found (Supplementary Data S1–S4). When applying a background consensus threshold of 80%, significantly fewer taxon-specific peptides were obtained with relative numbers between 48.8% and 90.4% for SABV and LUJV, respectively (Figure 4). Overall, the mean decrease in the ratio of all species is 16.6% and the strongest ratio decrease can be found for MACV (25.3%), SABV (28.2%), CHAV (26.2%), and JUNV (26.0%). These four species are all New World arenaviruses and part of the clade B (see Table 2). The close relationship of these four virus species (as shown in the phylogenetic tree in Figure 5) causes high numbers of shared peptides which explains the decline in taxon-specific peptides. The Old World arenavirus LUJV shows the highest taxon-specific peptide ratio after homologous matching (90.4%) and even after homologous analysis against all virus proteomes (79.2%). This illustrates that LUJV has the lowest sequence similarity with the other arenaviruses. The low similarity can be explained by the isolated geographical distribution of LUJV in Southern Africa [32]. In 2008, an outbreak of LUJV led to a high case fatality rate of 80% (4/5 cases), and a follow-up analysis of its genome confirmed that LUJV is a novel virus species being only distantly related to known arenaviruses and groups genetically closer to Old World viruses not associated with VHF [33].

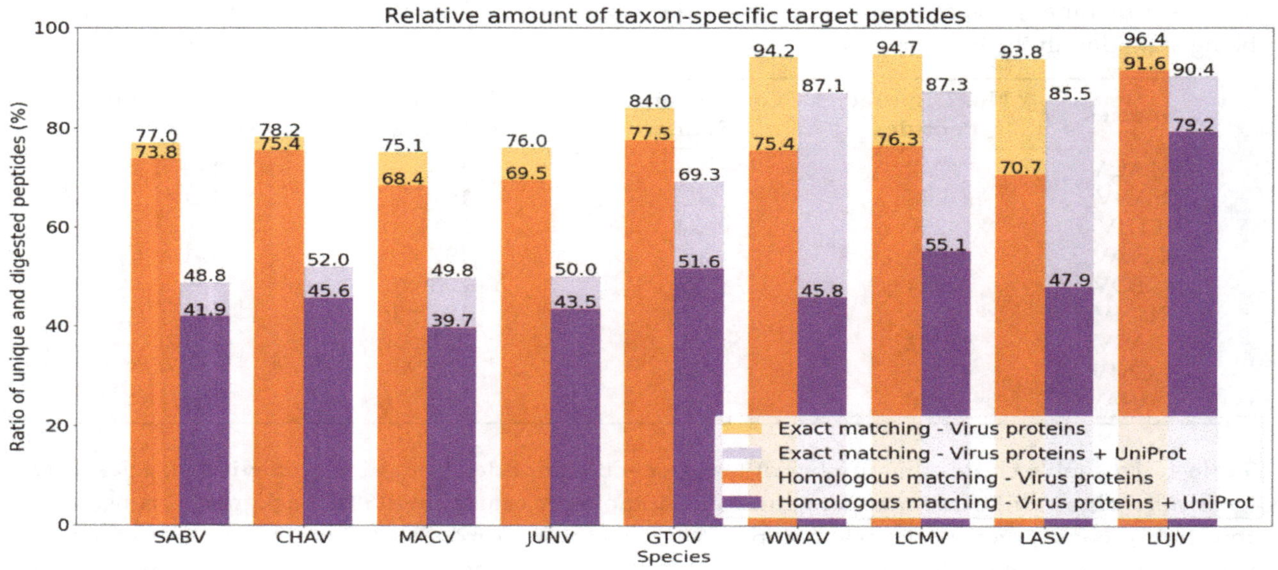

Figure 4. Relative amount of taxon-specific target peptides from nine arenavirus species proteomes. The ratio of unique to in silico-digested peptides is shown for exact (lighter colors) and homologous (darker colors) matching mode with a background consensus threshold of 80%. Orange bars show the results for the database consisting of four virus proteins for each arenavirus species. Purple bars indicate results that were generated when adding protein sequences from all reviewed virus proteomes (from UniProt/Swiss-Prot) as additional background.

Next, we assessed the protein sequence coverage on the basis of Purple-selected unique peptides for all four arenavirus proteins (RNA-directed RNA polymerase L; Nucleoprotein N; Pre-glycoprotein polyprotein GP complex GLYC; RING finger protein Z). We evaluated two different backgrounds here: (i) a small background with the arenavirus proteomes (containing the four proteins) of the remaining eight non-target species and (ii) a large background containing all arenavirus proteomes combined with all reviewed virus proteomes from UniProt/Swiss-Prot (see Section 2.3.2).

The analysis of the protein sequence coverage shows that L, GLYC and Z are relatively well covered by the taxon-specific peptides across all nine species for the small background (Figure 6). Nucleoprotein NCAP has the highest variability in protein coverage with an interquartile range (IQR) of 35.22% on the small background, suggesting that NCAP is the best-conserved protein among the considered arenavirus species. When taking a closer look at the results of the larger background analysis with all reviewed virus proteins, it can be found that the coverage decreases for all four proteins. The NCAP protein shows the lowest median in protein coverage (20.18%). This shows that NCAP has the lowest sequence consensus of taxon-specific peptides with other virus proteomes, indicating that it is the best-conserved of the four proteins. Indeed, the other three proteins (L, GLYC, and Z) have above 40% sequence coverage, thus more taxon-specific peptides can be obtained from these proteins. This analysis shows that, depending on the use case, it may make sense to investigate individual proteins instead of whole proteomes. For example, proteins with low sequence coverage based on taxon-specific peptides may be excluded.

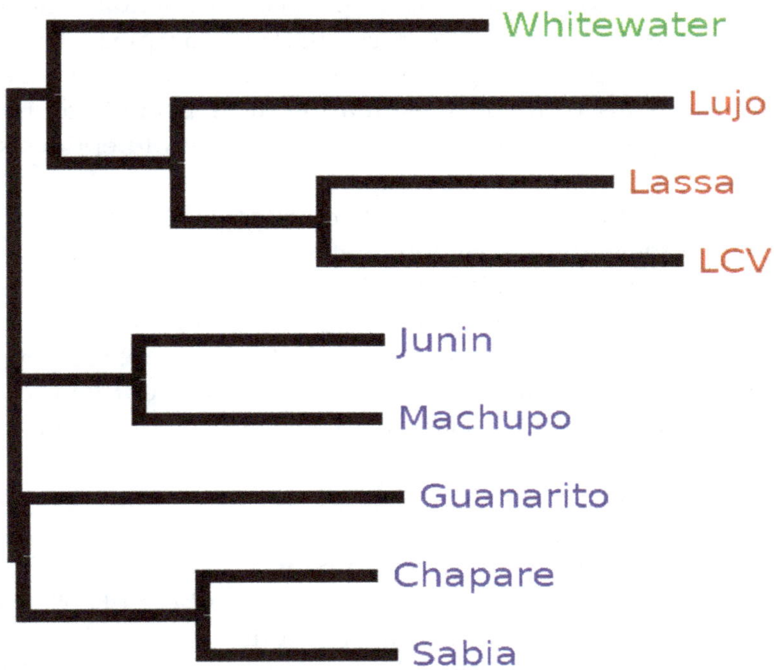

Figure 5. Phylogenetic tree of the pre-glycoprotein polyprotein GP complex (GLYC) of nine arenaviruses. The Whitewater strain is the only New World clade A/rec arenavirus (green). Lujo (LUJV), Lassa (LASV), and Lymphocytic choriomeningitis (LCV) are geographical Old World arenaviruses (red). Junin (JUNV), Machupo (MACV), Guanarito (GTOV), Chapare (CHAV), and Sabia (SABV) are members of the New World arenaviruses clade B (blue). The neighbor-joining tree without distance corrections was created using CLUSTAL Omega [34] for the multiple sequence alignment and the tree visualization software FigTree (http://tree.bio.ed.ac.uk/software/figtree/).

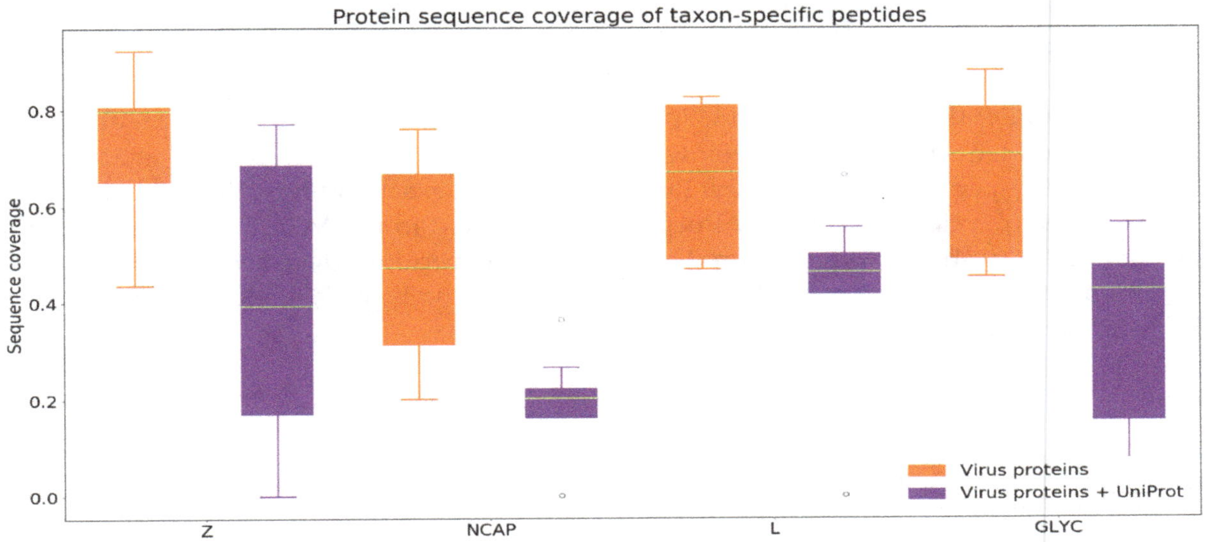

Figure 6. Protein sequence coverage of taxon-specific peptides selected by Purple on proteins for nine arenavirus species proteomes. The four proteins of the arenavirus proteomes are RNA-directed RNA polymerase L (L), nucleoprotein N (NCAP), pre-glycoprotein polyprotein GP complex (GLYC), and RING finger protein Z (Z). The coverage of selected peptides is displayed for homologous matching when applying a background consensus threshold of 80%.

3.2. Evaluating Species-Level Classification Based on Detected Peptides from Viral Shotgun Proteomics Measurements

To evaluate the peptide selection method in Purple on experimental data, we used representative MS/MS datasets derived from human cowpox virus (CPXV) samples. The main goal was to test whether peptides identified in a typical shotgun proteomics experiment can be used for differentiating viruses at the species level. We also aimed for estimating the expected accuracy gain for taxonomic classification when using a targeted proteomics assay on the basis of peptides suggested by Purple.

In a pre-analysis, we performed a Purple run using CPXV as target proteome to select species-specific peptides. For the peptide selection process, 18 reviewed (from UniProt/Swiss-Prot) and 208 unreviewed (from UniProt/TrEMBL) CPXV-specific protein sequences were used as target database, which is part of the PRIDE project (see Section 2.3.4). We used this combined database consisting of reviewed and unreviewed protein sequences because the available reviewed protein sequences for the Brighton Red strain yielded to a very limited number of peptide identifications during the database search (Supplementary Table S2). All available virus proteomes (a total of 16,846 sequences) and all reviewed human proteins were taken as background. These proteomes were obtained from UniProt/Swiss-Prot (see Section 2.3 for database details).

The Purple run resulted in 1509 in silico-digested peptides after exact matching and 885 peptides after homologous matching (using a background consensus threshold of 80%). The distribution of the homologous background consensus shows a normal distribution below 50% (Supplementary Figure S2). 3986 peptides were discarded, because they were shared with other (i.e., non-CPXV) viral proteomes or the human proteome. The remaining 885 CPXV-specific peptides have a mean background consensus of 53.9%, which means that on average around half of the amino acids of each peptide are equal to residues of peptides in the background.

Next, we searched experimental MS/MS spectra from CPXV samples using the search algorithm MS-GF+ [31] against a CPXV and human sequence database for peptide identification (see Section 2.3). In this analysis, CPXV datasets from MS measurements of three technical replicates, each with ~19,000 MS/MS spectra, were evaluated. The database search resulted in 4028, 4125, and 3967 identified peptides per sample replicate with sequence duplicates removed. More than twice the amount of CPXV peptides were identified as human peptides in this sample before applying a FDR filtering. After applying an FDR threshold of 1%, 1067, 1028, and 1004 CPXV peptides were identified (Table 5). Subsequently, the identified peptides (below 1% FDR threshold) were compared against the set of taxon-specific CPXV peptides suggested by Purple using both exact and homologous matching mode. Between 83 and 94 peptides selected by Purple were detected in the MS/MS experiments (without applying any FDR threshold). When filtered by 1% FDR, the peptides decreased to numbers between 78 and 84. Consequently, this analysis demonstrates that it would be possible to reliably identify CPXV for these three sample replicates.

Table 5. This table shows the number of peptides from the cowpox virus (CPXV) Brighton Red strain after (i) database search with duplicates removed (CPXV); (ii) database search with duplicates removed (human); (iii) intersection of peptides obtained from Purple and peptide identifications from database search; (iv) database search, duplicates removed and filtered by 1% FDR threshold; and (iiv) intersection of peptides suggested by Purple and peptide identifications from FDR-filtered database search. The CPXV Brighton Red strain was compared against the background of all reviewed virus proteomes and the reviewed human proteome. In addition, the second column specifies the sample replicate data that was used for the database search.

Strain	Replicate	No. Database Search (CPXV)	No. Database Search (HUMAN)	No. Intersection	No. Database Search Filtered	No. Intersection Filtered
Brighton Red	1	4028	10319	94	1067	**84**
Brighton Red	2	4125	10286	83	1028	**78**
Brighton Red	3	3967	10068	92	1004	**84**

When considering the results of all three replicates, it can be observed that 61 CPXV-specific peptides were detected without any applied FDR threshold (Figure 7A). Filtered by 1% FDR, 56 peptides across all replicates can be used to specifically identify the species within the sample as a member of CPXV (Figure 7B).

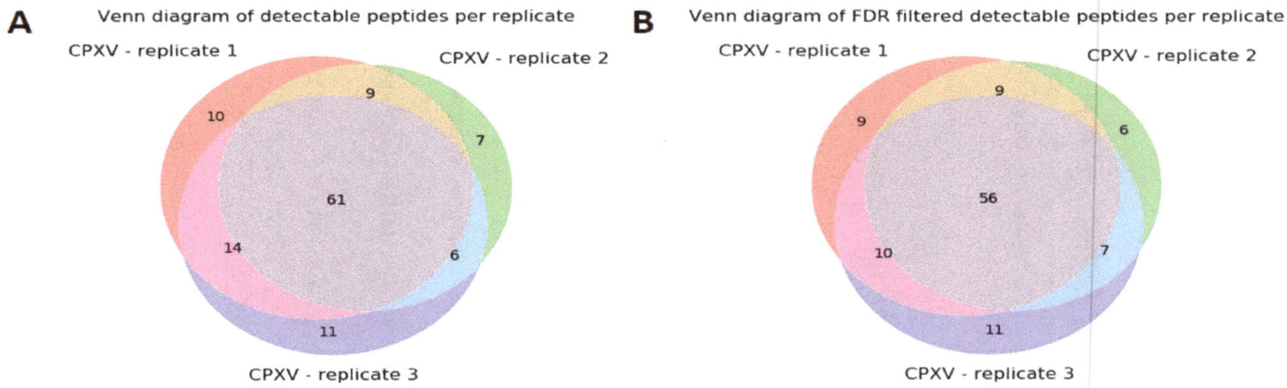

Figure 7. Intersection of detectable peptides of CPXV sample replicates. These Venn diagrams show the intersection of the detectable peptides in replicates 1–3. The subfigures depict the number of peptides without applying any false discovery rate (FDR) threshold (**A**) and filtered by 1% FDR (**B**).

When examining the peptides shared by the target and background proteomes, it can be found that the Cowpox virus shares ~3000 peptide sequences per strain with the Vaccinia virus strains and Variola virus strains (Figure 8). Other Orthopoxviruses were found as well, although the number of peptides is low, due to fewer proteins of these strains in the background database. The CPXV Brighton Red strain-specific peptides are small in number because most matches originate from the Cowpox virus species proteome without giving any details about a particular strain. Around 500 peptides were shared with the human proteome and were consequently discarded.

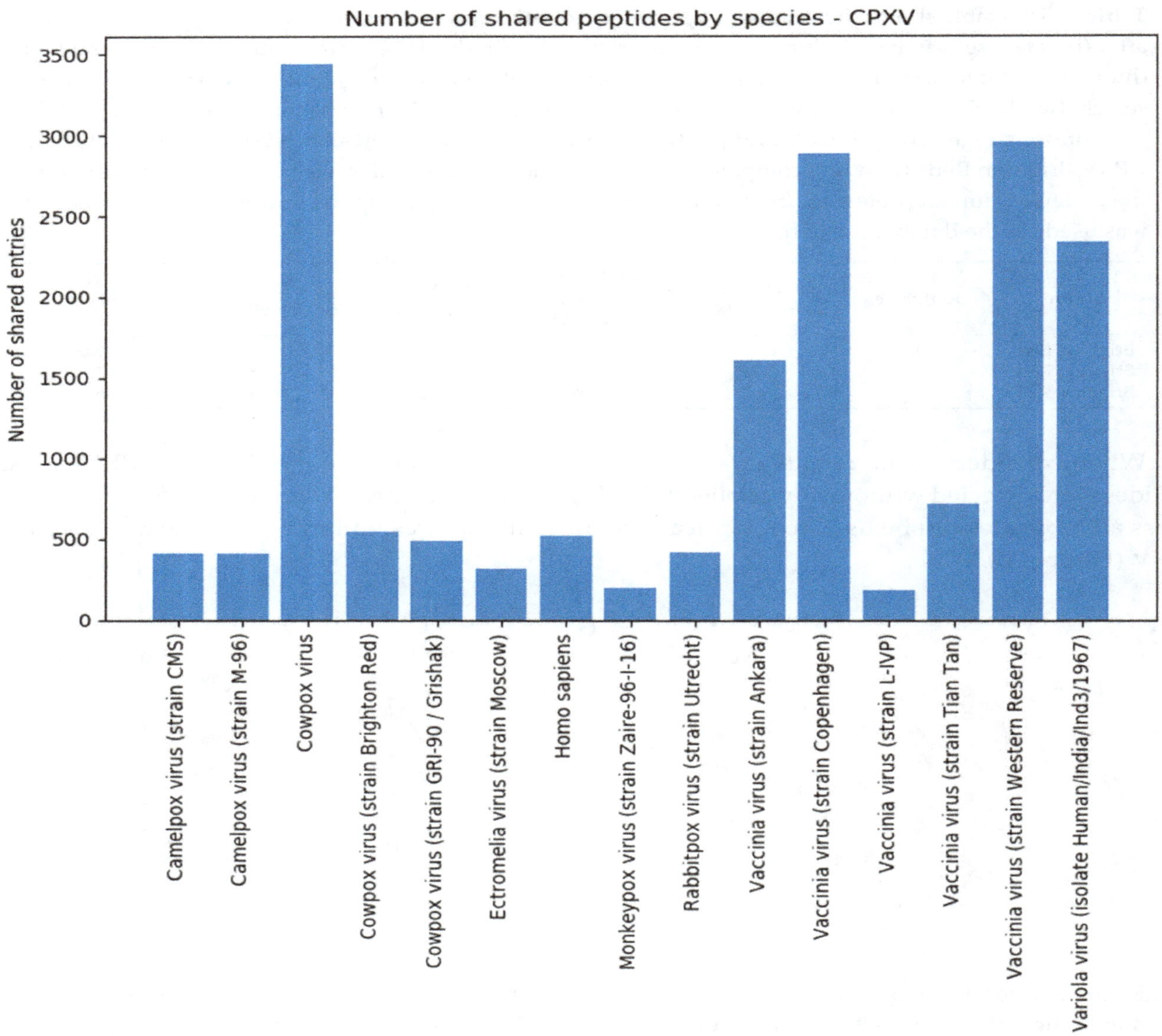

Figure 8. Number of shared CPXV peptides by species/strain assigned. This plot shows the number of shared peptides that Purple detected in the background for a species/species after the CPXV Brighton Red target analysis. All species that contribute less than 0.5% to the total amount of shared peptides were removed.

3.3. Comparison of Strain vs. Strain and Strain vs. All Virus Level Resolution

Next, we conducted a performance evaluation using two different, yet highly similar Vaccinia virus strains, namely VACV Copenhagen and VACV Western Reserve. The objective was to test whether Purple can retrieve strain-specific peptides that are then used in the targeted proteomics assay for accurate taxonomic classification. In this analysis, the target database contained sequences from one of the two VACV virus strains (either Copenhagen or Western reserve). Consequently, the background database contained the remaining VACV strain and all reviewed virus proteins available on UniProt. This procedure was repeated with the remaining VACV strains as target. The goal was to find strain-specific peptides to accurately detect the virus strain. We used a background consensus threshold of 80% to filter out homologous peptides. Afterwards, experimental data (see Section 2.3.3) was used to validate the results and to show if the selected strain-specific peptides are found in the acquired tandem mass spectrometry (MS/MS) data. For peptide identification, we used the software MS-GF+ [31] with an 1% FDR threshold (see Section 2.3.3).

In the case of VACV Copenhagen, Purple discarded 3848 peptides because a perfect sequence match was present in the background with a peptide of another strain or virus (Table 6). Equally, 3971 VACV Western Reserve peptides are marked as shared with the background and discarded. After exact matching, 498 and 341 strain-specific peptides could be obtained for VACV Copenhagen and VACV Western Reserve, respectively. The homologous matching removed additional 157 (VACV Copenhagen) and 172 (VACV Western Reserve) peptides from the set of unique peptides. The remaining 352 (VACV Copenhagen) and 169 (VACV Western Reserve) peptides can be used to uniquely identify the strain in a mixture of all reviewed virus proteins available on UniProt/Swiss-Prot.

Table 6. This table shows the number of taxon-specific peptides from the VACV Copenhagen and VACV Western Reserve strain after (i) *in silico* digest, (ii) exact matching, and (iii) homologous matching (80% background consensus threshold). Each target strain was compared against the background of the other strain and all reviewed virus proteomes. The second column provides the number of nonspecific peptides, i.e., the ones being shared with the background.

Species	No. Digested Peptides	No. Background Shared	No. Exact Matching	No. Homologous Matching
Copenhagen	4200	3848	498	**352**
Western Reserve	4140	3971	341	**169**

In addition, we categorized the shared peptides by virus species to check for close relationships in the background. For VACV Copenhagen, it can be observed that most peptide matches are found in the Vaccinia species (Figure 9), owing to a high protein sequence similarity of involved Vaccinia strains. Other contributing species are Camelpox virus, Cowpox virus, Monkeypox virus, Rabbitpox virus, and Ectromelia virus. All these viruses are, as expected, members of the orthopoxvirus genus. Similar findings could be observed for the results of the VACV Western Reserve strain (Supplementary Figure S1). Note here that Figure 9 shows the number of peptides and if a species is underrepresented in the databases, it will affect the outcome concerning the number of peptides that contribute to the shared peptides.

To evaluate the detectability of taxon-specific peptides for the given DDA experiments, we performed database searches for peptide identification using three different technical replicates of VACV Copenhagen. Without any FDR cut-off, we could identify between 60 and 66 strain-specific peptides selected by Purple (Table 7). However, when filtered by an FDR of 1% the number of peptides decreased drastically and only one or two taxon-specific peptides were confirmed in the shotgun proteomics data. It was possible to identify Replicate 1 and 2 as VACV Copenhagen by using the peptide sequence ILFWPYIEDELR. The number of peptides can be increased by switching to a targeted proteomics approach and by considering PTMs or by an improved homologous matching. The three technical replicates of the VACV Western Reserve strain resulted in fewer peptides in the intersection with the database search results (between 32 and 42), but when filtered by 1% FDR, the number of peptides was increased up to 11-fold (with nine to 11 peptides) in comparison to the VACV Copenhagen replicates. Six peptides were detected, and their sequences were identical among all three replicates.

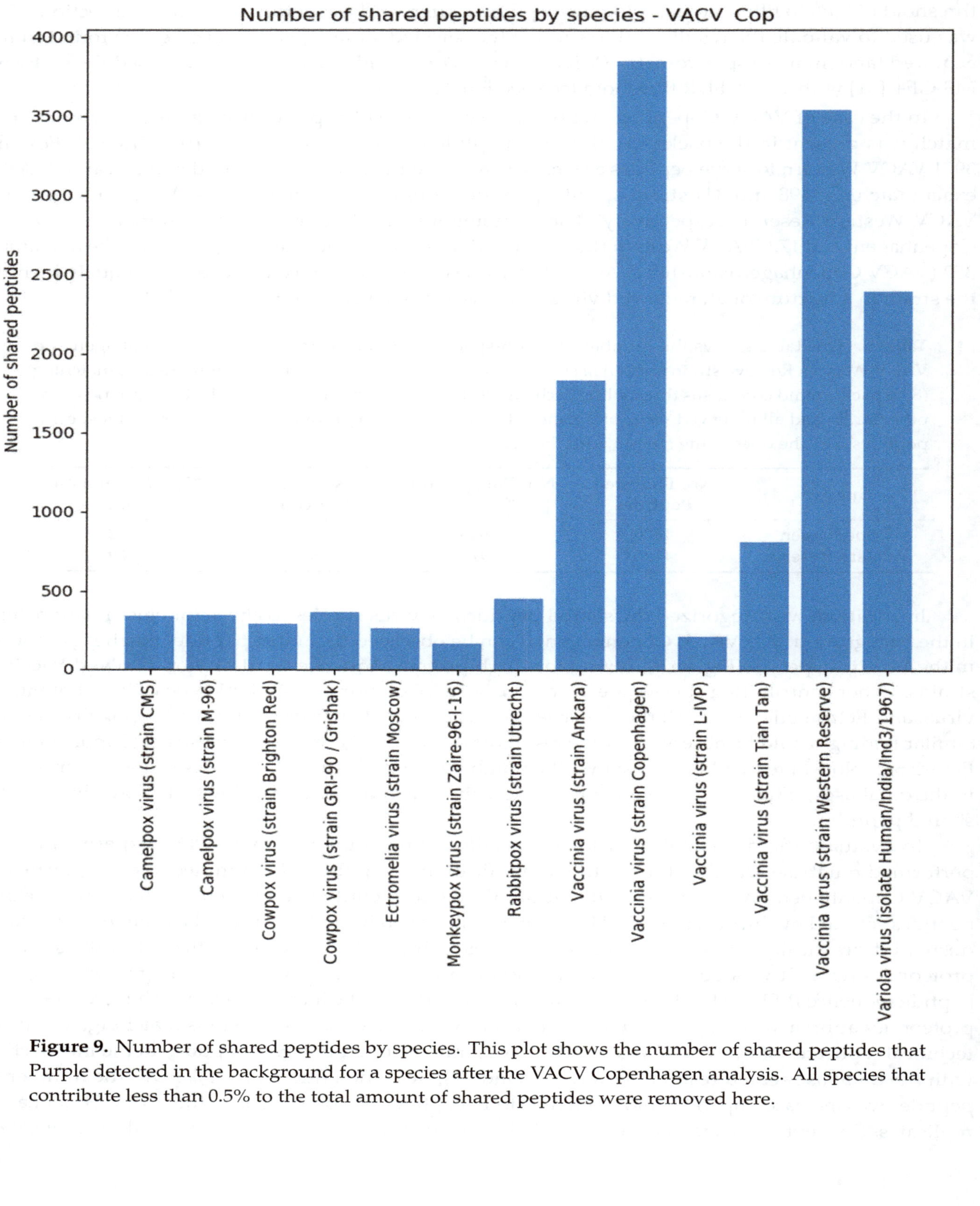

Figure 9. Number of shared peptides by species. This plot shows the number of shared peptides that Purple detected in the background for a species after the VACV Copenhagen analysis. All species that contribute less than 0.5% to the total amount of shared peptides were removed here.

Table 7. This table shows the number of peptides from VACV Copenhagen and VACV Western Reserve strain after (i) database search with duplicates removed; (ii) intersection of peptides obtained by Purple and database search; (iii) database search, duplicates removed and filtering by FDR; and (iv) intersection of peptides obtained by Purple and filtered database search. Each target strain was compared against the background of the other strain and all reviewed virus proteomes. The second column specifies the replicate data that was used for the database search.

Strain	Replicate	No. Database Search	No. Intersection	No. Database Search Filtered	No. Intersection Filtered
Copenhagen	1	3585	66	825	**2**
Copenhagen	2	3507	62	800	**1**
Copenhagen	3	3525	60	828	**1**
Western Reserve	1	3636	35	841	**9**
Western Reserve	2	3736	42	800	**11**
Western Reserve	3	3507	32	809	**9**

In conclusion, we were able to identify every strain in each sample with an applied FDR of 1%. For VACV Western Reserve, the number of peptides was higher than for the VACV Copenhagen strain. The number of detectable peptides could be increased by improving scoring and filtering or by switching from shotgun to targeted proteomics methods or by considering PTMs.

Figure 10 reveals a normal distributed homologous consensus in the interval from 10% to 50%. This is caused by random matches with background peptides and these peptides should be unique for the strain. We could not observe a distinct distribution above 50%. This could be improved by moving from identity to a similarity-based matching, as this would differentiate peptides with the same amount of matching consensus residuals.

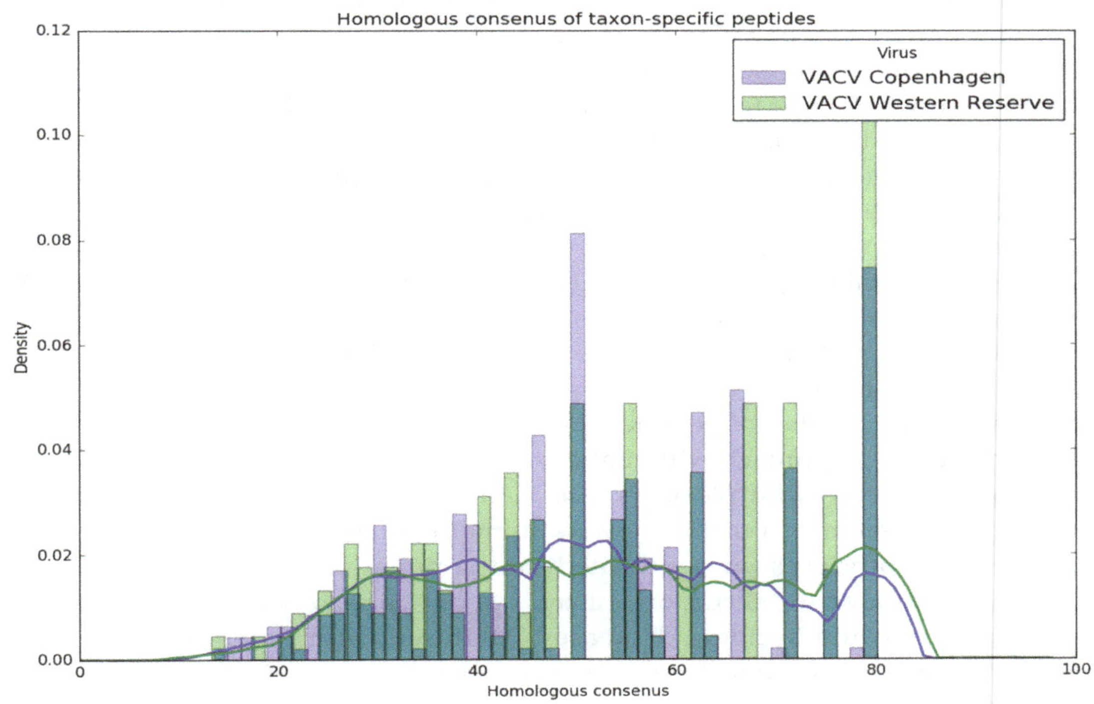

Figure 10. Histogram and density plot of homologous consensus. This histogram shows the distribution of the homologous consensus for the VACV Copenhagen (blue) and Western Reserve (green) strains. Additionally, the kernel density was calculated utilizing the Epanechnikov kernel and a Silverman bandwidth estimation.

4. Discussion

The main goal of our developed Purple software is to provide taxon-specific peptides for a targeted proteomics assay. These targeted assays can be used in a diagnostic setting to identify a virus species/strain or even a whole virus family in a sample in sensitive and time-efficient manner. In this work, we validated the software in three different benchmarking experiments.

Purple enabled us to retrieve taxon-specific peptides to distinguish between arenavirus species proteomes that are very similar in their sequences (see Section 3.1). Accordingly, we observed a comparable decrease in the ratio of unique to *in silico*-digested peptides for New and Old World arenaviruses based on differences between their proteomes (Figure 4). This effect could also be recognized also on the clade level for the New World viruses.

The data analysis of CPXV (see Section 3.2) resulted in 56 taxon-specific peptides (Figure 7). These peptides were present in each MS/MS sample replicate and can be used to uniquely identify CPXV in a mixed biological sample, although its proteome is very similar to other Orthopoxvirus species and strains (Figure 8). By changing to a Brighton Red strain-specific target database, a reliable determination of the strain would be possible as well. This underlines that Purple relies on a correct and complete database to yield to the best possible results. Missing or incorrectly assigned protein sequences could result in incorrect selected unique peptides or discarded ones. Furthermore, although many spectra in the shotgun proteomics experiment were assigned to human peptides, this does not present a limitation for the targeted proteomics approach, because unique virus peptides selected by Purple can be detected using a targeted (e.g., PRM-based) assay in specific and sensitive manner; for example, in a recently published study [35], a PRM-based assay was used to identify dengue virus species directly from clinical serum samples. Nevertheless, to validate the resulting set of peptides, it would be recommended to test them on other CPXV samples and to check if the peptides are detectable in these samples likewise. In addition, the selected background database might be incomplete, e.g., when proteome references were missed to be included for the Purple analysis. In this case, it is useful to validate Purple-selected peptides using secondary tools such as Unipept [36] for resolving the taxonomic origin of any tryptic peptide based on the complete UniProt database. Furthermore, false negatives may result from issues during sample preparation or poor instrument performance. Therefore, these parameters need to be controlled in diagnostic PRM assays, e.g., by using internal standards and running further quality control samples.

It can be crucial in virus infection scenarios to accurately distinguish between specific strains. To cover these cases, we examined the strain-level resolution of our tool using data of VACV Copenhagen and VACV Western Reserve strains (see Section 3.3). Purple was able to find a reliable amount of strain-specific peptides (Table 7). The intersection between the Purple-selected peptides and the peptide identification from the database search showed that it is possible to detect these peptides. In general, strain-level identification was possible even for an applied FDR threshold of 1%, however, it became apparent that the shotgun proteomics approach becomes limited due to the spurious numbers of identified peptides. The number of peptides could be increased by adjusting the FDR filtering or by using a targeted proteomics approach with higher sensitivity.

In comparison to other tools, Purple offers several advantages, such as cross-platform compatibility on multiple operating systems. Purple allows a homology-based analysis of multiple proteome databases at once and produces an aggregated and summarized export on various levels. In addition, Purple is not limited to specific organisms, but can be used with general UniProt databases, also including eukaryotic and bacterial databases. High sequence similarity between strains and horizontal gene transfer may complicate taxon-specific classification for bacterial samples. However, Purple could help to overcome complications and can be helpful for creating targeted assays for bacterial detection as well. The graphical user interface and compatibility with all UniProt databases enables researchers without bioinformatics background to find taxon-specific peptides in an easy and straightforward manner.

A potential improvement to the software would be to move from a sequence identity-based metric based on the Hamming distance to similarity-based matching for the homologous matching mode. In this case, amino acid substitutions are not weighted equally, for example by using a PAM or BLOSUM matrix [37]. This similarity-based metric might allow a more accurate homologous matching in Purple. For example, an approach based on a structural alignment as introduced by Ogata et al. [38] might be useful. Further potential improvements with useful features in Purple include adding plots for better data exploration and a tabular view for inspecting the results (that are currently exportable as text files to spreadsheet software).

In summary, the most promising application of Purple is to select taxon-specific peptides for creating tailored SRM or PRM assays with high sensitivity and specificity. This application will allow for new time- and cost-efficient diagnostic methods in healthcare and further biological applications. It could even be used to identify multiple organisms in a single sample in the context of targeted metaproteomics [39].

Purple is available for download on our GitLab website (https://gitlab.com/rki_bioinformatics), by using the Python package manager pip (https://pypi.org/project/purple-bio/) or via the Bioconda channel (https://anaconda.org/bioconda/purple-bio) [40]. The software is available as graphical user interface version, Python package and command line version for Windows, Linux, and MacOS. In addition, user support, tutorials, and the documentation manual can be found on the GitLab webpages.

Supplementary Materials:
Table S1: Genome sequence similarities of cowpox virus; Table S2: Number of peptides from CPXV Brighton Red strain processing; Figure S1: Number of shared peptides by species for VACV Western Reserve; Figure S2: Histogram and density plot of homologous consensus—CPXV; Data S1: MSA of the pre-glycoprotein polyprotein GP complex (GPC gene); Data S2: MSA of nucleocapsid protein (N gene); Data S3: MSA of RNA-directed RNA polymerase L (L gene); Data S4: MSA of RING finger protein Z (Z gene).

Author Contributions: Conceptualization: J.D., B.Y.R., and T.M.; methodology: J.D., J.L., P.H., and T.M.; software: F.H., J.L., and P.H.; validation: F.H. and J.L.; formal analysis: F.H., J.L., and T.M..; investigation: F.H. and J.L..; resources: A.N., B.Y.R., J.D., and M.G.; data curation: F.H.; writing—original draft preparation: F.H. and J.L.; writing—review and editing: A.N., B.Y.R., J.D., M.G., P.H., and T.M.; visualization: F.H.; supervision: B.Y.R. and T.M.; project administration: T.M.

References

1. Singhal, N.; Kumar, M.; Kanaujia, P.K.; Virdi, J.S. MALDI-TOF mass spectrometry: An emerging technology for microbial identification and diagnosis. *Front. Microbiol.* **2015**, *6*, 791. [CrossRef] [PubMed]
2. Ebhardt, H.A.; Root, A.; Sander, C.; Aebersold, R. Applications of targeted proteomics in systems biology and translational medicine. *Proteomics* **2015**, *15*, 3193–3208. [CrossRef] [PubMed]
3. Deutsch, E.W. The PeptideAtlas project. *Methods Mol. Biol.* **2010**, *604*, 285–296. [PubMed]
4. Picotti, P.; Aebersold, R. Selected reaction monitoring-based proteomics: Workflows, potential, pitfalls and future directions. *Nat. Methods* **2012**, *9*, 555–566. [CrossRef] [PubMed]
5. Peterson, A.C.; Russell, J.D.; Bailey, D.J.; Westphall, M.S.; Coon, J.J. Parallel reaction monitoring for high resolution and high mass accuracy quantitative, targeted proteomics. *Mol. Cell. Proteom.* **2012**, *11*, 1475–1488. [CrossRef] [PubMed]
6. Borràs, E.; Sabidó, E. What is targeted proteomics? A concise revision of targeted acquisition and targeted data analysis in mass spectrometry. *Proteomics* **2017**, *17*, 17–18.
7. Karlsson, C.; Malmström, L.; Aebersold, R.; Malmström, J. Proteome-wide selected reaction monitoring assays for the human pathogen *Streptococcus pyogenes*. *Nat. Commun.* **2012**, *3*, 1301. [CrossRef]
8. Peters, J.S.; Calder, B.; Gonnelli, G.; Degroeve, S.; Rajaonarifara, E.; Mulder, N.; Blackburn, J.M. Identification of quantitative proteomic differences between *Mycobacterium tuberculosis* lineages with altered virulence. *Front. Microbiol.* **2016**, *7*, 813. [CrossRef]
9. MacLean, B.; Tomazela, D.M.; Shulman, N.; Chambers, M.; Finney, G.L.; Frewen, B.; MacCoss, M.J. Skyline: An. open source document editor for creating and analyzing targeted proteomics experiments. *Bioinformatics* **2010**, *26*, 966–968. [CrossRef]
10. Mohammed, Y.; Domański, D.; Jackson, A.M.; Smith, D.S.; Deelder, A.M.; Palmblad, M.; Borchers, C.H. PeptidePicker: A scientific workflow with web interface for selecting appropriate peptides for targeted proteomics experiments. *J. Proteom.* **2014**, *106*, 151–161. [CrossRef]

11. Demeure, K.; Duriez, E.; Domon, B.; Niclou, S.P. PeptideManager: A peptide selection tool for targeted proteomic studies involving mixed samples from different species. *Front. Genet.* **2014**, *5*, 305. [CrossRef] [PubMed]
12. Zauber, H.; Kirchner, M.; Selbach, M. Picky: A simple online PRM and SRM method designer for targeted proteomics. *Nat. Methods* **2018**, *15*, 156–157. [CrossRef] [PubMed]
13. Zolg, D.P.; Wilhelm, M.; Schnatbaum, K.; Zerweck, J.; Knaute, T.; Delanghe, B.; Yu, P. Building ProteomeTools based on a complete synthetic human proteome. *Nat. Methods* **2017**, *14*, 259–262. [CrossRef] [PubMed]
14. Zolg, D.P.; Wilhelm, M.; Schmidt, T.; Medard, G.; Zerweck, J.; Knaute, T.; Kuster, B. ProteomeTools: Systematic characterization of 21 post-translational protein modifications by liquid chromatography tandem mass spectrometry (LC-MS/MS) using synthetic peptides. *Mol. Cell. Proteom.* **2018**, *17*, 1850–1863. [CrossRef] [PubMed]
15. Mesuere, B.; Van der Jeugt, F.; Devreese, B.; Vandamme, P.; Dawyndt, P. The unique peptidome: Taxon-specific tryptic peptides as biomarkers for targeted metaproteomics. *Proteomics* **2016**, *16*, 2313–2318. [CrossRef] [PubMed]
16. Mallick, P.; Schirle, M.; Chen, S.S.; Flory, M.R.; Lee, H.; Martin, D.; Kuster, B. Computational prediction of proteotypic peptides for quantitative proteomics. *Nat. Biotechnol.* **2007**, *25*, 125–131. [CrossRef] [PubMed]
17. Eyers, C.E.; Lawless, C.; Wedge, D.C.; Lau, K.W.; Gaskell, S.J.; Hubbard, S.J. CONSeQuence: Prediction of reference peptides for absolute quantitative proteomics using consensus machine learning approaches. *Mol. Cell. Proteom.* **2011**, *10*, M110-003384. [CrossRef]
18. Qeli, E.; Omasits, U.; Goetze, S.; Stekhoven, D.J.; Frey, J.E.; Basler, K.; Ahrens, C.H. Improved prediction of peptide detectability for targeted proteomics using a rank-based algorithm and organism-specific data. *J. Proteom.* **2014**, *108*, 269–283. [CrossRef]
19. Tang, H.; Arnold, R.J.; Alves, P.; Xun, Z.; Clemmer, D.E.; Novotny, M.V.; Radivojac, P. A computational approach toward label-free protein quantification using predicted peptide detectability. *Bioinformatics* **2006**, *22*, e481–e488. [CrossRef]
20. Sanders, W.S.; Bridges, S.M.; McCarthy, F.M.; Nanduri, B.; Burgess, S.C. Prediction of peptides observable by mass spectrometry applied at the experimental set level. *BMC Bioinform.* **2007**, *8*, S23. [CrossRef]
21. Fusaro, V.A.; Mani, D.R.; Mesirov, J.P.; Carr, S.A. Prediction of high-responding peptides for targeted protein assays by mass spectrometry. *Nat. Biotechnol.* **2009**, *27*, 190–198. [CrossRef] [PubMed]
22. Cock, P.J.; Antao, T.; Chang, J.T.; Chapman, B.A.; Cox, C.J.; Dalke, A.; De Hoon, M.J. Biopython: Freely available python tools for computational molecular biology and bioinformatics. *Bioinformatics* **2009**, *25*, 1422–1423. [CrossRef] [PubMed]
23. Olayiwola, J.O.; Bakarey, A.S. Epidemiological trends of Lassa fever outbreaks and insights for future control in Nigeria. *Int. J. Trop. Dis. Heal.* **2017**, *24*, 1–14. [CrossRef]
24. Doellinger, J.; Schaade, L.; Nitsche, A. Comparison of the cowpox virus and Vaccinia virus mature virion proteome: Analysis of the Species- and strain-specific proteome. *PLoS ONE* **2015**, *10*, e0141527. [CrossRef] [PubMed]
25. Schrick, L.; Tausch, S.H.; Dabrowski, P.W.; Damaso, C.R.; Esparza, J.; Nitsche, A. An early American smallpox vaccine based on horsepox. *N. Engl. J. Med.* **2017**, *377*, 1491–1492. [CrossRef] [PubMed]
26. Esparza, J.; Schrick, L.; Damaso, C.R.; Nitsche, A. Equination (inoculation of horsepox): An. early alternative to vaccination (inoculation of cowpox) and the potential role of horsepox virus in the origin of the smallpox vaccine. *Vaccine* **2017**, *35*, 7222–7230. [CrossRef] [PubMed]
27. Altschul, S.F.; Madden, T.L.; Schäffer, A.A.; Zhang, J.; Zhang, Z.; Miller, W.; Lipman, D.J. Gapped BLAST and PSI-BLAST: A new generation of protein database search programs. *Nucleic Acids Res.* **1997**, *25*, 3389–3402. [CrossRef]
28. Jacobs, B.L.; Langland, J.O.; Kibler, K.V.; Denzler, K.L.; White, S.D.; Holechek, S.A.; Baskin, C.R. Vaccinia virus vaccines: Past, present and future. *Antivir. Res.* **2009**, *84*, 1–13. [CrossRef]
29. Apweiler, R.; Bairoch, A.; Wu, C.H.; Barker, W.C.; Boeckmann, B.; Ferro, S.; Martin, M.J. UniProt: The Universal protein knowledgebase. *Nucleic Acids Res.* **2004**, *32*, D115–D119. [CrossRef]
30. Kessner, D.; Chambers, M.; Burke, R.; Agus, D.; Mallick, P. ProteoWizard: Open source software for rapid proteomics tools development. *Bioinformatics* **2008**, *24*, 2534–2536. [CrossRef]

31. Kim, S.; Pevzner, P.A. MS-GF+ makes progress towards a universal database search tool for proteomics. *Nat. Commun.* **2014**, *5*, 5277. [CrossRef] [PubMed]
32. Fehling, S.; Lennartz, F.; Strecker, T. Multifunctional nature of the arenavirus RING finger protein Z. *Viruses* **2012**, *4*, 2973–3011. [CrossRef] [PubMed]
33. Briese, T.; Paweska, J.T.; McMullan, L.K.; Hutchison, S.K.; Street, C.; Palacios, G.; Nichol, S.T. Genetic detection and characterization of Lujo virus, a new hemorrhagic fever-associated arenavirus from southern Africa. *PLoS Pathog.* **2009**, *5*, e1000455. [CrossRef] [PubMed]
34. Chojnacki, S.; Cowley, A.; Lee, J.; Foix, A.; Lopez, R. Programmatic access to bioinformatics tools from EMBL-EBI update: 2017. *Nucleic Acids Res.* **2017**, *45*, W550–W553. [CrossRef] [PubMed]
35. Wee, S.; Alli-Shaik, A.; Kek, R.; Swa, H.L.; Tien, W.P.; Lim, V.W.; Gunaratne, J. Multiplex targeted mass spectrometry assay for one-shot flavivirus diagnosis. *Proc. Natl. Acad. Sci. USA* **2019**, *116*, 6754–6759. [CrossRef]
36. Mesuere, B.; Devreese, B.; Debyser, G.; Aerts, M.; Vandamme, P.; Dawyndt, P. Unipept: Tryptic peptide-based biodiversity analysis of metaproteome samples. *J. Proteome Res.* **2012**, *11*, 5773–5780. [CrossRef]
37. Henikoff, S.; Henikoff, J.G. Amino acid substitution matrices from protein blocks. *Proc. Natl. Acad. Sci. USA* **1992**, *89*, 10915–10919. [CrossRef] [PubMed]
38. Ogata, K.; Ohya, M.; Umeyama, H. Amino acid similarity matrix for homology modeling derived from structural alignment and optimized by the Monte Carlo method. *J. Mol. Gr. Model.* **1998**, *16*, 178–189. [CrossRef]
39. Saito, M.A.; Dorsk, A.; Post, A.F.; McIlvin, M.R.; Rappé, M.S.; DiTullio, G.R.; Moran, D.M. Needles in the blue sea: Sub-species specificity in targeted protein biomarker analyses within the vast oceanic microbial metaproteome. *Proteomics* **2015**, *15*, 3521–3531. [CrossRef]
40. Dale, R.; Grüning, B.; Sjödin, A.; Rowe, J.; Chapman, B.A.; Tomkins-Tinch, C.H.; Köster, J. Bioconda: A sustainable and comprehensive software distribution for the life sciences. bioRxiv 2017, bioRxiv:207092. *bioRxiv* **2017**. bioRxiv:207092.

10

Secretory Nanoparticles of *Neospora caninum* Profilin-Fused with the Transmembrane Domain of GP64 from Silkworm Hemolymph

Hamizah Suhaimi [1], Rikito Hiramatsu [2], Jian Xu [3], Tatsuya Kato [1,2,3] and Enoch Y. Park [1,2,3,*]

[1] Laboratory of Biotechnology, Department of Bioscience, Graduate School of Science and Technology, Shizuoka University, 836 Ohya, Suruga-ku, Shizuoka 422-8529, Japan; noor.hamizah.binsuhaimi.16@shizuoka.ac.jp (H.S.); kato.tatsuya@shizuoka.ac.jp (T.K.)

[2] Laboratory of Biotechnology, Department of Applied Biological Chemistry, College of Agriculture, Graduate School of Integrated Science and Technology, Shizuoka University, 836 Ohya Suruga-ku, Shizuoka 422-8529, Japan; hiramatsu.rikito.17@shizuoka.ac.jp

[3] Laboratory of Biotechnology, Research Institute of Green Science and Technology, Shizuoka University, 836 Ohya Suruga-ku, Shizuoka 422-8529, Japan; xu.jian@shizuoka.ac.jp

* Correspondence: park.enoch@shizuoka.ac.jp

Abstract: Neosporosis, which is caused by *Neospora caninum*, is a well-known disease in the veterinary field. Infections in pregnant cattle lead to abortion via transplacental (congenitally from mother to fetus) transmission. In this study, a *N. caninum* profilin (NcPROF), was expressed in silkworm larvae by recombinant *Bombyx mori* nucleopolyhedrovirus (BmNPV) bacmid and was purified from the hemolymph. Three NcPROF constructs were investigated, native NcPROF fused with an N-terminal PA tag (PA-NcPROF), PA-NcPROF fused with the signal sequence of bombyxin from *B. mori* (bx-PA-NcPROF), and bx-PA-NcPROF with additional C-terminal transmembrane and cytoplasmic domains of GP64 from BmNPV (bx-PA-NcPROF-GP64TM). All recombinant proteins were observed extra- and intracellularly in cultured Bm5 cells and silkworm larvae. The bx-PA-NcPROF-GP64TM was partly abnormally secreted, even though it has the transmembrane domain, and only it was pelleted by ultracentrifugation, but PA-NcPROF and bx-PA-NcPROF were not. Additionally, bx-PA-NcPROF-GP64TM was successfully purified from silkworm hemolymph by anti-PA agarose beads while PA-NcPROF and bx-PA-NcPROF were not. The purified bx-PA-NcPROF-GP64TM protein bound to its receptor, mouse Toll-like receptor 11 (TLR-11), and formed unique nanoparticles. These results suggest that profilin fused with GP64TM was secreted as a nanoparticle with binding affinity to its receptor and this nanoparticle formation is advantageous for the development of vaccines to *N. caninum*.

Keywords: BmNPV bacmid; nanobiomaterials; *Neospora caninum*; *Neospora caninum* profilin; neosporosis; silkworm expression system

---

1. Introduction

Neosporosis is a disease caused by apicomplexan parasites such as *Neospora caninum* and *Neospora hugheshi* [1]. Considering its morphology, *N. caninum* is quite similar to *Toxoplasma gondii*, so neosporosis was misdiagnosed as *T. gondii* infection until 1988 [2]. Although the morphologies are quite similar, these parasites are biologically different and are considered as zoonotic compared with toxoplasmosis where it affects human, sheep and many other warm-blooded animals [3]. Neosporosis infection occurs from unsporulated oocysts in feces followed by ingestion by the cattle, transformation from the tachyzoites stage to the sporozoites stage and leading to neosporosis in the cattle [4]. Recrudescence of the infection occurs in pregnant cattle, where this parasite is transmitted from the placenta to the unborn fetus and may lead to abortion [5,6].

N. caninum, a parasite of the phylum Apicomplexa, has been known as the etiologic agent of neosporosis disease [5]. Apicomplexa contains specific organelles, micronemes, rhoptries, and dense granules. Most of the specific antigen proteins are secreted from each organelle and are located at the apical end of the *N. caninum* parasite and have been extensively studied as recombinant vaccine candidates against *N. caninum* infection [7,8]. In our previous study, we demonstrated the capability of baculoviruses displaying *N. caninum*-derived antigens, such as surface antigen 1 (NcSAG1), SAG1-related sequence 2 (NcSRS2) and microneme protein 3, as an alternative method to control neosporosis because the combination of three antigens could induce T-cell activation and interferon gamma (IFN-γ) production and suppress *N. caninum* infection in mice [9].

Profilin of *N. caninum* (NcPROF) is recognized and conserved as a vaccine candidate with high potential against neospora infections [10]. Interestingly, profilin is known as a small actin-binding protein located at the apical end of *N. caninum* tachyzoites and is essential for invasion of the host cell by regulating the polymerization and depolymerization of actin filaments [11]. Furthermore, Jenkins et al. [12] and Mansilla et al. [13] revealed that *T. gondii* profilin binds to Toll-like receptor 11 (TLR11) in mice and is responsible for activating dendritic cells and stimulating the release of cytokines such as interleukin 12 (IL-12) and IFN-γ. According to Innes et al. [14], some cytokines help in controlling the infection by inhibiting parasite multiplication.

In this study, to produce recombinant NcPROF, the *Bombyx mori* nucleopolyhedrovirus (BmNPV) bacmid-based silkworm expression system was used. The BmNPV bacmid-based silkworm expression system contributes to several advantages for recombinant protein expression, such as low cost, ease of treatment and high safety [15]. Previous studies of recombinant NcPROF have not dealt with the expression fusion of the transmembrane and cytoplasmic domains of GP64 from BmNPV. Therefore, to evaluate the expression and purification of fusion native NcPROF fused with an N-terminal PA tag, fused with the signal sequence of bombyxin from *B. mori* (bx-PA-NcPROF) fused with the transmembrane and cytoplasmic domains of GP64 from BmNPV (bx-PA-NcPROF-GP64TM), NcPROF was expressed in two other NcPROF constructs, one fused with a PA tag (PA-NcPROF) and PA-NcPROF fused with the signal sequence of bombyxin from *B. mori* (bx-PA-NcPROF). The bx-PA-NcPROF-GP64TM was successfully purified from silkworm hemolymph and the binding of bx-PA-NcPROF-GP64TM with recombinant mouse TLR11 (mTLR11), and the morphological analysis of the nanoparticles were reported.

2. Materials and Methods

2.1. Construction of the Recombinant BmNPV Bacmid Containing NcPROF Constructs

The NcPROF gene (GenBank accession no. BK006901.1) was used to express recombinant NcPROF. The bx-PA (Peptide tag, GVAMPGAEDDVV)-NcPROF-GP64TM gene was synthesized by GENEWIZ Japan (Saitama, Japan) with the bombyxin signal peptide (bx) sequence (NCBI reference sequence no. NP_001103771.1) at its N-terminus and transmembrane cytoplasmic domains of GP64 from BmNPV (GP64TM) (GenBank accession no. BAF32568.1) at its C-terminus. The synthetic sequence of bx-PA-NcPROF-GP64TM was digested using *Eco*RI (NEB, Tokyo, Japan) and *Not*I (NEB, Tokyo, Japan) and was cloned into pFastBac1 (ThermoFisher Scientific. K.K., Tokyo, Japan), resulting in the construct pFast/bx-PA-NcPROF-GP64TM. The gene coding bx-PA-NcPROF was amplified by polymerase chain reaction (PCR) using (bx-PA-NcPROF) with primer sets (Table 1). The amplified gene was digested by *Eco*RI and *Xho*I and was cloned into pFastbac1, resulting in the construct pFast/bx-PA-NcPROF. Additionally, pFast/PA-NcPROF and pFast/bx-PA-NcPROF were generated by PCR amplification with primers listed in Table 1 and then underwent self-ligation. Subsequently, the constructed pFast/bx-PA-NcPROF-GP64TM, pFast/bx-PA-NcPROF, and pFast/bx-PA-NcPROF were transformed into *Escherichia coli* BmDH10Bac, respectively, as described previously [15]. The white colonies were identified as positive transformants containing each recombinant BmNPV bacmid.

Table 1. Primers used in this study.

Name	5′ to 3′
pFastBac1	
Forward	5′-TATTCCGGATTATTCATACC-3′
Reverse	5′-ACAAATGTGGTATGGCTGATT-3′
NcPROF	
Forward	5′-GGACACAATCGGAGAGGACG-3′
Reverse	5′-GTGCACACATGGTGATGTCG-3′
pUC/M13	
Forward	5′-CCCAGTCACGACGTTGTAAAACG-3′
Reverse	5′-AGCGGATAACAATTTCACACAGG-3′
PA-NcPROF	
Forward	5′-GGCGTTGCCATGCCAGGTGC-3′
Reverse	5′-CATGAATTCCGCGCGCTTCG-3′
Bx-PA-NcPROF	
Forward	5′-GCGCGGAATTCATGAAGATACTCCTTGCT-3′
Reverse	5′-GCATGCCTCGAGTTAATAGCCAGACTGGTGAAGGTACTCG-3′

2.2. Expression of Recombinant NcPROF in Silkworm Larvae and Bm5 Cells

Each recombinant BmNPV bacmid DNA (10 or 20 μg) was mixed with 0.1% chitosan (ratio of amino group to phosphate group 2) and 2% (*w*/*v*) of 2-(*N*-morpholino) ethanesulfonic acid (MES) buffer [16]. The mixture of each recombinant BmNPV bacmid DNA was incubated at room temperature (RT) for approximately 45 min before injection into the silkworm larvae. Subsequently, approximately 50 μL of the mixture was injected into a fifth instar silkworm larva (Ehime Sansyu, Ehime, Japan), and the larvae were reared at 25 °C and 65 ± 5% relative humidity with the artificial diet Silkmate S2 (Nosan, Yokohama, Japan) for 6–7 day. The collected hemolymph was diluted with 10-fold of phosphate-buffered saline (PBS, pH 7.4) and was injected again into fifth-instar silkworm larva and reared for 4–5 d. The larval hemolymph was collected through cutting the caudal leg, was mixed with 5 μL of 200 mM 1-phenyl-2-thiourea, and then was centrifuged at 10,000× *g* for 10 min at 4 °C. The larval fat body was dissected in PBS mixed with 0.1% Triton-X 100 before sonication on ice with an interval time of 15 s until the solution was clear and then it was centrifuged at 10,000× *g* for 30 min at 4 °C. For the expression of NcPROF in Bm5 cells, diluted hemolymph containing recombinant BmNPVs was added to the cultured Bm5 cells in a 6-well plate at 27 °C in Sf-900II medium (ThermoFisher Scientific K.K.) supplemented with 1% antibiotic-antimycotic (ThermoFisher Scientific K.K.) and 10% FBS (Gibco, Tokyo, Japan). The culture supernatant and infected-Bm5 cells were collected at 3 d post infection. The culture supernatant (culture media) was collected after centrifugation at 10,000× *g* for 10 min at 4 °C together with the same solution as the larval hemolymph. Infected-Bm5 cells were mixed with the same solution and were subjected to the same treatment as the larval fat body, followed by separation of cell lysate soluble and insoluble fractions after centrifugation. The supernatants of the larval hemolymph, fat body, cell medium, cell lysate soluble, and insoluble were immediately frozen at −80 °C until further analysis.

2.3. Purification of bx-PA-NcPROF-GP64TM from Silkworm Larvae

bx-PA-NcPROF-GP64TM was purified from silkworm larval hemolymph using anti-PA tag affinity chromatography (WAKO Pure Chemical Industries, Osaka, Japan). Ten-times the diluted supernatant of the hemolymph was mixed with anti-PA tag affinity beads and incubated at 4 °C for 24 h with gentle agitation. Next, the beads were washed 10 times with Tris-buffered saline (TBS, pH 7.6), and the bound proteins were eluted with 0.1 M glycine-HCl (pH 3) and immediately neutralized with 1.5 M Tris-HCl (pH 8.0). The elution was conducted at RT in a stepwise manner.

2.4. Ultracentrifugation Analysis of bx-PA-NcPROF-GP64TM, bx-PA-NcPROF, and PA-NcPROF

Two milliliters of crude hemolymph was mixed with 1 mL of PBS (pH 6.2). This mixture was centrifuged at 100,000× *g* for 90 min at 4 °C, and the pellet was suspended in 1 mL of PBS (pH 6.2). This suspension was sonicated to dissolve the pellet, and the suspension was collected and subjected to western blotting.

2.5. Transmission Electron Microscopy Observation of bx-PA-NcPROF-GP64TM

Purified bx-PA-rNcPROF-GP64TM was analyzed through negative staining. The bx-PA-rNcPROF-GP64TM (20 μL) drop was loaded onto the surface of film 200 mesh copper grid (Nisshin Em Co. Ltd., Tokyo, Japan) within 30 s at RT. Next, the grid was washed 3 times with PBS and negatively stained with phosphotungstic acid (2% of *v*/*v*). To investigate the surface of the bx-PA-rNcPROF-GP64TM nanoparticle, the sample on the grid was blocked in 2% (*v*/*v*) bovine serum albumin (BSA) for approximately 5 min after washing with PBS 3 times. Next, the grid was incubated for 1 h at RT, and then it was loaded onto the surface drop of rat anti-PA tag monoclonal antibody (NZ-1, 1:50 in PBS) (WAKO Pure Chemical Industries). After 1 h of incubation, the grid was washed 6 times and was loaded onto the surface with goat anti-rat immunoglobulin G (IgG) (H+L)-conjugated with 12 nm gold beads (1:50 in PBS) (Jackson ImmnunoResearch Inc., West Grove, PA, USA) drops and was incubated for 1 h at RT. Finally, the grid was washed 6 times with PBS, followed negative staining with phosphotungstic acid (2% *v*/*v*). Images were acquired with a transmission electron microscope (TEM; JEM-2100F; JEOL, Ltd., Tokyo, Japan) operated at 100 kV.

2.6. Sodium Dodecyl Sulfate-Polyacrylamide Gel Electrophoresis (SDS-PAGE) and Western Blotting

The collected larval hemolymph and extract of a fat body were used to confirm the expression of recombinant proteins through SDS-PAGE (Bio-Rad, Hercules, CA, USA). The proteins were transferred onto a polyvinylidene fluoride (PVDF) membrane using a trans-blot SD semidry transfer cell (Bio-Rad). Next, 5% (*w*/*v*) skimmed milk in Tris-buffered saline containing 0.1% (*v*/*v*) Tween 20 (TBST) was used to block the PVDF membrane for 1 h. Thereafter, the membrane was incubated with rat anti-PA tag monoclonal antibody (NZ-1, 0.1 μg/mL) (Wako Pure Chemical Industries, Ltd.), followed by incubation with horseradish peroxidase (HRP)-conjugated goat anti-rat IgG (H+L) (1:10,000) (Bios Antibodies Inc., Woburn, MA, USA) and development for 1 min with Immobilon western chemiluminescence HRP substrate (Merck Millipore, Burlington, MA, USA). Stained proteins were detected using a molecular imager VersaDoc MP imaging systems (Bio-rad).

2.7. Binding Assay of bx-PA-NcPROF-GP64TM with Mouse TLR11

The binding assay for purified bx-PA-NcPROF-GP64TM against recombinant mTLR11 Fc chimera (R&D Systems, Minneapolis, MN, USA) was carried out by enzyme-linked immunosorbent assay (ELISA). The binding assay was performed in 96-well plates coated with 50 μL/well of mTLR11 (100 ng/well) at 4 °C overnight. The plates were blocked with 100 μL/well of 2% skimmed milk in PBS for 2 h at RT. After 2 h, the plates were washed 3 times with PBS with 0.1% Tween 20 (PBST). The purified bx-PA-NcPROF-GP64TM with different concentrations (100, 300 and 500 ng/well) with the 1 mM dithiothreitol (DTT) treatment and 100 ng without 1 mM DTT treatment was diluted in blocking buffer in triplicate, followed by incubation for 2 h at RT. The plate was washed with PBST and incubated for 2 h at RT with 50 μL/well of rat anti-PA tag monoclonal antibody (NZ-1, 1:1000 dilution). Subsequently, the plates were washed again with PBST and were incubated for 2 h at RT with 50 μL/well of HRP-conjugated goat anti-rat IgG (H+L) secondary antibody (1:5000 dilution). Next, the plates were washed with PBST and developed by incubating with 50 μL/well of 3,3′,5,5′-tetramethylbenzidine (TMBZ) (Dojindo Co. Ltd., Kanagawa, Japan) for 20 min, followed by stopping the reaction with

100 μL/well of stop solution (10% H_2SO_4). Thereafter, 1 mM DTT and BSA (100 ng/well) were used as negative controls. Finally, the absorbance of the mixture was measured using a microplate reader (Model 680, Bio-Rad, Hercules, CA, USA) at 450 nm.

3. Results and Discussion

3.1. Expression of bx-PA-NcPROF-GP64TM, bx-PA-NcPROF, and PA-NcPROF in Silkworms

In this study, three NcPROF constructs were generated. PA-NcPROF has a PA tag sequence at its N-terminus (Figure 1A), whereas bx-PA-NcPROF has the signal sequence of bombyxin from *B. mori* at its N-terminus (Figure 1B). Meanwhile, bx-PA-NcPROF-GP64TM has the transmembrane and cytoplasmic domains of GP64 from BmNPV at the C-terminus of bx-PA-NcPROF (Figure 1C).

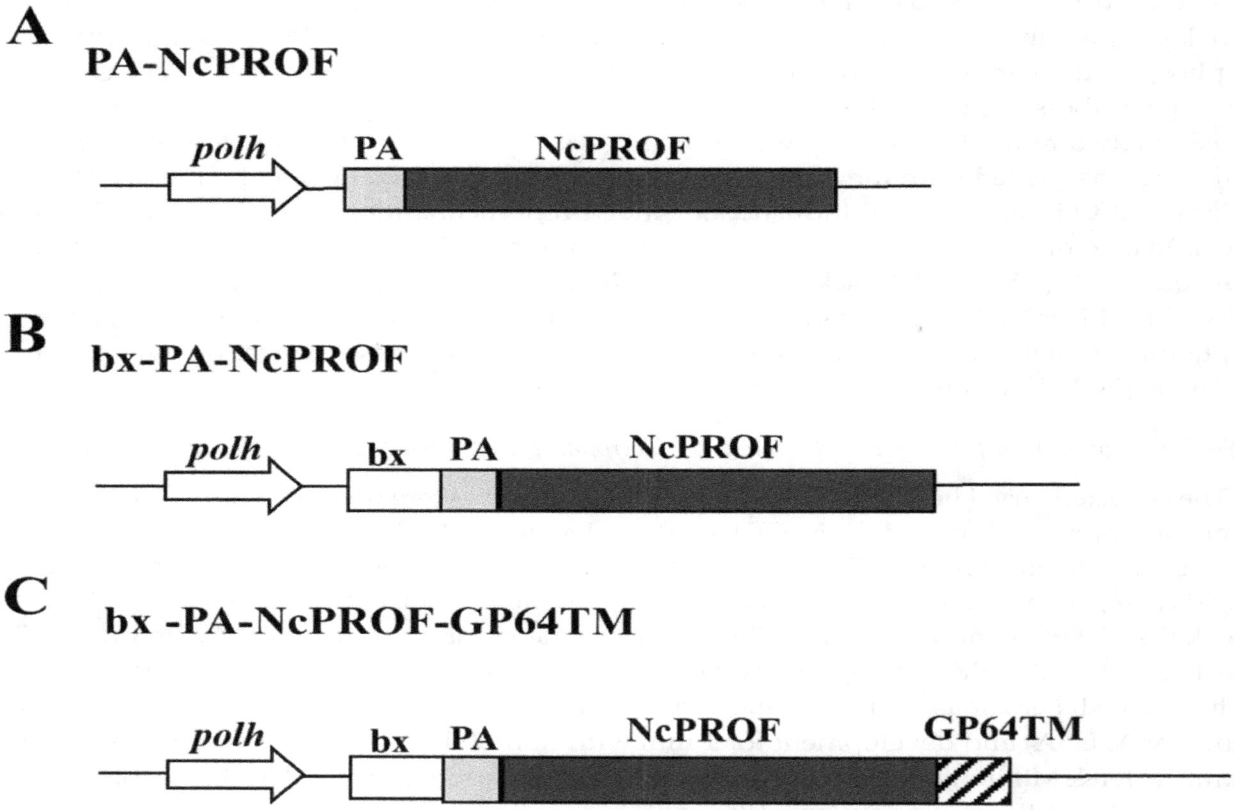

Figure 1. Constructs of native NcPROF fused with an N-terminal PA tag (PA-NcPROF) (**A**), PA-NcPROF fused with the signal sequence of bombyxin from *B. mori* (bx-PA-NcPROF) (**B**) and bx-PA-NcPROF with additional C-terminal transmembrane and cytoplasmic domains of GP64 from *Bombyx mori* nucleopolyhedrovirus (BmNPV) (bx-PA-NcPROF-GP64TM) (**C**). *polh*, polyhedrin promoter; bx, bombyxin signal sequence; GP64TM, transmembrane and cytoplasmic domains of GP64 from BmNPV; PA, PA-tag sequence.

The bombyxin signal peptide (bx) is responsible for the secretion of recombinant proteins to the hemolymph of silkworm larvae [17,18]. The GP64TM were used as a fusion partner with NcPROF as a target protein to be incorporated into the cell membrane [19]. All NcPROF constructs were observed in both the fat body and hemolymph when they were expressed in silkworm larvae (Figure 2A). PA-NcPROF, bx-PA-NcPROF, and bx-PA-NcPROF-GP64TM comprise 176, 195, and 226 amino acids, respectively, and their estimated molecular weights are 19, 21, and 25 kDa, respectively. All NcPROFs were expressed and verified intra- and extracellularly in Bm5 cells (Figure 2B). The molecular weight of PA-NcPROF was approximately 25 kDa by western blotting.

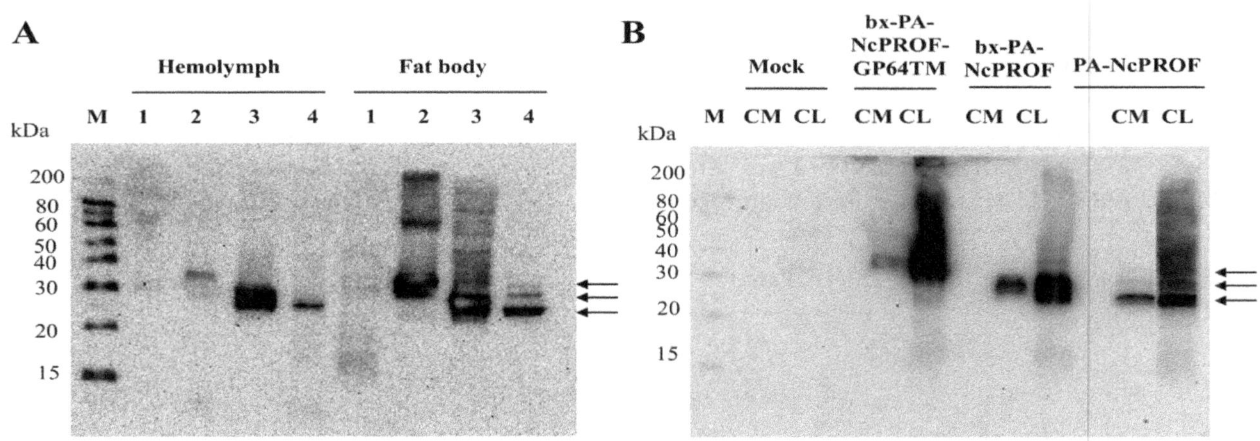

Figure 2. Western blot analysis of bx-NcPROF-GP64TM, bx-NcPROF and NcPROF expressed in silkworm larvae (**A**) and Bm5 cells (**B**). Recombinant BmNPV/bx-PA-NcPROF-GP64TM, BmNPV/bx-PA-NcPROF or BmNPV/PA-NcPROF were injected into the silkworms, followed by harvesting at 5 d post injection. Bm5 cells were harvested at 3 d post infection of the indicated recombinant BmNPV. Lanes M, 1, 2, 3 and 4 in (**A**) denote the molecular marker, mock, bx-PA-NcPROF-GP64TM, bx-PA-NcPROF, and PA-NcPROF, respectively. Lanes M, CM and CL in (**B**) denote the molecular marker, protein samples from cell medium and cell lysate, respectively. Upper, middle and lower arrows in (**A**,**B**) indicate bx-NcPROF-GP64TM, bx-NcPROF and NcPROF, respectively.

The molecular weight of native NcPROF in this study was similar to that of full-length profilin expressed in *E. coli* (22 kDa) [12]. This finding suggests that PA-NcPROF expressed in silkworm larvae may be not modified posttranslationally. bx-PA-NcPROF was observed in the fat body and hemolymph at 25–30 kDa as a large obscure band that appeared as two bands. The molecular weight of the lower band was the same as that of PA-NcPROF and that of the upper band was larger than that of PA-NcPROF. This finding suggests that bx-PA-NcPROF may be partially modified posttranslationally after the cleavage of the bombyxin signal peptide. In fact, N128 was predicted as an *N*-glycosylation site estimated by NetNGlyc 1.0 Server (http://www.cbs.dtu.dk/services/NetNGlyc/). The bx-PA-NcPROF-GP64TM product was observed at 30–40 kDa by western blotting. Interestingly, bx-PA-NcPROF-GP64TM was observed in the hemolymph and culture medium although it was fused with the transmembrane domain of GP64. In our previous report, *N. caninum*-derived antigens NcSAG1 and NcSRS2 fused with the transmembrane and cytoplasmic domains of GP64 were not observed in hemolymph [9]. In this study, bx-PA-NcPROF-GP64TM was secreted into hemolymph, in part, and the culture medium from Bm5 cells. It has been reported that profilins of *T. gondii* and *Babesia canis* were partially secreted even though no signal peptide is predicted in these genes [20,21]. In this study, NcPROF was also secreted regardless of the addition of the signal sequence. To investigate *N*-glycosylation of the *N*-glycan of three NcPROFs, PNGase F treatment was carried out (Figure S1). Molecular weight of each NcPROF was not changed between before and after the PNGase F treatment, indicating no *N*-glycan was attached into the three NcPROFs. These results suggest that the two bands of bx-PA-NcPROF did not come from its *N*-glycosylation. Hence, it showed that the bx-PA-NcPROF and bx-PA-NcPROF-GP64TM were not posttranslationally modified with an *N*-glycan in endoplasmic reticulum even though these proteins have the bx signal peptide. In addition, the discrepancy of the molecular weight of each PROF to its estimated that was not caused by the *N*-glycosylation. In nature, estimated pI of this NcPROF is around 4.89, which may have influence on the mobility of NcPROFs on the SDS-PAGE gel [22]. In addition, the negative charge of acidic residues may create repulsion and this repulsion caused the anomalous migration of protein in SDS-polyacrylamide gels [23]. This may be the reason why the molecular weight of bx-PA-NcPROF-GP64TM was increased.

The expressed NcPROFs in the Bm5 cell culture broth were further investigated by ultracentrifugation (Figure 3). It showed that PA-NcPROF and bx-PA-NcPROF were not precipitated by ultracentrifugation

(100,000× g), while bx-PA-NcPROF-GP64TM was pelleted. Normally, transmembrane proteins are anchored in the membrane fraction of cells and are not easily secreted. However, glycoproteins from some viruses are secreted and are partially incorporated into extracellular vesicles when they are overexpressed [24,25]. Moreover, the transmembrane and cytoplasmic domains of GP64 from BmNPV facilitated the display of recombinant proteins on BmNPV particles. These findings suggest that bx-PA-NcPROF-GP64TM was secreted as nanoparticles or by display on the envelope of BmNPV particles.

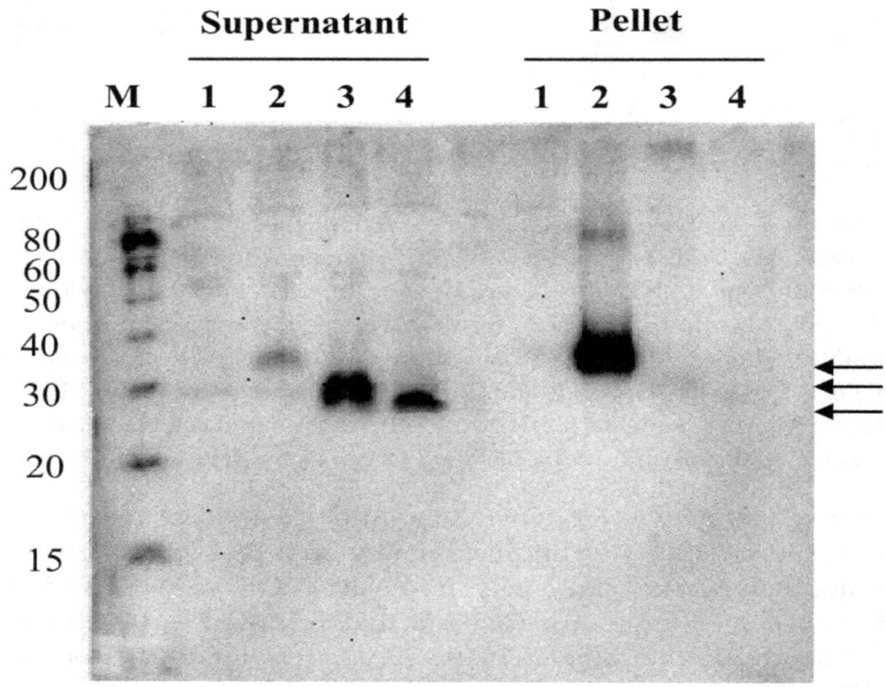

Figure 3. Western blots of soluble and insoluble fractions from the Bm5 cell culture supernatant. The culture supernatant containing bx-PA-NcPROF-GP64TM, bx-PA-NcPROF or PA-NcPROF was centrifuged at 100,000× g, and the pellet was resuspended in 1 mL of PBS. Lanes M, 1, 2, 3 and 4 denote the molecular marker, mock, bx-PA-NcPROF-GP64TM, bx-PA-NcPROF and PA-NcPROF, respectively. Upper, middle and lower arrows indicate bx-NcPROF-GP64TM, bx-NcPROF and NcPROF, respectively.

bx-PA-NcPROF-GP64TM was then purified from silkworm hemolymph using anti-PA tag agarose. As shown in Figure 4A, bx-PA-NcPROF-GP64TM was successfully purified as a single band in SDS-PAGE and was further verified by western blotting (Figure 4B). This finding indicates that bx-PA-NcPROF-GP64TM does not display on the surface of BmNPV particles because no band was observed in the purified sample except for bx-PA-NcPROF-GP64TM. These results suggested that bx-PA-NcPROF-GP64TM forms only some nanoparticles and can be secreted into hemolymph and the culture broth. The yield of purified NcPROF-GP64TM was approximately 14 μg from 3 mL of larval hemolymph.

Interestingly, PA-NcPROF and bx-PA-NcPROF could not be purified from hemolymph using the same protocols as bx-PA-NcPROF-GP64TM. From the results, it was demonstrated that, without modification of the transmembrane region at the C-terminus for both constructs, bx-PA-NcPROF was only slightly purified and PA-NcPROF was almost impossible to be purified from hemolymph (Figure S2) although a portion of bx-PA-NcPROF was secreted into hemolymph compared with bx-PA-NcPROF-GP64TM (Figure 2A). However, when 0.5% (w/v) nonyl phenoxypolyethoxylethanol (NP-40) was added, both bx-PA-NcPROF and PA-NcPROF could be purified from hemolymph and showed positive bands at 30 kDa and 25 kDa, respectively (Figure S2), suggesting that bx-PA-NcPROF and PA-NcPROF were aggregated abnormally and the aggregates were dissolved with NP-40.

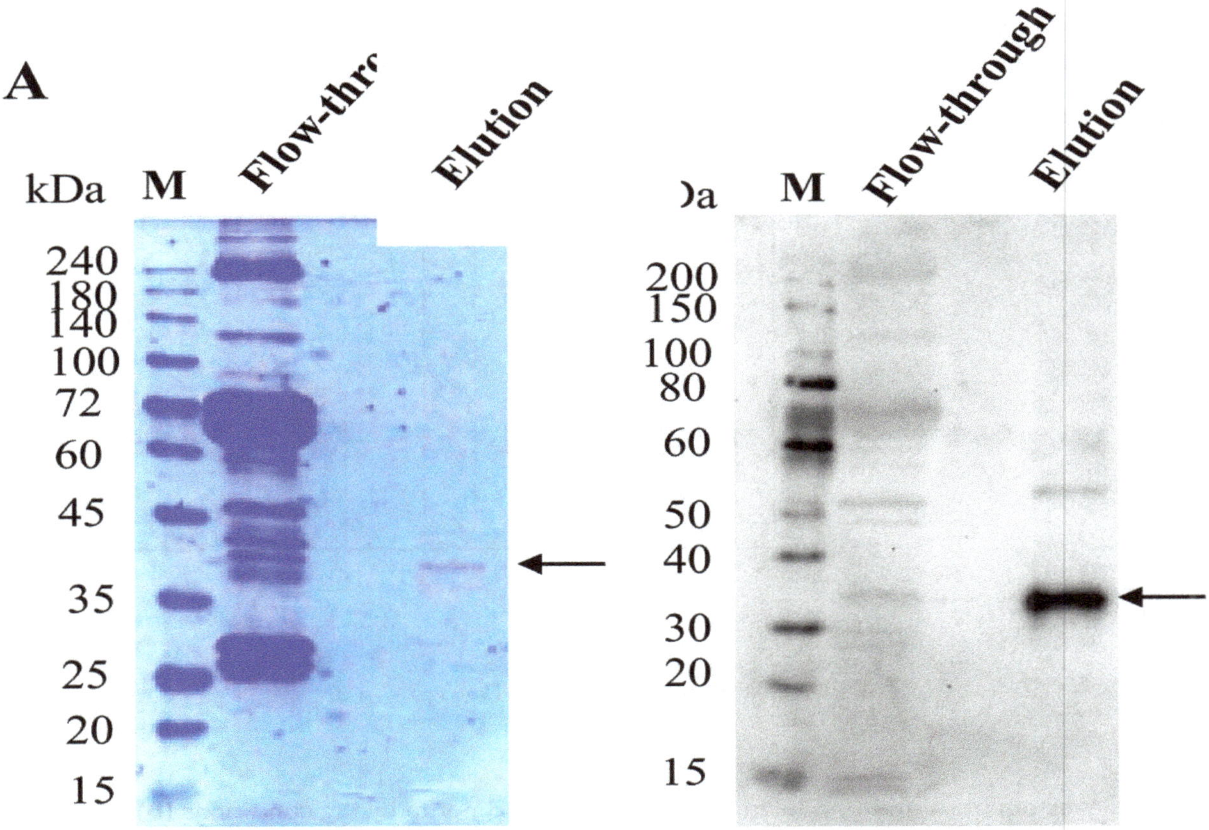

Figure 4. (**A**) Sodium dodecyl sulfate-polyacrylamide gel electrophoresis (SDS-PAGE) and western blot under reducing conditions. (**B**) SDS-PAGE of purified bx-PA-NcPROF-GP64TM from hemolymph using anti-PA tag affinity chromatography. M indicates the molecular weight marker; arrows indicate the recombinant bx-PA-NcPROF-GP64TM band.

3.2. *Binding Assay of bx-NcPROF-GP64TM with mTLR11*

Recently, it was reported that TLR is responsible for parasite recognition and the induction of cytokines in *T. gondii* [26]. NcPROF induces limited protection and the T-cell response in mice [13]. In this study, NcPROF expressed in silkworms was investigated to determine whether it binds to recombinant mTLR11. The binding of purified bx-NcPROF-GP64TM to mTLR11 was not clearly observed compared with that of the negative control. Because bx-PA-NcPROF-GP64TM formed multimers under nonreducing conditions, DTT was then added to purified bx-PA-NcPROF-GP64TM before ELISA (Figure 5A,B). As shown in Figure 5C, the specific binding was enhanced by an increment of almost 72.7% after DTT treatment and simultaneously increased the binding ability when different concentrations (100, 300, 500 ng/well) were challenged. Hence, it suggests that bx-PA-NcPROF-GP64TM expressed in silkworm larvae is functional. These results indicate that multimer formation might prevent bx-PA-NcPROF-GP64TM from binding to mTLR11. Similar to the results of a previous study, Hedhli et al. [27], showed that PROF of *T. gondii* secreted into *Drosophila* S2 cell broth with the signal sequence of insect binding immunoglobulin protein (BiP) enhanced the cellular and humoral responses in mice after DTT treatment. These results indicate that bx-PA-NcPROF-GP64TM was secreted through the secretory pathway in the host and irregular disulfide bonds were formed in the endoplasmic reticulum. In nature, NcPROF does not have the signal peptide at its N-terminus and does not enter the endoplasmic reticulum [20,28].

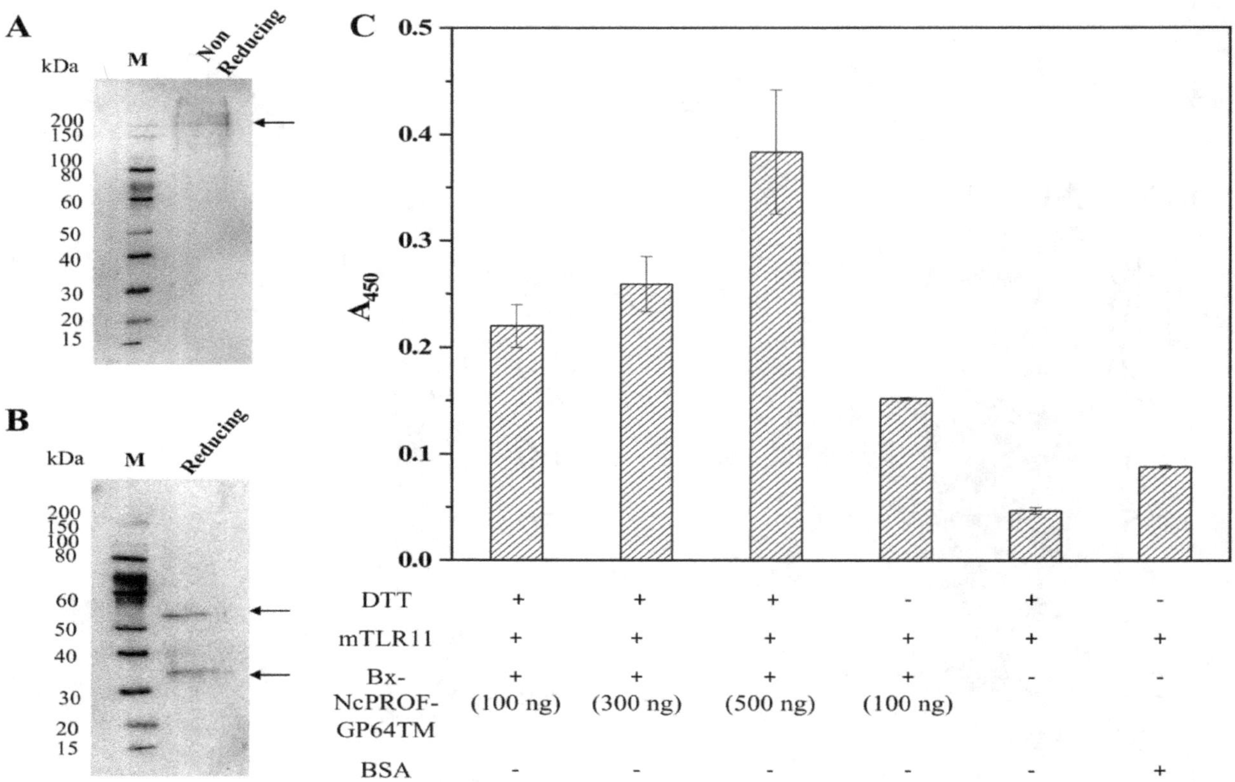

Figure 5. Western blots under reducing conditions (**A**) and nonreducing conditions (**B**) of purified bx-PA-NcPROF-GP64TM. (**C**) enzyme-linked immunosorbent assay (ELISA) analysis of the binding of purified of bx-NcPROF-GP64TM to recombinant mTLR11 Fc chimera. Purified bx-NcPROF-GP64TM (100 ng, 300 ng, 500 ng) with the 1 mM dithiothreitol (DTT) treatment and 100 ng of bx-NcPROF-GP64T without the 1 mM DTT treatment were assayed. Bovine serum albumin (BSA) was used as a negative control. The absorbance was detected at 450 nm and presented as means ±0.001~0.05 standard deviation with coefficient of variance ±0.7%~15%.

3.3. Morphology of bx-PA-NcPROF-GP64TM Nanoparticles

The morphology of purified bx-PA-NcPROF-GP64TM was observed using TEM. Nanoparticles with a diameter of 30 nm were observed in purified bx-PA-NcPROF-GP64TM (Figure 6A). To further analyze the purified bx-PA-NcPROF-GP64TM, immuno-TEM was conducted. The recombinant proteins were probed with rat anti-PA tag (NZ-1) and goat anti-rat IgG (H+L) conjugated with 12-nm gold beads. As shown in Figure 6B, bx-PA-NcPROF-GP64TM demonstrated a spherical morphology with a diameter of approximately 30 nm. These results indicate that bx-PA-NcPROF-GP64TM was secreted as nanoparticles (Figure 3).

Vesicular stomatitis virus G glycoprotein (VSV-G) could be recovered from the culture supernatant when VSV-G alone was expressed in mammalian cells [24,25]. Similarly, the S protein of severe acute respiratory syndrome-associated coronavirus was incorporated into exosomes when its transmembrane and cytoplasmic domains were replaced with those of VSV-G [29]. Extracellular vesicles containing virus-encoded glycoprotein were secreted into the culture supernatant when mammalian cells were infected with the modified *Vaccinia* virus Ankara strain [30]. Additionally, glycoprotein 64 (AcGP64) of *Autographa californica* multiple nucleopolyhedrovirus was secreted into the culture medium when AcGP64 was expressed in mammalian cells [31]. In this study, it is possible that the bx-PA-NcPROF-GP64TM protein formed nanoparticles when it was expressed in silkworm larvae and Bm5 cells. In our previous paper, antigens of *N. caninum* fused with the transmembrane and cytoplasmic domains of BmGP64 were displayed on the surface of BmNPV particles, but few of these fusion proteins were secreted into silkworm hemolymph [9]. These studies suggest that the secretion of recombinant fusion proteins with the transmembrane and cytoplasmic domains of BmGP64 depends on the properties of recombinant proteins of interest. Currently, nanoparticles were only observed in purified bx-PA-NcPROF-GP64TM but not in the other two constructs. Additionally, when purified bx-PA-NcPROF-GP64TM was treated with 1% TritonX-100, some of this protein moved to the fraction at low sucrose concentration during sucrose density gradient centrifugation (Figure S3), implying that this fusion protein forms nanoparticles. We are investigating whether these nanoparticles are aggregates or not and how these nanoparticles form.

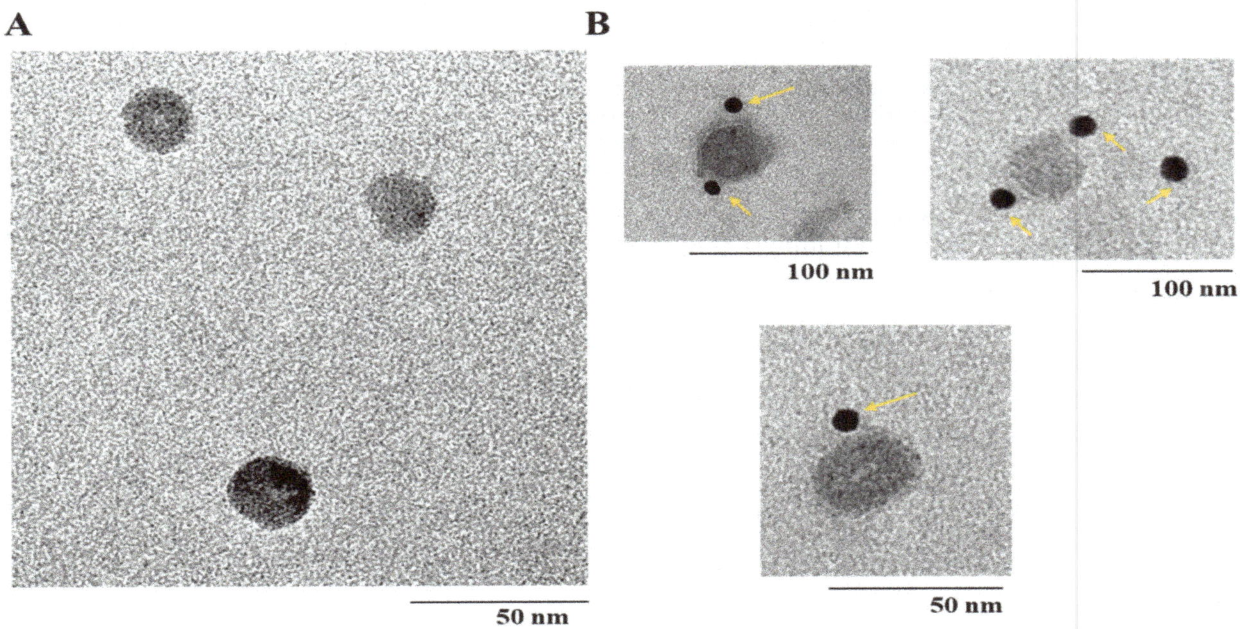

Figure 6. (**A**) Transmission electron microscope (TEM) image of bx-PA-NcPROF-GP64TM. (**B**) Immuno-TEM using gold-labeled antibody. The arrows in (**B**) anti-PA tag-conjugated-gold nanoparticles, respectively.

Apicomplexan profilin, including NcPROF, is a promising protein target with an adjuvant activity as a vaccine candidate [10]. Previous research from Gause et al. [32], stated that delivery of the subunit antigen using particle-based delivery systems can lead to significant improvement in immunogenicity because these systems have now been enhanced by engineering through the physiochemical properties of the particle to promote an immune response. Hence, it was advantageous to purify bx-PA-NcPROF-GP64TM because it naturally forms unique functional nanoparticles without modification through its physiochemical properties to deliver subunit antigens through a particle-based delivery system. Antigen-displaying nanoparticles have been normally prepared by the co-expression of viral structural proteins and antigens or chemical conjugation and genetic fusion of antigens to virus-like particles [33,34]. In this study, NcPROF-displaying nanoparticles were prepared in insect cells and silkworm larvae only by expression of bx-PA-NcPROF-GP64TM, which has the bombyxin signal sequence and transmembrane and cytoplasmic domains of BmGP64 at its N-terminus and C-terminus, respectively.

4. Conclusions

bx-PA-NcPROF-GP64TM was secreted partly into the silkworm larval hemolymph and culture broth of Bm5 cells, although this protein has the transmembrane domain of BmGP64. NcPROF was purified from silkworm hemolymph as a single band only when the bombyxin signal sequence and transmembrane and cytoplasmic domains of BmGP64 were fused with the N- and C-termini, respectively. The purified bx-PA-NcPROF-GP64TM formed unique nanoparticles and was bound to mTLR11.

Supplementary Materials: Figure S1: Analysis of *N*-Glycan attached to crude of hemolymph NcPROF (bx-PA-NcPROF-GP64TM, bx-PA-NcPROF, PA-NcPROF). Western blotting with anti-PA tag antibody deglycosylated with PNGase F (incubation at 37 °C for overnight) in denature condition. Arrows show the molecular weight of NcPROF (bx-PA-NcPROF-GP64TM, bx-PA-NcPROF, PA-NcPROF). Figure S2: Western blots of purified bx-PA-NcPROF (A) and PA-NcPROF (B) from hemolymph with and without treatment with 0.5% of *w/v* Np40 using anti-PA tag affinity chromatography. M indicates molecular weight marker; arrows show the recombinant bx-PA-NcPROF and PA-NcPROF; Figure S3: Western blotting analysis of incorporation level of bx-PA-NcPROF-GP64TM, bx-PA-NcPROF and PA-NcPROF with and without treatment of 1% Triton X-100.

Author Contributions: H.S. performed the whole experiments and wrote the first draft of the MS. R.H. and J.X. provided a helpful discussion and helped H.S. to carry out some biological in vitro experiments. T.K., J.X. and E.Y.P. designed and wrote the critical discussion, and prepared the final version of the MS. E.Y.P. supervised the experiments. All authors have given their approval to the final version of the manuscript.

Acknowledgments: We extend our deep appreciation to Jeawook Lee and Kenshin Takemura for their help and guidance for acquiring the TEM and immuno-TEM images.

Abbreviations

BmNPV	*Bombyx mori* nucleopolyhedrovirus
bx	Bombyxin signal peptide from *Bombyx mori*
PA	Peptide tag (GVAMPGAEDDVV)
bx-PA-NcPROF	PA-NCPROF fused with bx
bx-PA-NcPROF-GP64TM	bx-PA-NcPROF with additional C-terminal GP64TM
GP64TM	C-terminal transmembrane and cytoplasmic domains of GP64 from BmNPV
mTLR11	Recombinant mouse TLR11
NcPROF	Neospora caninum profilin
PA-NcPROF	N-terminal PA tag with *N. caninum* profilin
TLR11	Toll-like receptor 11

References

1. Buxton, D.; McAllister, M.M.; Dubey, J.P. The comparative pathogenesis of neosporosis. *Trends Parasitol.* **2002**, *18*, 546–552. [CrossRef] [PubMed]
2. Dubey, J.P.; Schares, G.; Ortega-Mora, L.M. Epidemiology and control of neosporosis and *Neospora caninum*. *Clin. Microbiol. Rev.* **2007**, *20*, 323–367. [CrossRef] [PubMed]
3. Dubey, J.P.; Schares, G. Neosporosis in animals-the last five years. *Vet. Parasitol.* **2011**, *180*, 90–108. [CrossRef]
4. Innes, E.A.; Andrianarivo, A.G.; Björkman, C.; Williams, D.J.L.; Conrad, P.A. Immune responses to *Neospora caninum* and prospects for vaccination. *Trends Parasitol.* **2002**, *18*, 497–504. [CrossRef]
5. Donahoe, S.L.; Lindsay, S.A.; Krockenberger, M.; Phalen, D.; Šlapeta, J. A review of neosporosis and pathologic findings of *Neospora caninum* infection in wildlife. *Int. J. Parasitol. Parasit. Wildl.* **2015**, *4*, 216–238. [CrossRef] [PubMed]
6. Reichel, M.P.; Ellis, J.T. *Neospora caninum*-How close are we to development of an efficacious vaccine that prevents abortion in cattle? *Int. J. Parasitol.* **2009**, *39*, 1173–1187. [CrossRef] [PubMed]
7. Jin, C.; Yu, L.; Wang, Y.; Hu, S.; Zhang, S. Evaluation of *Neospora caninum* truncated dense granule protein 2 for serodiagnosis by enzyme-linked immunosorbent assay in dogs. *Exp. Parasitol.* **2015**, *157*, 88–91. [CrossRef]
8. Lv, Q.; Xing, S.; Gong, P.; Chang, L.; Bian, Z.; Wang, L.; Zhang, X.; Li, J. A 78 kDa host cell invasion protein of *Neospora caninum* as a potential vaccine candidate. *Exp. Parasitol.* **2015**, *148*, 56–65. [CrossRef] [PubMed]
9. Kato, T.; Otsuki, T.; Yoshimoto, M.; Itagaki, K.; Kohsaka, T.; Matsumoto, Y.; Ike, K.; Park, E.Y. Bombyx mori nucleopolyhedrovirus displaying *Neospora caninum* antigens as a vaccine candidate against *N. caninum* infection in mice. *Mol. Biotechnol.* **2015**, *57*, 145–154. [CrossRef] [PubMed]

10. Mansilla, F.C.; Capozzo, A.V. Apicomplexan profilin in vaccine development applied to bovine neosporosis. *Exp. Parasitol.* **2017**, *183*, 64–68. [CrossRef]
11. Mansilla, F.C.; Quintana, M.E.; Cardoso, N.P.; Capozzo, A.V. Fusion of foreign T-cell epitopes and addition of TLR agonists enhance immunity against *Neospora caninum* profilin in cattle. *Parasit. Immunol.* **2016**, *38*, 663–669. [CrossRef]
12. Jenkins, M.C.; Tuo, W.; Feng, X.; Cao, L.; Murphy, C.; Fetterer, R. *Neospora caninum*: cloning and expression of a gene coding for cytokine-inducing profilin. *Exp. Parasitol.* **2010**, *125*, 357–362. [CrossRef]
13. Mansilla, F.C.; Quintana, M.E.; Langellotti, C.; Wilda, M.; Martinez, A.; Fonzo, A.; Moore, D.P.; Cardosa, N.; Capozzo, A.V. Immunization with *Neospora caninum* profilin induces limited protection and a regulatory T-cell response in mice. *Exp. Parasitol.* **2016**, *160*, 1–10. [CrossRef] [PubMed]
14. Innes, E.A.; Wright, S.; Bartley, P.; Maley, S.; Macaldowie, C.; Esteban-Redondo, I.; Buxton, D. The host-parasite relationship in bovine neosporosis. *Vet. Immunol. Immunopathol.* **2005**, *108*, 29–36. [CrossRef]
15. Motohashi, T.; Shimojima, T.; Fukagawa, T.; Maenaka, K.; Park, E.Y. Efficient large-scale protein production of larvea and pupae of silkworm by *Bombyx mori* nuclear polyhedrosis virus bacmid system. *Biochem. Biophys. Res. Commun.* **2005**, *326*, 564–569. [CrossRef] [PubMed]
16. Kato, T.; Sugioka, S.; Itagaki, K.; Park, E.Y. Gene transduction in mammalian cells using *Bombyx mori* nucleopolyhedrovirus assisted by glycoprotein 64 of *Autographa californica* multiple nucleopolyhedrovirus. *Sci. Rep.* **2016**, *6*, 1–9. [CrossRef] [PubMed]
17. Otsuki, T.; Dong, J.; Kato, T.; Park, E.Y. Expression, purification and antigenicity of *Neospora caninum*-antigens using silkworm larvae targeting for subunit vaccines. *Vet. Parasitol.* **2013**, *192*, 284–287. [CrossRef] [PubMed]
18. Southey, B.R.; Sweedler, J.V.; Rodriguez-Zas, S.L. Prediction of neuropeptide cleavage sites in insects. *Bioinformatics* **2008**, *24*, 815–825. [CrossRef] [PubMed]
19. Oomens, A.G.P.; Wertz, G.W. The Baculovirus GP64 protein mediates highly stable infectivity of a human respiratory syncytial virus lacking its homologous transmembrane glycoproteins. *J. Virol.* **2004**, *78*, 124–135. [CrossRef]
20. Plattner, F.; Yarovinsky, F.; Romero, S.; Didry, D.; Carlier, M.F.; Sher, A.; Soldati-Favre, D. *Toxoplasma* profilin is essential for host cell invasion and TLR11-dependent induction of an interleukin-12 response. *Cell Host Microb.* **2008**, *3*, 77–87. [CrossRef]
21. Eichenberger, R.M.; Ramakrishnan, C.; Russo, G.; Deplazes, P.; Hehl, A.B. Genome-wide analysis of gene expression and protein secretion of *Babesia canis* during virulent infection identifies potential pathogenicity factors. *Sci. Rep.* **2017**, *7*, 3357. [CrossRef]
22. Graur, D. The evolution of electrophoretic mobility of proteins. *J. Theor. Biol.* **1986**, *118*, 443–469. [CrossRef]
23. Shi, Q.; Jackowski, G. One-Dimensional Polyacrylamide Gel Electrophoresis. In *Gel Electrophoresis of Proteins. A Practical Approach*, 3rd ed.; Hames, B.D., Ed.; Oxford University Press: New York, NY, USA, 1998; pp. 31–33.
24. Abe, A.; Miyanohara, A.; Friedmann, T. Enhanced gene transfer with fusogenic liposomes containing vesicular stomitis virus G glycoproattein. *J. Virol.* **1998**, *72*, 6159–6163.
25. Okimoto, T.; Friedmann, T.; Miyanohara, A. VSV-G envelope glycoprotein forms complexes with plasmid DNA and MLV retrovirus-like particles in cell free conditions and enhances DNA transfection. *Mol. Ther.* **2001**, *4*, 232–238. [CrossRef]
26. Yarovinsky, F. Innate immunity to *Toxoplasma gondii* infection. *Nat. Rev. Immunol.* **2014**, *14*, 109–121. [CrossRef]
27. Hedhli, D.; Moiré, N.; Akbar, H.; Laurent, F.; Héraut, B.; Dimier-Poisson, I.; Mévélec, M.N. The antigen-specific response to *Toxoplasma gondii* profilin, a TLR11/12 ligand, depends on its intrinsic adjuvant properties. *Med. Microbiol. Immunol.* **2016**, *205*, 345–352. [CrossRef]
28. Hiasa, J.; Nishimura, M.; Itamoto, K.; Xuan, X.; Inokuma, H.; Nishikawa, Y. Enzyme-linked immunosorbent assays based on *Neospora caninum* dense granule protein 7 and profilin for estimating the stage of neosporosis. *Clin. Vaccine Immunol.* **2012**, *19*, 411–417. [CrossRef]
29. Kuate, S.; Cinatl, J.; Doerr, H.W.; Uberla, K. Exosomal vaccines containing the S protein of the SARS coronavirus induce high levels of neutralizing antibodies. *Virology* **2007**, *362*, 26–37. [CrossRef]
30. Spehner, D.; Drillien, R. Extracellular vesicles containing virus-encoded membrane protein are a byproduct of infection with modified vaccinia virus Ankara. *Virus Res.* **2008**, *137*, 129–136. [CrossRef]
31. Guibinga, G.H.; Song, S.; Loring, J.; Friedmann, T. Characterization of the gene delivery properties of boculoviral-based virosomal vectors. *J. Virol. Methods* **2008**, *148*, 277–282. [CrossRef]

32. Gause, K.T.; Wheatley, A.K.; Cui, J.; Tan, Y.; Kent, S.J.; Caruso, F. Immunological principles guiding the rational design of particles for vaccine delivery. *ACS Nano* **2017**, *11*, 54–68. [CrossRef]
33. Koudelka, K.J.; Pitek, A.S.; Manchester, M.; Steinmetz, N.F. Virus-based nanoparticles as versatile nanomachines. *Annu. Rev. Virol.* **2015**, *2*, 379–401. [CrossRef]
34. Yan, D.; Wei, Y.Q.; Guo, H.C.; Sun, S.Q. The application of virus-like particles as vaccines and biological vehicles. *Appl. Microbiol. Biotechnol.* **2015**, *99*, 10415–10432. [CrossRef]

11

RNAseq Analysis Reveals Virus Diversity within Hawaiian Apiary Insect Communities

Laura E. Brettell [1,2,*], Declan C. Schroeder [3,4] and Stephen J. Martin [2]

1. Hawkesbury Institute for the Environment, Western Sydney University; Locked bag 1797, Penrith 2751, NSW, Australia
2. School of Environment and life Sciences, University of Salford, Manchester, M5 4WT, UK; s.j.martin@salford.ac.uk
3. School of Biological Sciences, University of Reading, Reading RG6 6LA, UK; dcschroe@umn.edu
4. Veterinary Population Medicine, College of Veterinary Medicine, University of Minnesota, St Paul, MN 55108, USA

* Correspondence: l.brettell@westernsydney.edu.au

Abstract: Deformed wing virus (DWV) is the most abundant viral pathogen of honey bees and has been associated with large-scale colony losses. DWV and other bee-associated RNA viruses are generalists capable of infecting diverse hosts. Here, we used RNAseq analysis to test the hypothesis that due to the frequency of interactions, a range of apiary pest species would become infected with DWV and/or other honey bee-associated viruses. We confirmed that DWV-A was the most prevalent virus in the apiary, with genetically similar sequences circulating in the apiary pests, suggesting frequent inter-species transmission. In addition, different proportions of the three DWV master variants as indicated by BLAST analysis and genome coverage plots revealed interesting DWV-species groupings. We also observed that new genomic recombinants were formed by the DWV master variants, which are likely adapted to replicate in different host species. Species groupings also applied when considering other viruses, many of which were widespread in the apiaries. In social wasps, samples were grouped further by site, which potentially also influenced viral load. Thus, the apiary invertebrate community has the potential to act as reservoirs of honey bee-associated viruses, highlighting the importance of considering the wider community in the apiary when considering honey bee health.

Keywords: RNAseq; honey bees; deformed wing virus; quasispecies; apiary pests; recombination

---

1. Introduction

Across the globe, emerging infectious diseases (EIDs) pose a significant threat to biodiversity and health [1]. This has been clearly demonstrated in recent years both by the recent catastrophic decline of amphibians caused by pathogenic fungi [2] and by the cases of large-scale honey bee colony losses, a major factor in which is the spread of pathogenic viruses [3].

EIDs often occur as a consequence of human-mediated translocations of infected hosts and/or parasites and due to the close proximity of wild and domesticated hosts [1]. As such the honey bee, which over the last century has been spread across the globe by humans for honey production and pollination services [4] and shares complex communities with a wide range of insect taxa [5–8], is a prime candidate to facilitate the spread of EIDs into new insect hosts. Pollinators and other insects with which they share environments are of particular interest due to their value in terms of economy (including pollination services) and biodiversity and are currently already experiencing a number of pressures, e.g., from habitat loss and pesticides [9–11]. The combination of multiple pressures can have the additional effect of decreasing immunity and thus potentially increasing susceptibility to pathogens [12], although further studies are needed in this area [13].

Most host switching results in a dead end or a limited low-level outbreak; however, on rare occasions the transmission can result in sustained outbreaks or major epidemics. This can happen when there is sufficient increased exposure or the evolution of new variants in the original or new host allows successful replication and efficient spread between members of the new host species [14]. RNA viruses with their high mutation rates, diverse sequences, and often very high population sizes [15] are thus prime candidates for emergence or re-emergence in novel hosts. This is of particular concern when considering honey bee populations, which host abundant and diverse RNA viruses, are frequently transported en masse, and come into frequent contact with other arthropods both in the hive and when foraging.

Deformed wing virus (DWV) is now well known to be one of the major factors responsible for honey bee colony losses across the world [3,4,16]. This single virus has come to dominate the virome of honey bee populations due to its spread and amplification in the host being aided by the ectoparasitic mite *Varroa destructor* (referred to as varroa from now on) [17,18]. Additionally, the new mite–bee transmission route serves to reduce viral genotypic diversity and select for the amplification of virulent strains [16,17]. Although initially described as a honey bee virus, it has since become apparent that DWV is a generalist insect virus capable of infecting 64 species from eight orders of arthropods [19]. However, the extent of the generality and capacity for emergence in new hosts is still the subject of contention [20]. Worryingly, recent studies have found that viral pathogens circulating in managed pollinators may be driving infections in wild species [21,22], and in the Hawaiian system, it has recently been shown using RT-PCR-based studies that the presence of varroa in honey bee populations has resulted in a dramatic increase in the prevalence of DWV in species of wasps (*Polistes* sp.) and solitary bees (*Ceratina smaragdula*) [23]. Furthermore, in the yellowjacket wasp (*Vespula pensylvanica*) on the Big Island, Hawaii, DWV prevalence has increased along with a decrease in strain diversity [24], mirroring the situation seen in honey bees [17].

Although gene targeted RT-PCR and RT-qPCR have provided very useful insights into the virome of insect communities, RNAseq allows entire genomes of RNA viruses to be sequenced easily and thus provides much deeper anaylsis. More specifically, using oligo dT-derived RNAseq on field collected samples in Hawaiian honey bee apiaries in which varroa is established, we aimed to identify the extent to which +ssRNA viruses (the overwhelming majority of viruses found associated with honey bees are +ss polyadenylated RNA viruses [10]) were present in the wider arthropod community and to characterize any DWV master variant genomes present in the different species. Common pest species were targeted as they would have the most frequent contact with honey bees that are known to harbor high DWV loads in the study locations [17,25]. We hypothesized that the frequency of interactions would result in common DWV genotypes circling in the apiary environment due to repeated viral transmission events between species. Furthermore, we aimed to determine whether other viruses commonly found in honey bees are also found in apiary pests, and if so, whether certain viruses are associated with particular hosts. The pest species investigated were varroa mites (*Varroa destuctor*), small hive beetles (*Aethina tumida*) yellowjacket wasps (*Vespula pensylvanica*), and two species of ant, big headed ants (*Pheidole megacephala*) and ghost ants (*Tapinoma melanocephalum*)—all widely distributed invasive pests known to interact directly with honey bees in Hawaii.

2. Materials and Methods

2.1. Site and Species Selection

In November and December 2012, opportunistic sampling was carried out in managed apiaries on the islands of Oahu and the Big Island, Hawaii (Figure 1), and where found, common apiary pests (yellowjacket wasps, small hive beetles, big headed ants, and ghost ants) were collected from within brood boxes or at hive entrances. Additionally, reference samples of asymptomatic honey bees

(50–100 individuals per hive) were also collected. Varroa populations had been established for at least 3–5 years at all locations sampled, and honey bee populations were known to harbor high DWV levels [17,25]. All samples were kept on ice for transportation to the laboratory, where they were stored at –80 °C. Additional samples of small hive beetles had been collected in the same way earlier in the year from colonies in an advanced stage of collapse, i.e., "slime out phase".

2.2. RNA Extraction and Next Generation Sequencing

Pools of 30 asymptomatic honey bees were taken and checked for the presence of varroa mites, which, where found, were removed and stored separately. The pools of whole bees were then ground in liquid nitrogen using a sterile pestle and mortar to produce a fine homogenous powder, 30 mg of which was used for RNA extraction. Varroa mites that had been removed from bees were homogenized in pools of 10 using a mini pestle in a 1.5 mL eppendorf tube. Small hive beetles were homogenized either individually or in pools of six. Yellowjacket wasps were homogenized individually, and all ants were crushed in pools of 20–40 due to their small size. RNA extractions proceeded for all samples in the same way using the RNeasy mini kit (Qiagen, Manchetser, UK) following the manufacturers' instructions, eluting in 30 μL nuclease free water followed by DNase treatment using DNase I (Promega, Southampton, UK). Initial screening for the presence of DWV was then carried out using RT-PCR as per [17,26]. A selection of positive samples was then chosen to be analyzed in greater depth using total RNA sequencing (RNAseq). The list of samples used for RNAseq analysis along with the species name, site from which they were sampled, and symbol used to denote them in subsequent figures is given in Table S1. These samples were transported to The BBSRC Earlham Institute (Norwich), where cDNA libraries were prepared using oligo (dT) priming. Resulting libraries were then run on the HiSeq 2000 (The Earlham Institute, Norwich). Raw data were deposited in the National Centre for Biotechnology Information (NCBI) Sequence Read Archive (SRA) under BioProject accession number: PRJNA531527. Sample V_des_2 was previously deposited in the European Nucleotide Archive (ENA) as sample "V_S48" under the Study Accession PRJEB8112.

2.3. Bioinformatic Analysis

Initially, quality control of generated reads was performed using FASTQC (http://www.bioinformatics.babraham.ac.uk/projects/fastqc/). Then, a pipeline initially described by [25] was applied to identify DWV-like reads in individual samples. Briefly, this involved taking all reads that passed QC and performing a nucleotide BLAST against a custom database (E value of 10^{-5}) containing DWV types A, B. and C (type A: NC_004830.2 and NC_005876.1 (Kakugo virus), type B: AY251269.2 (VDV-1), and type C: CEND01000001.1). Custom perl, sed, and awk scripts were used to parse the data, take the top BLAST hit, remove empty lines, and remove reads which did not map to the database. BLAST top hit analysis was then used to quantify the numbers of reads belonging to each of the three master variants by identifying which of the reference genomes each read matched to most closely. This was used to produce pie charts showing the proportions of DWV type A, B, and C reads in each sample.

To assemble contigs representing the full diversity of DWV sequences present in each sample, DWV-like reads were identified by BLAST (read1 files for each sample in fasta format), and the corresponding read 2 files were selected. These were then used to generate de novo assemblies using VICUNA, an assembler specifically designed to accommodate the highly variable sequence data typical of RNA viruses [27].

To investigate DWV diversity within the apiary insect community, all VICUNA-generated contigs >300 bp for each sample were imported into Geneious (Version 7.04, Biomatters), and the map to reference tool was then used with a MUSCLE alignment to competitively map contigs to the DWV-A (NC_004830.2), -B (AY251269.2), and -C (CEND01000001.1) reference genomes (mapping ambiguous reads to all). The resulting alignments containing all the DWV contigs were then trimmed to use a 507 bp region of the *RdRp* gene (nucleotide positions 8016–8522 on NC_004830.2, 7989–8495 on AY251269.2, and 7999–8505 on CEND01000001.1 [4]). Alignments were visually inspected, and contigs showing recombination breakpoints were removed so as to comply with the assumptions of the phylogenetic construction, and a Baysian phylogenetic analysis was constructed using the MrBayes v3.2.6 plugin in Geneious. [28], using a GTR substitution model with gamma rate variation. We ran four chains for 1.1×10^6 MCMC generations, sampling trees every 200 generations, and a consensus tree was then created using a 10% burn in and 50% support threshold.

Furthermore, to understand the diversity of DWV-A in Hawaiian apiaries in a global context, a second phylogeny was generated using a 410 bp region of the *RdRp* gene, as used by [4] in their global DWV-A phylogeny along with 10 of their DWV-A sequences of geographically diverse origins from both honey bees and varroa mites (Table S2). DWV-A contigs generated in this study were aligned with the 410 bp sequences from [4] in Geneious (MUSCLE alignment), and this time, no recombinant contigs were observed in the alignment. A Bayesian phylogeny was then created using the same parameters as described above. The corresponding region of the DWV-C reference genome (CEND01000001.1) was used as the outgroup.

To determine whether recombinants were dominating in the samples, competitive alignment plots were created to look at the DWV reference genomes to which reads preferentially mapped across the length of the genome. This used all reads passing QC (read 1, in fasta format) and the "Map to Reference" tool in Geneious, using DWV types A, B and C (NC_004830.2, AY251269.2 and CEND01000001.1) reference genomes, discarding all ambiguous reads. Recombinants were observed to be present if preferential coverage switched from one master variant to another along the length of the genome. To provide additional evidence for the presence of recombinants, two samples were chosen for which competitive alignment plots had revealed recombinants to be present (V_pen_8 and A_tum_5), and their de novo assembled DWV contigs were aligned using MUSCLE and visually inspected to identify recombinant contigs.

Finally, to investigate whether other honey bee-associated viruses were circulating in the apiary community, a custom BLAST database was created using DWV types A, B, and C along with seven common honey bee-associated viruses and an additional two viruses (Moku and Milolii viruses [29,30]) previously identified from RNAseq data generated from Hawaiian apiary insects (data from individual insects were also used in this study) (Table S3), and BLASTn was used to identify reads belonging to each virus (E value 10^{-5}). Read counts were expressed as reads per kilobase million (RPKM).

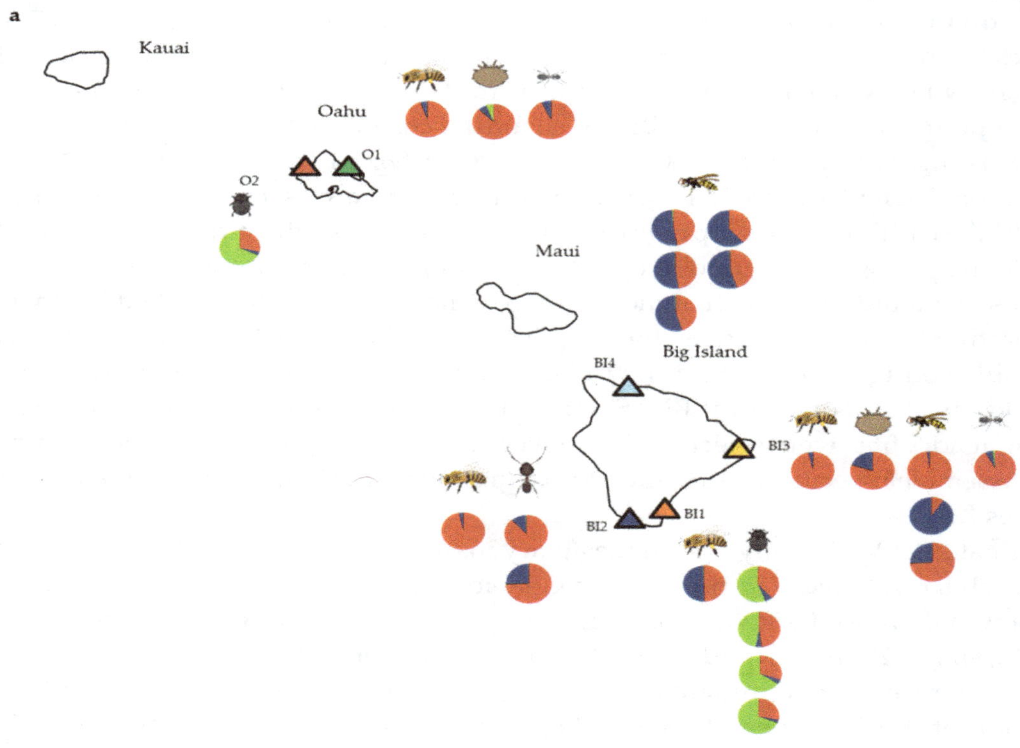

Figure 1. *Cont.*

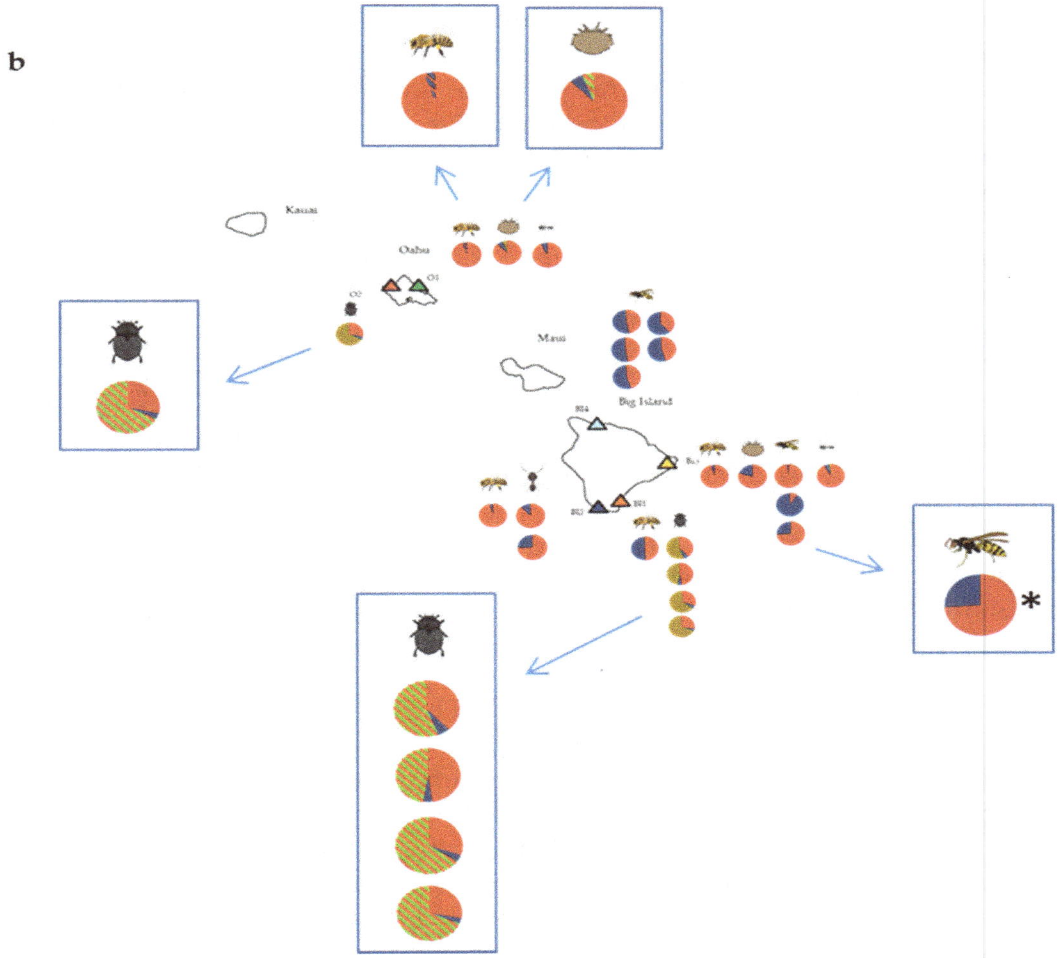

Figure 1. Proportions of deformed wing virus (DWV)-like reads in individual samples from each species mapping to DWV type A (red), type B (blue), and type C (green) along with the site from which they were collected. (**a**) Results using BLAST top hit analysis only (read numbers provided in Table S4); (**b**) where results were amended after considering additional Geneious mapping analysis (Figure S1). Green and red stripes represent that DWV-A/DWV-C recombinants were present (rather that true full-length DWV-C,), red and blue stripes denote a DWV-A/DWV-B recombinant (no full-length DWV-B), and the asterisk represents an additional sample identified as containing recombinant(s). Results for all other samples remain unchanged. Insect images are from BioRender [23,31].

2.4. Statistical Analysis

Redundancy analysis (RDA) was carried out in R (version 3.3.2) to investigate whether the response variables of read count data (RPKM) for each virus were associated with explanatory variables of insect species and site. Dummy variables were randomly assigned as varroa for the species category and B4 for site.

3. Results

3.1. Deformed Wing Virus

RNAseq data was generated from samples of honey bees (A_mel_1-A_mel_4), small hive beetles (A_tum_1-A_tum_5), yellowjacket wasps (V_pen_1-V_pen_8), big headed ants (P_meg_1 and P_meg_2) and ghost ants (T_mel_1 and T_mel_2). As expected from our previous work [17,25], BLAST top hit analysis revealed DWV to dominate sequence reads in all honey bee and varroa mite samples, reaching as high as 91.25% of total reads in the varroa sample V_des_2 from apiary O1. DWV was ubiquitous, being detected in all ant, wasp and beetle samples, although read counts were highly variable (Table S4). The proportions of DWV-types A, B and C present also differed considerably and appeared to group by species (Figure 1). Of the DWV reads, DWV-A reads dominated the majority of honey bee, varroa and ant samples; however the small hive beetle samples were all dominated by type C and the wasps usually contained relatively similar proportions of types A and B. Interestingly, the wasp samples showed further separation by location.

DWV genome coverage plots were generated for each sample, and all alignments showed the typical 3′ bias resulting from the oligo (dT) priming method of cDNA synthesis in library preparation [32]. Similarly, the samples containing type C all showed a spike in the Helicase region caused by the presence of a poly-(A) region of the genome, irrespective of location (Figure 2; Figure S1). Plots showed that the majority of samples were dominated by either full-length DWV-A or DWV-B sequences (or both), as demonstrated by the consistent coverage depth across the genome, and were consistent with the results of the BLAST top hit analysis (Figure 1a). Again, the yellowjacket wasp data appeared to separate by site, with samples V_pen_1–V_pen_5 from site B4 all showing identical coverage plots, with coverage being restricted to the 3′ end indicative of low virus levels in the samples and showing almost identical coverage for DWV-A and DWV-B. The samples V_pens_6 to V_pen_8 from site B3, however, were more variable and contained greater coverage depths, again, in keeping with the BLAST top hit analysis.

Although the majority of samples were dominated by full-length genomes, recombinants were detected in this study as evidenced by the competitive alignment plots (Figure 2) and associated alignments of assembled contigs (Figure S1). The dominant master variants present were the same using both BLASTn and mapping to DWV reference genomes (Geneious). However, the mapping (competitive alignment plots) revealed additional information regarding the presence of recombinants. These additional data were therefore used to create amended pie charts showing variant proportions (Figure 1b) for samples where recombinants were detected. Interestingly all beetles were dominated by a DWV-A/DWV-C recombinant (Figure 1b, Figure 2b, Figures S1 and S2) with a breakpoint in the 5′ UTR immediately upstream of the open reading frame, albeit at low levels, irrespective of location. These beetles also contained lower level full-length DWV-A, as evidenced by type A coverage across the whole genome. The same recombinant was also present in the honey bee sample A_mel_4 (Figure S1), which was collected from a different location from any of the beetles, although the recombinant did not dominate in the honey bee sample, which was instead dominated by full-length DWV-A. Finally, one yellowjacket wasp sample, V_pen_8, showed a distinct coverage pattern not seen in any other sample in this study (Figure 2c). This sample showed three recombination breakpoints—one in the 5′ UTR and one at either end of the helicase gene—and showed coverage of both DWV-A and DWV-B across the full length of the genome. As such, these data cannot confirm the makeup or proportions of recombinant(s) and full-length genomes in this sample.

A Bayesian phylogeny created using a 507 bp fragment of the *RdRp* gene assembled from all samples showed sequences to cluster according to master variant (DWV-A, B, and C), as expected (Figure 3). The origin of all the DWV-C sequences came from a single DWV-C variant, as all the sequences obtained from this region were almost identical to one another and to the reference genome. Within the DWV-A and DWV-B clades, samples did not cluster by either species or location and showed often very similar sequences, indicating that common variants are circling in the apiary.

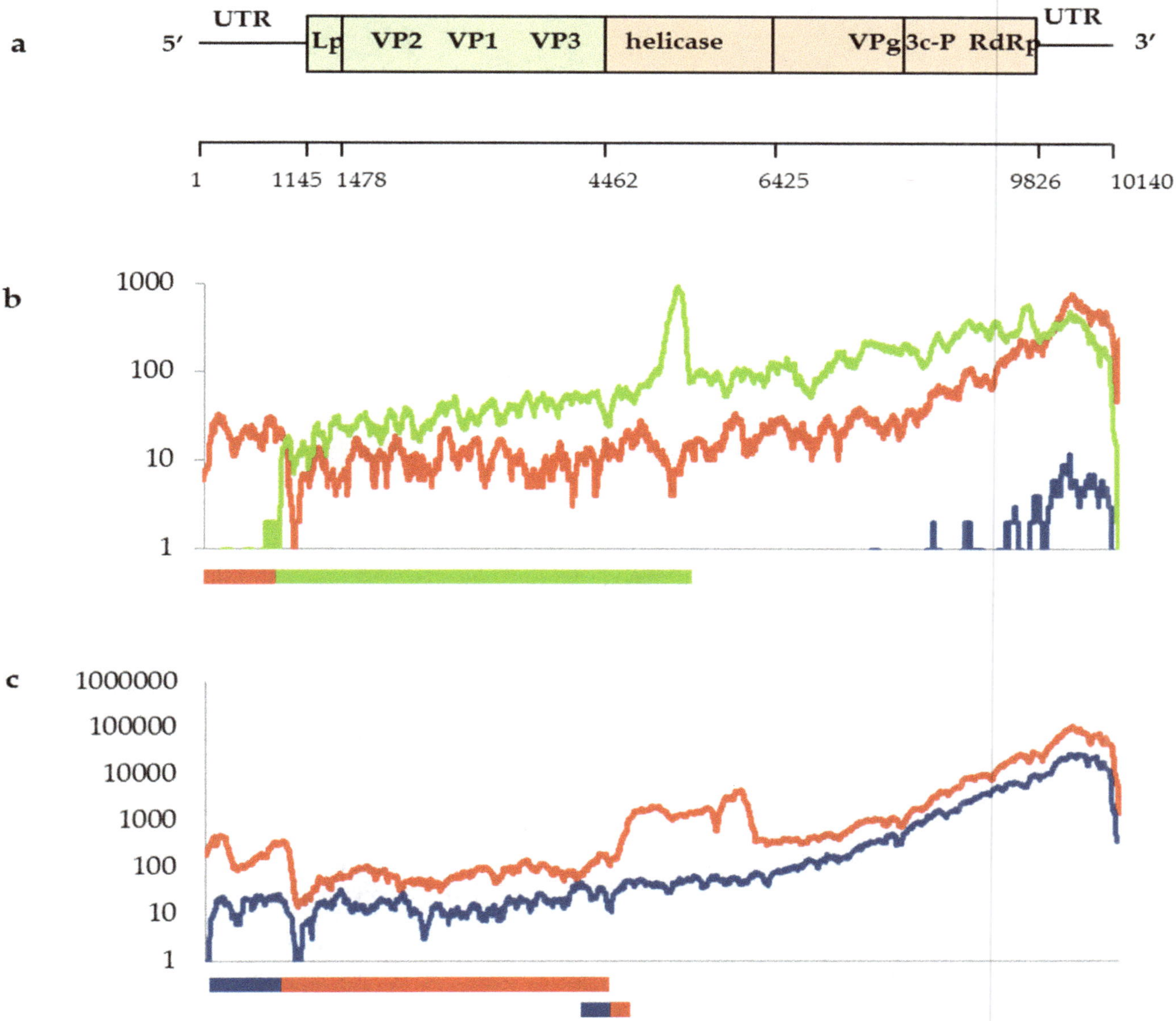

Figure 2. DWV genome coverage plots showing read depth of DWV-A (red), DWV-B (blue), and DWV-C (green), across the length of the genome, along with representations of VICUNA-assembled contigs showing recombination breakpoints; (**a**) genome organization with the structural region in green and the non-structural region in orange, showing nucleotide positions below (adapted from [33,34]); (**b**) small hive beetle sample A_tum_5; (**c**) yellowjacket wasp sample V_pen_8. Plots for all samples are shown in Figure S1, and nucleotide alignments showing the recombinant contigs for samples A_tum_5 and V_pens_8 are shown in Figure S2.

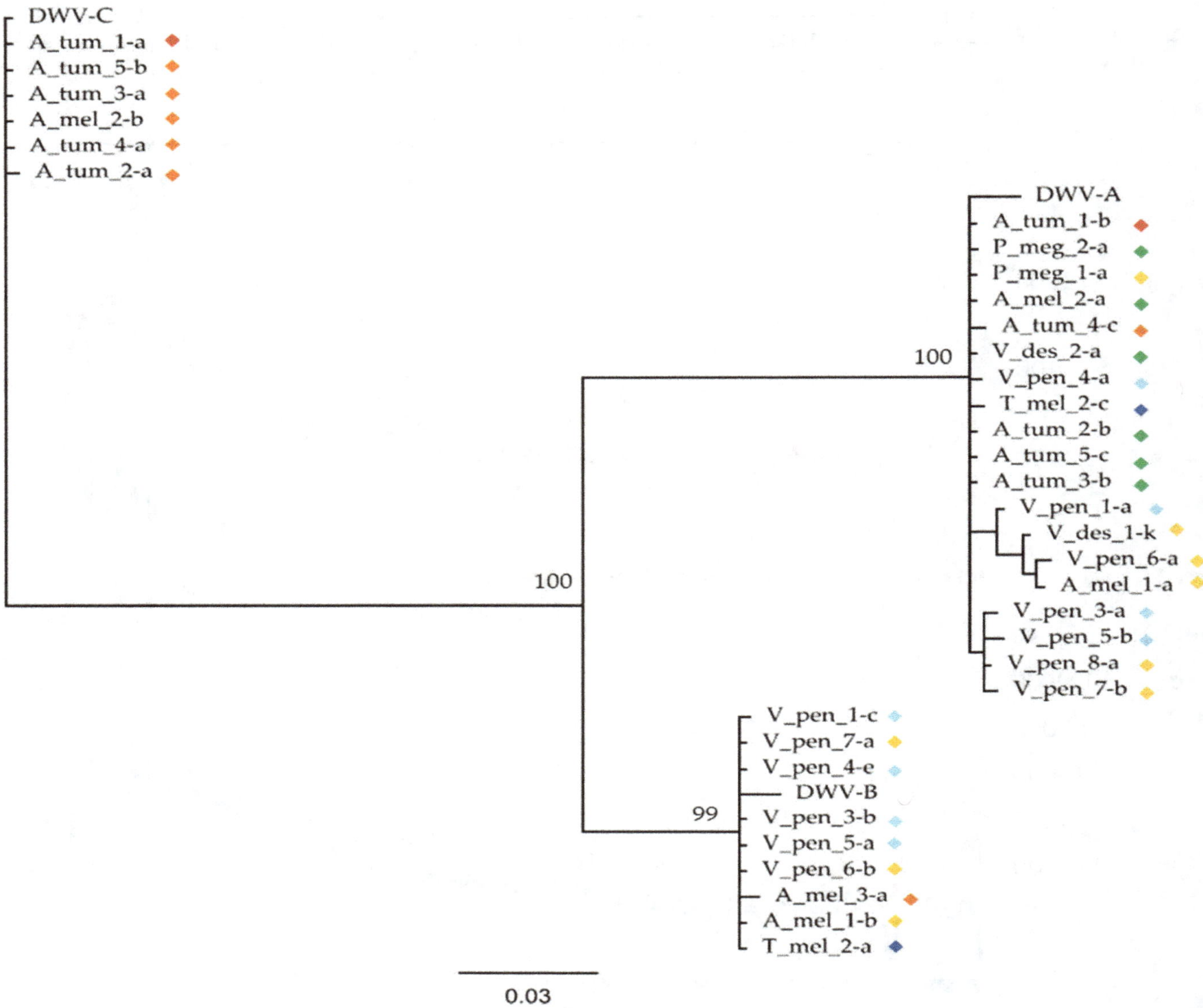

Figure 3. A Bayesian phylogeny of a 507 bp region of the DWV *RdRp* gene using trimmed contigs assembled for each sample individually using VICUNA, along with the corresponding regions from the DWV-A, B, and C reference genomes (NC_004830.2, AY251269.2, and CEND01000001.1). Colored diamonds represent the sample collection location: O1 = green, O2 = red, B1 = orange, B2 = dark blue, B3 = yellow, and B4 = light blue. Sample name suffixes represent the contig name. The bar represents the number of nucleotide substitutions per site, and the consensus support (%) is shown for selected branches.

To investigate how the DWV-A sequences in the Hawaiian apiary insects compared with DWV-A sequences on a broader scale, we constructed a phylogeny using the same 410 bp *RdRp* region used by [4] in their global phylogeny, along with some of their samples (Figure 4). This revealed that all samples in the current study fell within those detected elsewhere and clustered together with little sequence variation in this region of the genome.

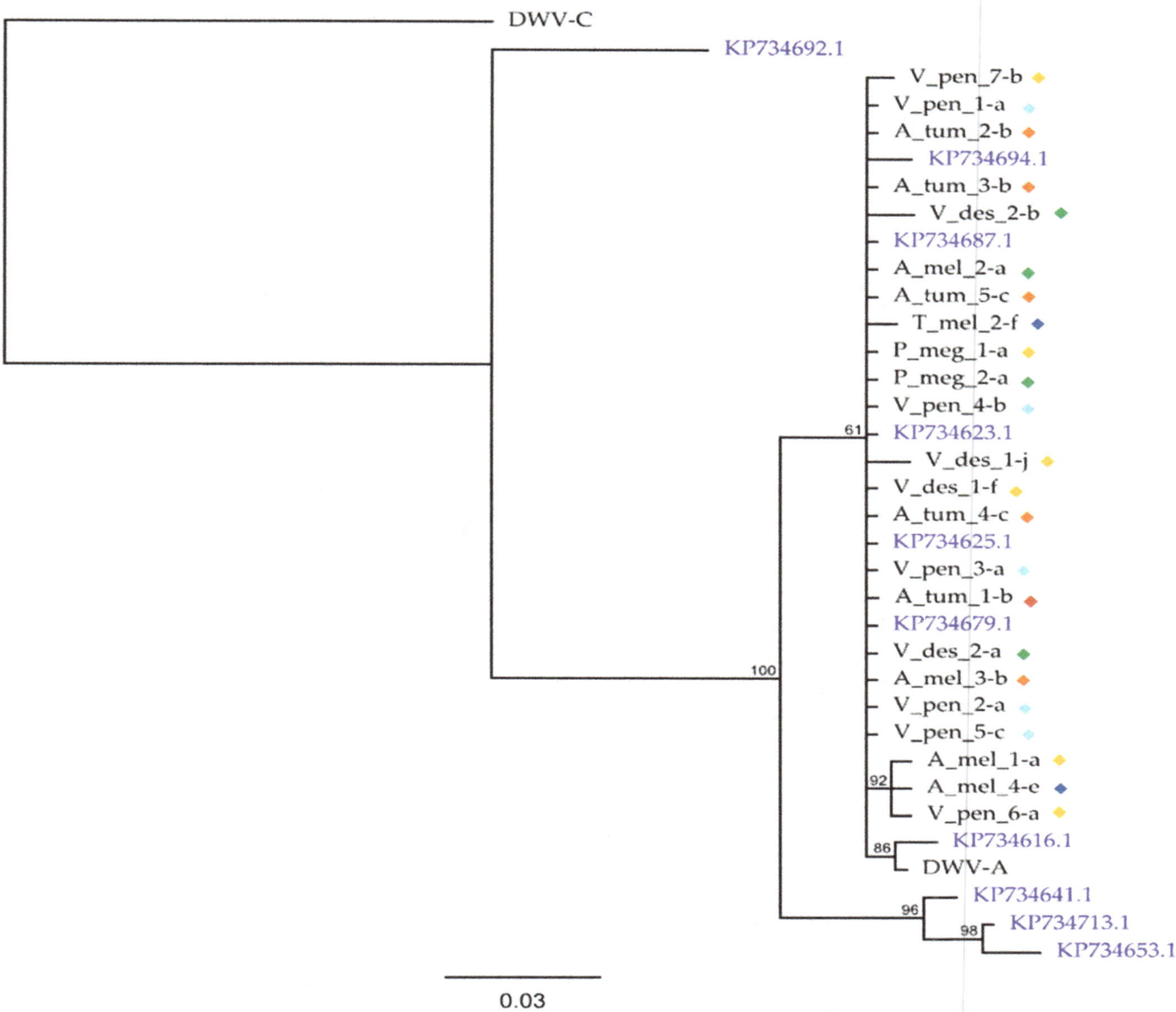

Figure 4. A Bayesian phylogeny of a 410 bp region of the DWV-A *RdRp* gene using trimmed contigs assembled for each sample individually using VICUNA, along with 10 additional sequences, as featured in the global DWV-A phylogeny of [4] (shown in blue text). Colored diamonds represent the site from which the sample was collected as follows: O1 = green, O2 = red, B1 = orange, B2 = dark blue, B3 = yellow, and B4 = light blue. Sample name suffixes represent the contig name. DWV-C is used as the outgroup, and the bar represents the number of nucleotide substitutions per site. Consensus support values (%) are shown.

3.2. *Other Honey Bee-Associated Viruses*

In addition to generally grouping in terms of DWV types A, B, and C proportions and coverage, geographical and taxa groupings were also observed when considering the other honey bee-associated viruses they harbored. BLASTn was used to identify reads belonging to eight of the most common honey bee viruses and two additional viruses known from our previous work to be prevalent in Hawaiian insects. As expected, honey bees (with the exception of sample A_mel_3) and varroa were dominated by DWV reads, as were big headed ants. We found ghost ants to be dominated by Milolii virus, yellowjacket wasps to be dominated by Moku virus, and small hive beetles to contain relatively few virus reads (Figure 5). The wasp samples further separated into two groups, with samples V_pen_1–V_pen_5 (from site B4) containing high numbers of Moku virus reads and also consistent amounts of SBPV, whereas V_pen_6–V_pen_8 (from site B3) showed increased DWV, more variable Moku virus, and no notable SBPV. The honey bees were the only samples to contain any notable BQCV.

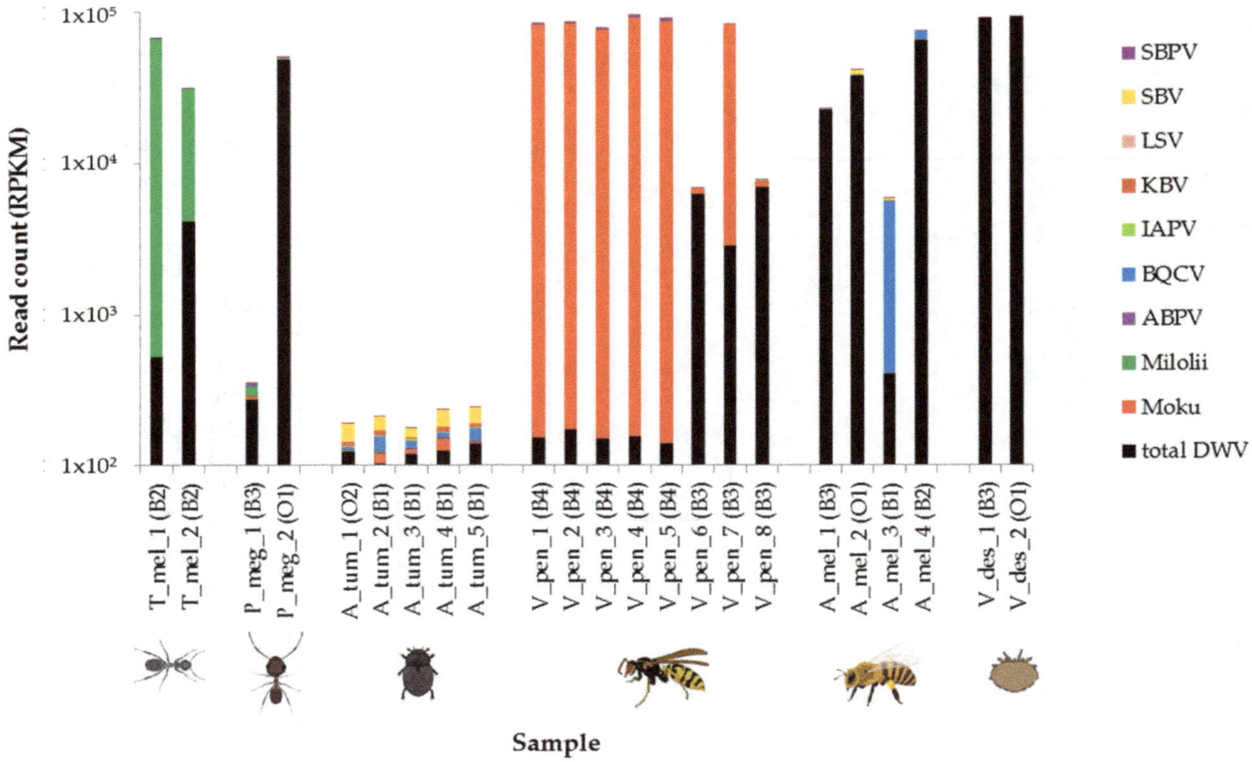

Figure 5. Histograms represent numbers of reads mapping to each of 10 commonly found insect viruses (slow bee paralysis virus: SBPV, sacbrood virus; SBV, Lake Sinai virus; LSV, Kashmir bee virus; KBV, Israeli acute paralysis virus; IAPV, black queen cell virus; BQCV, acute bee paralysis virus; ABPV, Milolii virus, Moku virus and deformed wing virus; DWV) as determined using BLASTn. Read numbers are expressed as reads per kilobase million (RPKM) and are shown on a log-10 scale starting 1×10^2. The samples are positioned according to taxa: ghost ants (T_mel_1–2), big headed ants (P_meg_1–2), small hive beetles (A_tum_1–5), yellowjacket wasps (V_pens_1–8), honey bees (A_mel_1–4), and varroa (V_des_1–2). Apiary locations from where the samples were taken are given after each sample name.

Redundancy analysis was conducted ($p < 0.001$) to further investigate the variation in viral read count data (log-10) for each sample in the context of species and site explanatory variables. RDA1 and RDA2 together explained 54.8% of the variation, and all variables (sites and species) were significantly different from the dummy variables ($p < 0.05$) (Figure 6). The RDA plot shows groupings by site (colors), as well as demonstrating separation by species, especially "honey bee", "small hive beetle", and "yellowjacket wasp".

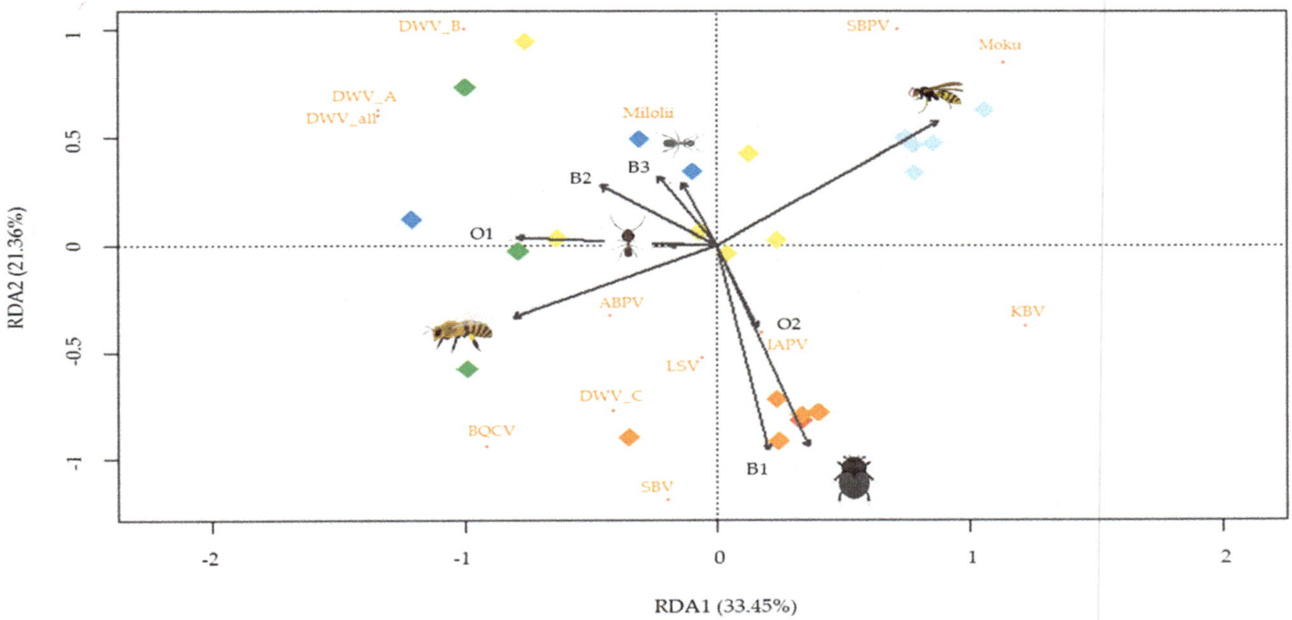

Figure 6. Redundancy Analysis (RDA) plot showing the variation in virus read count data (RPKM log-10) for each sample correlated with environmental variables: site and species. Dummy variables were randomly assigned to varroa (species) and B4 (site). Sites are colored: O1 = green, O2 = red, B1 = orange, B2 = dark blue, B3 = yellow, and B4 = light blue.

4. Discussion

This study revealed that common apiary pests have the potential to act as reservoirs for, or could be impacted by, a number of honey bee-associated viruses. Although recent studies have highlighted the ability of a number of viruses initially described as honey bee pathogens to infect a range of taxa and in some cases cause pathogenic effects (reviewed by [19]), this is the first study to assess the viral burden of taxonomically diverse common apiary pests.

Of 10 common +ssRNA viruses surveyed, DWV was the most common virus in honey bees and varroa, as expected, because it is the most prevalent virus in honey bees across the world [19] and was the most prevalent virus in the apiary with full-genome coverage achieved from samples of each species. DWV-A was the most common variant detected, although DWV-B was also widespread, correlating with the recent finding by [35] that DWV-B dominance is replacing DWV-A on the mainland United States of America (USA). Although the BLAST top hit analysis was reliable when considering full-length virus genomes, such as when detecting a number of different viruses (Figure 5), or when only full length master variants are present, e.g., considering DWV diversity in all ant samples (Figure 1 and Figure S2), methods that identify all reads independently, irrespective of genome location, mean that it is impossible to discriminate between full genomes and recombinant forms. This is evidenced by the beetle data that BLAST top hit analysis suggested was dominated by DWV-C; however closer inspection using competitive alignments revealed this to, in fact, be a DWV-A/DWV-C recombinant. Thus, the two analyses (BLAST top hit results and competitive alignment plots) gave consistent results when only full-length genomes were present but differed when recombinants were detected.

Interestingly, beetles sampled from across two locations were the only ones to be dominated by a DWV-A/DWV-C recombinant. These all had very low read counts but were very consistent. All beetle samples were collected earlier in the year from collapsing honey bee colonies suffering from extremely heavy small hive beetle infestations (slime out conditions), and unfortunately, no corresponding bee samples were taken at this time. As such it may be possible that the same recombinant was present at high levels in the dying bees at that time, and the high amounts of virus in the hives resulted in passive contamination of the beetles, which would explain why DWV is only very rarely detected in beetles [36]. In this case, contamination during library preparation or during the Illumina run cannot be ruled out, because it is a particular issue for RNAseq analysis. This can occur with invertebrate viruses due to the extreme dominance of viral reads in heavily infected samples. These beetle samples had previously tested positive for DWV using RT-PCR prior to selection for RNAseq analysis and a recent RT-PCR-based study showed that stingless bees (*Melipona subnitida*) from Brazil also harbored DWV-C but at higher viral levels [37], confirming this DWV variant also has the ability to infect diverse hosts.

Investigating viral reads generated by oligo (dT), RNAseq is not the only method to assess whether samples are harboring true infections [38,39]. Nonetheless, this method will reveal high read counts, full-genome coverage, and assembled full genomes, suggesting that DWV is replicating in at least some samples from each species. Furthermore, the unique dominance of recombinant(s) in the wasp sample V_pens_8 that was not seen in any other sample provides further evidence that there are specific DWV variants/recombinants present in the quasispecies that are able to successfully replicate in different species.

Considering both the phylogenies together, it appears that common DWV variants are circling in the apiary insect community, suggesting frequent interspecies transmission events. The nature of these transmission events remains unclear, with trophylaxis (between small hive beetles) [20], fecal–oral routes and predation on bees, and consumption of contaminated hive materials, pollen, and nectar all being implicated as routes by which viruses can spread both between and within species [40].

The DWV-A phylogeny, which in the original publication by [4] only contained honey bee- and varroa-derived sequences (Figure 5) showed all sequences from this study to fall within the greater diversity and all cluster within one clade. When considering the phylogeny of DWV sequences from the current study only (Figure 4), again all DWV-A clustered together as was also true of DWV-B sequences. The limitations both in terms of sampling and in using only one gene region means that is not possible to draw strong conclusions on the evolutionary history of these sequences. Furthermore, when considering the read count data and genome coverage plots for wasps only, they appear to separate by location with samples V_pen_1–5 from B4 all showing almost identical results in terms of their DWV variant composition, whereas V_pen_6–8 are very different (Figure 1 and Figure S1). These findings are in keeping with the results of [24] who found, using RT-PCR and Sanger sequencing, that DWV sequences in yellowjacket wasps (also from Hawaii) tracked strain changes observed in the honey bees [17] on which they were feeding, i.e., DWV diversity decreased in honey bees as varroa became established on Oahu and then on the Big Island, and this change was detected in yellowjackets on the Big Island. Although the current study is limited to eight individuals across two locations, it is clear that DWV is present in the wasp samples, that sequences are diverse, and that the presence and potential level of DWV variants are affected by location.

This is the first study of arthropods in the wider apiary community to consider the newly described Moku [29] and Milolii [30] viruses. Although these viruses only dominated in the samples in which they were first described, low numbers of reads were also found in other varroa, beetle, wasp, and ant samples. The wider +ssRNA virus detection study revealed that all of the common honey bee-associated viruses tested for were present at some level in the Hawaiian apiary insects, although IAPV, KBV, ABPV, and LSV were only present in very low levels (as detected through BLAST), and therefore, true infections cannot be confirmed in any sample. It is interesting to note that, similar to the findings of the DWV data, when considering the other honey bee-associated virus data, species and location groupings were seen. In the case of species groupings, this may be related to different species having different virus susceptibilities and, in the case of site groupings, may be more related to variation in particular virus levels at different sites.

5. Conclusions

Although this pilot study has limitations, namely unbalanced sampling design, we have shown that several common honey bee-associated +ssRNA viruses are common in taxonomically diverse apiary pests. We showed that DWV was the most prevalent virus and that DWV infections grouped between species in terms of dominant variants and recombinants, but that common variants (predominantly type A) were circling between all species, suggesting repeated transmission events between species. Within the wasps, DWV was further separated by location. Species also grouped in terms of which other honey bee-associated viruses they harbored, i.e., particular viruses are associated with particular hosts. Therefore, this study highlights the need to consider the wider arthropod community as potential reservoirs of viral pathogens in the apiary.

Supplementary Materials:
Figure S1: DWV genome coverage plots for individual samples created using Geneious. Read depths are shown on a log-10 scale and represent DWV-A (red), -B (blue), and -C(yellow) along the ~10.1 kb genomes. Figure S2: DWV alignments (MUSCLE) created using Geneious showing de novo assembled contigs from samples A_tum-5 and V_pens-8, which contain recombination breakpoints. (a) Contig A_tum-5-a aligned with DWV-A (NC_004830.2) and DWV-C (CEND01000001.1) reference genomes, (b) contig V_pen_8-b aligned with DWV-A (NC_004830.2) and DWV-B (AY251269.2), and (c) a second contig from sample V_pen_8; V_pen_8-t also aligned with DWV-A (NC_004830.2) and DWV-B (AY251269.2). All alignments show disagreements with the assembled contigs highlighted in black, and recombinant contigs are shaded red where they map most closely to DWV-A, blue to DWV-B, and green to DWV-C. Table S1: Samples used in this study. Sample names are given along with the site from which they were sampled, species name, and the symbol used to denote them in Figures 1 and 5. Table S2: DWV-A *RdRp* sequences originally from [4] and used in this study in the construction of the DWV-A phylogeny in Figure 5. Table S3: Viruses commonly found in bees used for BLAST analysis along with accession numbers. Table S4: Numbers of reads mapping to DWV types A, B, and C using Blast top hit analysis for each sample, along with total numbers of reads passing QC (read1.fasta).

Author Contributions: Conceptualization, L.E.B. and S.J.M.; methodology, L.E.B. and S.J.M.; software, L.E.B.; validation, L.E.B.; formal analysis, L.E.B. and S.J.M.; investigation, L.E.B., S.J.M. and D.C.S; resources, S.J.M. and D.C.S.; data curation, L.E.B.; original draft preparation, L.E.B.; review and editing, L.E.B., S.J.M. and D.C.S.; visualization, L.E.B.; S.J.M. and D.C.S.; and funding acquisition, S.J.M.

Acknowledgments: We would like to think E. Villolobos and S. Nikaido for their assistance in organizing and carrying out the sampling, C. Ahrens for assistance with statistical analyses using R, J. Finch for assistance with the phylogenetic reconstructions and all the beekeepers in Hawaii who provided access to their bees and apiaries. We also would like to thank the anonymous reviewers, whose comments helped to improve this manuscript.

References

1. Daszak, P.; Cunningham, A.A.; Hyatt, A.D. Emerging infectious diseases of wildlife–threats to biodiversity and human health. *Science* **2000**, *287*, 443. [CrossRef] [PubMed]
2. Blaustein, A.R.; Johnson, P.T.J. Conservation biology: When an infection turns lethal. *Nature* **2010**, *465*, 881–882.
3. Schroeder, D.C.; Martin, S.J. Deformed wing virus: The main suspect in unexplained honeybee deaths worldwide. *Virulence* **2012**, *3*, 589–591. [CrossRef] [PubMed]
4. Wilfert, L.; Long, G.; Leggett, H.C.; Schmid-Hempel, P.; Butlin, R.; Martin, S.J.; Boots, M. Deformed wing virus is a recent global epidemic in honeybees driven by Varroa mites. *Science* **2016**, *351*, 594–597. [CrossRef]
5. Potts, S.G.; Vulliamy, B.; Dafni, A.; Ne'eman, G.; Willmer, P. Linking bees and flowers: How do floral communities structure pollinator communities? *Ecology* **2003**, *84*, 2628–2642.
6. Steffan-Dewenter, I.; Tscharntke, T. Resource overlap and possible competition between honeybees and wild bees in central Europe. *Oecologia* **2000**, *122*, 288–296. [CrossRef]
7. Steffan-Dewenter, I.; Tscharntke, T. Insect communities and biotic interactions on fragmented calcareous grasslands—a mini review. *Biol. Cons.* **2002**, *104*, 275–284.
8. Thomson, D. Competitive interactions between the invasive European honeybee and native bumble bees. *Ecology* **2004**, *85*, 458–470. [CrossRef]
9. Potts, S.G.; Biesmeijer, J.C.; Kremen, C.; Neumann, P.; Schweiger, O.; Kunin, W.E. Global pollinator declines: Trends, impacts and drivers. *Trends Ecol. Evol.* **2010**, *25*, 345–353. [CrossRef]
10. Brown, M.J.; Dicks, L.V.; Paxton, R.J.; Baldock, K.C.; Barron, A.B.; Chauzat, M.P.; Freitas, B.M.; Goulson, D.; Jepsen, S.; Kremen, C.; et al. A horizon scan of future threats and opportunities for pollinators and pollination. *PeerJ* **2016**, *4*, e2249. [CrossRef] [PubMed]
11. Vanbergen, A.J. Threats to an ecosystem service: Pressures on pollinators. *Front. Ecol. Environ.* **2013**, *11*, 251–259. [CrossRef]
12. Cornman, R.S.; Tarpy, D.R.; Chen, Y.; Jeffreys, L.; Lopez, D.; Pettis, J.S.; Evans, J.D. Pathogen webs in collapsing honeybee colonies. *PLoS ONE* **2012**. [CrossRef]

13. Collison, E.; Hird, H.; Cresswell, J.; Tyler, C. Interactive effects of pesticide exposure and pathogen infection on bee health—A critical analysis. *Biol. Rev. Camb. Philos. Soc.* **2016**, *91*, 1006–1019. [CrossRef]
14. Parrish, C.R.; Holmes, E.C.; Morens, D.M.; Park, E.C.; Burke, D.S.; Calisher, C.H.; Laughlin, C.A.; Saif, L.J.; Daszak, P. Cross-species virus transmission and the emergence of new epidemic diseases. *Microbiol. Mol. Biol. Rev.* **2008**, *72*, 457–470. [CrossRef] [PubMed]
15. Domingo, E.J.; Holland, J.J. RNA virus mutations and fitness for survival. *Annu. Rev. Microbiol.* **1997**, *51*, 151–178. [CrossRef] [PubMed]
16. Ryabov, E.V.; Wood, G.R.; Fannon, J.M.; Moore, J.D.; Bull, J.C.; Chandler, D.; Mead, A.; Burroughs, N.; Evans, D.J. A virulent strain of Deformed wing virus (DWV) of honeybees (*Apis mellifera*) prevails after *Varroa destructor*-mediated, or in vitro, transmission. *PLoS Pathog.* **2014**, *10*, 1–21. [CrossRef] [PubMed]
17. Martin, S.J.; Highfield, A.C.; Brettell, L.; Villalobos, E.M.; Budge, G.E.; Powell, M.; Nikaido, S.; Schroeder, D.C. Global honeybee viral landscape altered by a parasitic mite. *Science* **2012**, *336*, 1304–1306. [CrossRef]
18. Mondet, F.; de Miranda, J.R.; Kretzschmar, A.; Le Conte, Y.; Mercer, A.R. On the front line: Quantitative virus dynamics in honeybee (*Apis mellifera* L.) colonies along a new expansion front of the parasite *Varroa destructor*. *PLoS Pathog.* **2014**, *10*, e1004323. [CrossRef]
19. Martin, S.J.; Brettell, L.E. Deformed wing virus in Honeybees and Other Insects. *Annu. Rev. Virol.* **2019**, *6*. [CrossRef]
20. Eyer, M.; Chen, Y.P.; Schäfer, M.O.; Pettis, J.; Neumann, P. Small hive beetle, *Aethina tumida*, as a potential biological vector of honeybee viruses. *Apidologie* **2009**, *40*, 419–428. [CrossRef]
21. Fürst, M.A.; McMahon, D.P.; Osborne, J.L.; Paxton, R.J.; Brown, M.J.F. Disease associations between honeybees and bumblebees as a threat to wild pollinators. *Nature* **2014**, *506*, 364–366. [CrossRef]
22. McMahon, D.P.; Fürst, M.A.; Caspar, J.; Theodorou, P.; Brown, M.J.F.; Paxton, R.J. A sting in the spit: Widespread cross-infection of multiple RNA viruses across wild and managed bees. *J. Anim. Ecol.* **2015**, *84*, 615–624. [CrossRef]
23. Santamaria, J.; Villalobos, E.M.; Brettell, L.E.; Nikaido, S.; Graham, J.R.; Martin, S. Evidence of Varroa-mediated deformed wing virus spillover in Hawaii. *J. Invertebr. Pathol.* **2018**, *151*, 126–130. [CrossRef]
24. Loope, K.J.; Baty, J.W.; Lester, P.J.; Wilson Rankin, E.E. Pathogen shifts in a honeybee predator following the arrival of the Varroa mite. *Proc. R. Soc. B* **2019**, *286*. [CrossRef]
25. Mordecai, G.J.; Brettell, L.E.; Martin, S.J.; Dixon, D.; Jones, I.M.; Schroeder, D.C. Superinfection exclusion and the long-term survival of honeybees in Varroa-infested colonies. *Isme J.* **2016**, *10*, 1182–1191. [CrossRef]
26. Brettell, L.E.; University of Salford, Manchester, UK. Results of screening for Deformed wing virus in common apiary pest insects using RT-PCR. 2012.
27. Yang, X.; Charlebois, P.; Gnerre, S.; Coole, M.G.; Lennon, N.J.; Levin, J.Z.; Qu, J.; Ryan, E.M.; Zody, M.C.; Henn, M.R. De novo assembly of highly diverse viral populations. *BMC Genomics* **2012**, *13*, 475. [CrossRef]
28. Huelsenbeck, J.P.; Ronquist, F. MRBAYES: Bayesian inference of phylogenetic trees. *Bioinformatics* **2001**, *17*, 754–755. [CrossRef]
29. Mordecai, G.J.; Brettell, L.E.; Pachori, P.; Villalobos, E.M.; Martin, S.J.; Jones, I.M.; Schroeder, D.C. Moku virus; a new Iflavirus found in wasps, honeybees and Varroa. *Sci. Rep.* **2016**, *6*, 34983. [CrossRef]
30. Brettell, L.E.; Mordecai, G.; Pachori, P.; Martin, S. Novel RNA virus genome discovered in Ghost ants (*Tapinoma melanocephalum*) from Hawaii. *Genome Announc.* **2017**, *5*, e00669-17. [CrossRef]
31. Villalobos, E.M. The mite that jumped, the bee that traveled, the disease that followed. *Science* **2016**, *351*, 554–556. [CrossRef]
32. Brooks, E.M.; Sheflin, L.G.; Spaulding, S.W. Secondary structure in the 3′UTR of EGF and the choice of reverse transcriptases affect the detection of message diversity by RT-PCR. *Biotechniques* **1995**, *19*, 806–812.
33. Lanzi, G.; de Miranda, J.R.; Boniotti, M.B.; Cameron, C.E.; Lavazza, A.; Capucci, L.; Carmazine, S.M.; Rossi, C. Molecular and Biological Characterization of Deformed Wing Virus of Honeybees (*Apis mellifera* L.). *J. Virol.* **2006**, *80*, 4998–5009. [CrossRef]
34. Wood, G.R.; Burroughs, N.J.; Evans, D.J.; Ryabov, E.V. Error correction and diversity analysis of population mixtures determined by NGS. *PeerJ* **2014**, *2*, e645. [CrossRef] [PubMed]
35. Kevill, J.L.; de Souza, F.; Sharples, C.; Schroeder, D.; Martin, S.J. DWV-A lethal to honey bees (*Apis mellifera*): A colony level survey of DWV variants (A, B & C) in England, Wales and 32 states across the US. *Viruses* **2019**. (submitted).
36. Villalobos, E.M.; (University of Hawaii, Hawaii, USA). Personal communication, 2017.

37. De Souza, F.S.; Kevill, J.L.; Correia-Oliveira, M.E.; de Carvalho, C.A.; Martin, S.J. Occurrence of deformed wing virus variants in the stingless bee *Melipona subnitida* and honey bee *Apis mellifera* populations in Brazil. *J. Gen. Virol.* **2019**, *100*, 289–294. [CrossRef]
38. Boncristiani, H.F.; Di Prisco, G.; Pettis, J.S.; Hamilton, M.; Chen, Y.P. Molecular approaches to the analysis of deformed wing virus replication and pathogenesis in the honey bee, *Apis mellifera*. *Virol. J.* **2009**, *6*, 221. [CrossRef]
39. Fung, E.; Hill, K.; Hogendoorn, K.; Glatz, R.V.; Napier, K.R.; Bellgard, M.I.; Barrero, R.A. *De novo* assembly of honey bee RNA viral genomes by tapping into the innate insect antiviral response pathway. *J. Invertebr. Pathol.* **2018**, *152*, 38–47. [CrossRef]
40. Singh, R.; Levitt, A.L.; Rajotte, E.G.; Holmes, E.C.; Ostiguy, N.; Lipkin, W.I.; Toth, A.L.; Cox-Foster, D.L. RNA viruses in Hymenopteran pollinators: Evidence of inter-taxa virus transmission via pollen and potential impact on non-*Apis* hymenopteran species. *PLoS ONE* **2010**, *5*, e14357. [CrossRef]

12

Evaluation of Sequencing Library Preparation Protocols for Viral Metagenomic Analysis from Pristine Aquifer Groundwaters

René Kallies [1,*], Martin Hölzer [2,3], Rodolfo Brizola Toscan [1], Ulisses Nunes da Rocha [1], John Anders [1,4], Manja Marz [2,3,5] and Antonis Chatzinotas [1,5]

1. Helmholtz Centre for Environmental Research - UFZ, Department of Environmental Microbiology, 04318 Leipzig, Germany; rodolfo.toscan@ufz.de (R.B.T.); ulisses.rocha@ufz.de (U.N.d.R.); johnanders@posteo.de (J.A.); antonis.chatzinotas@ufz.de (A.C.)
2. Friedrich Schiller University Jena, RNA Bioinformatics and High-Throughput Analysis, 07743 Jena, Germany; martin.hoelzer@uni-jena.de (M.H.); manja@uni-jena.de (M.M.)
3. European Virus Bioinformatics Center, 07743 Jena, Germany
4. Bioinformatics Group, Department of Computer Science and Interdisciplinary Center for Bioinformatics, University Leipzig, 04081 Leipzig, Germany
5. German Centre for Integrative Biodiversity Research (iDiv) Halle-Jena-Leipzig, 04103 Leipzig, Germany

* Correspondence: rene.kallies@ufz.de

Abstract: Viral ecology of terrestrial habitats is yet-to be extensively explored, in particular the terrestrial subsurface. One problem in obtaining viral sequences from groundwater aquifer samples is the relatively low amount of virus particles. As a result, the amount of extracted DNA may not be sufficient for direct sequencing of such samples. Here we compared three DNA amplification methods to enrich viral DNA from three pristine limestone aquifer assemblages of the Hainich Critical Zone Exploratory to evaluate potential bias created by the different amplification methods as determined by viral metagenomics. Linker amplification shotgun libraries resulted in lowest redundancy among the sequencing reads and showed the highest diversity, while multiple displacement amplification produced the highest number of contigs with the longest average contig size, suggesting a combination of these two methods is suitable for the successful enrichment of viral DNA from pristine groundwater samples. In total, we identified 27,173, 5,886 and 32,613 viral contigs from the three samples from which 11.92 to 18.65% could be assigned to taxonomy using blast. Among these, members of the *Caudovirales* order were the most abundant group (52.20 to 69.12%) dominated by *Myoviridae* and *Siphoviridae*. Those, and the high number of unknown viral sequences, substantially expand the known virosphere.

Keywords: viral metagenome; groundwater; aquifer; AquaDiva; sequencing library preparation

---

1. Introduction

Groundwater systems are important compartments of the global hydrological cycle. They donate about 30% of all freshwater sources [1] and provide important ecosystem services. For example, purification and storage of water, active biodegradation of anthropogenic contaminants and nutrient recycling [2]. Many of these services are directly linked to the presence of microorganisms [2,3]. Studies in particular in marine systems have significantly contributed to a better understanding of viruses and their impacts on the mortality, diversity and genetic landscape of their microbial hosts [4–6]. However, only recently, and only in a limited number of surveys, has the potential role of viruses been explored in terrestrial subsurface systems [7–11].

In theory, metagenomics enables the identification and genomic characterisation of all (micro)organisms present in a sample, including viruses [12]. However, the proportion of viral sequences within a metagenome is usually far lower than for other organisms, leading to limitations in their detection. Especially, pristine aquifers are characterised by low microbial biomass and low abundances of virus particles [9,13,14], which might make their detection even more difficult. Size filtration or density-based enrichment methods are therefore widely used to concentrate virus particles from environmental samples [15,16]. However, a significant obstacle in applying metagenomics for pristine aquifers is the still too low amount of DNA required for the direct sequencing of such samples, making amplification techniques mandatory to further enrich viral nucleic acids. It is however widely known that DNA amplification is a source of bias that may lead to inaccurate conclusions after sequence analysis [17]. Three amplification techniques are commonly used to enrich low amounts of DNA [17,18], i.e., (i) linker amplification shotgun libraries (LASL) [19,20]; (ii) sequence-independent, single-primer amplification (SISPA) [21,22]; and (iii) multiple displacement amplification (MDA) [23,24]. Each method has its own potential source of bias. LASL relies on DNA fragmentation and subsequent linker ligation to blunt-end repaired DNA molecules prior to amplification, using primer oligos that bind to the linker sequences [19,25]. Linker ligation efficiency might be one source of bias, especially for very low amounts of DNA [26]. However, previous studies demonstrated that as little as a few pg to ng of DNA is sufficient for low amplification biases [26,27]. LASL may, in addition, be inefficient in recovering ssDNA viruses due to the double-stranded nature of linker DNA molecules [28] though this has recently been overcome with an adapted LASL protocol [29]. SISPA is built upon the use of pseudo-degenerated primer oligonucleotides, containing a stretch of random nucleotides at their 3′-end and a defined sequence at their 5′-end [21], and has successfully been applied to recover both RNA and DNA virus sequences [22,30]. It has, however, been reported that SISPA has a strong amplification bias resulting in an uneven sequencing read distribution and hence overrepresentation of some genome parts while other parts were completely uncovered. In addition, SISPA negatively affects the detection of low abundant genomes [31]. MDA works under isothermal conditions [32] with very low amounts of input DNA, random hexamer primer oligonucleotides and high fidelity as well as strand displacement functions of the phi29 polymerase [23]. Several sources of bias have been identified for phi29 amplification, including chimera formation [33], discontinuous amplification of linear DNA molecules [34] and preferential amplification of circular ssDNA molecules [35]. Recent studies evaluated different library preparation protocols using low input-DNA to assess the reconstruction of microbial communities from metagenomes [36,37]. Similar studies have been performed for the identification of virus sequences from, for example, seawater and human samples [17,35]. Despite these advances, to our knowledge no study has to date assessed and benchmarked sequence library preparation protocols for the identification of viral sequences from pristine aquifer groundwaters.

The Hainich Critical Zone Exploratory (Hainich CZE) in central Germany is an infrastructure designed to, among others, investigate the diversity, identity and abundance of microorganisms in the Hainich aquifers. In addition, analysis of metabolic potential and activities of microorganisms will be linked to physico-chemical parameters in spatial and temporal scales [38]. Here we sampled three carbonate-rock aquifer assemblages of the Hainich CZE, which represent a pristine and uncontaminated aquifer [38]. One problem in obtaining viral sequences from groundwater samples is the low amount of DNA (usually a pico- to few nanograms) that was extracted from isolated virus particles. The aim of this project is therefore two-fold. The first aim is to evaluate different DNA amplification techniques that may offer a sufficient amount of DNA for high-throughput sequencing. In addition, these methods should have a low amplification bias to reflect the natural diversity of the analysed samples. The second aim consisted of evaluating different viral sequence recovery tools to provide first insights on which viruses are present in the Hainich CZE groundwater aquifers.

2. Materials and Methods

2.1. Sample Collection

Groundwater samples were taken from three Hainich CZE aquifer wells in Thuringia, Germany, within the framework of the Collaborative Research Centre AquaDiva (http://www.aquadiva.uni-jena.de) (CRC 1076) [38]. The sampling site was located in the agriculturally used midslope and footslope regions of the Hainich low-mountain range. The three wells were drilled to depths of 50 m (H53), 65 m (H52) and 88 m (H51). H53 and H52 reflect anoxic conditions while oxic conditions prevailed for H51. A detailed description of hydrochemical and geostructural parameters can be found elsewhere [39,40].

Ten liters of groundwater (with approximately 2.3×10^5 (SD: 1.2×10^4) viral particles per milliliter) were collected from each well during a sampling campaign in May 2015. Water was filtered through 200 nm pore filters using a cross-flow system (Sartorius, Göttingen, Germany). Samples were then enriched for viral particles by filtration through 35 kDa filters using the same system. Approximately 60 mL were retained and further concentrated by ultracentrifugation at 22,000× g for 2 h and 4 °C. The viral particle containing pellet was resuspended in 500 μL TM buffer (50 mM Tris HCl, 10 mM Magnesium sulfate at pH 7.5). One volume of chloroform was given to the samples to remove microsized prokaryotes. The upper phase, intended for DNA extraction, was treated with DNase I to remove free DNA.

2.2. DNA Extraction, Library Construction and Sequencing

Viral DNA was extracted as described previously [20]. Viral DNA concentration was determined using the Qubit® dsDNA HS Assay Kit (Thermo Fisher Scientific, Waltham, MA, USA) resulting in total DNA amounts of 31.8 ng (H51), 5.4 ng (H52) and 25.9 ng (H53). DNA was divided into four parts to prepare four libraries for each sample. Non-amplified shotgun libraries (NASL): using a Covaris ultrasonicator, DNA was sheared to approx. 350 bp fragments and libraries were prepared with a TruSeq DNA PCR-Free Library Prep Kit (Illumina, San Diego, CA, USA) according to the manufacturer's instructions. Linker amplification shotgun libraries (LASL): DNA was sheared to approx. 350 bp fragments as mentioned above and LASL was performed with a NEBNext Ultra DNA Lib Prep Kit (New England Biolabs, Ipswich, MA, USA) as recommended by the manufacturer including 12 PCR cycles to enrich adaptor-ligated DNA. Single-primer amplification (SISPA): PCR was performed by ten cycles using random octamer primers that were linked to a specific primer sequence followed by amplification using a 1:9 mixture of random octamers and a primer targeting the specific primer sequence as described previously [41]. Multiple displacement amplification (MDA): DNA was subjected to phi29 amplification at 25 °C for 8 h using the illustra GenomiPhi V2 DNA Amplification Kit (Thermo Fisher Scientific) as described in the manual. PCR amplicons for the latter two libraries were purified using the Sureclean reagent (Bioline, Luckenwalde, Germany), fragmented as described above and libraries were prepared as described for NASL. Sequencing was performed on one lane of an Illumina HiSeq 2500 system to generate 100-bp paired-end reads.

2.3. Sequencing Read Processing and Assembly

PhiX contaminants were removed, SISPA primer sequences were clipped and raw sequencing reads were quality checked using Trimmomatic [42] and low-quality bases were trimmed from both ends. Reads were screened with a 4-base wide sliding window until the remaining sequences had a Phred-score of at least 15 and a minimum length of 36 nt. Sequencing read redundancy was identified by clustering at 90% sequencing read identity using CD-hit v.4.6 [43,44].

Sequencing reads were independently assembled for each sampling site and library preparation using metaSPAdes [45,46] and SOAPdenovo-Trans [47]. In addition, cross-assemblies were performed for each sampling site including all reads from LASL, SISPA and MDA libraries. We used the transcriptome assembler SOAPdenovo-Trans in addition to SPAdes because recent analyses revealed this assembly tool as very efficient for the assembly of RNA virus genomes [48]. Further analyses suggested this might be also true for the assembly of DNA virus genomes.

2.4. Viral Contig Recovery

Three viral sequence identification tools were used to recover viral contigs, i.e., VirSorter [49], VirFinder [50] and VrAP (https://www.rna.uni-jena.de/research/software/vrap-viral-assembly-pipeline/). VirSorter is based on the identification of viral hallmark genes, enrichment in hypothetical proteins and other viral signatures [49]. Only contigs identified as VirSorter categories 1 and 2 (higher confidence predictions) were retained for further analysis. VirFinder is a kmer based tool for the identification of viral contigs from metagenomes with improvements especially for the detection of short viral contigs [50]. Contigs with a p-value < 0.01 were used for further analysis. These two detection tools were completed by using VrAP, a novel de novo genome assembly pipeline especially designed for viruses. The pipeline is able to assemble complete genomes of viruses representing new strains and species, as well as prototypes of new genera and families. VrAP is based on the genome assembler SPAdes [45] combined with an additional read correction [51,52] and several filter steps. The pipeline classifies the contigs to distinguish host from viral sequences by annotation and open reading frame (ORF) density scores. By applying the ORF density method we were able to identify potential novel viruses without any sequence homology to known references (manuscript in preparation).

2.5. Virome Diversity Measures and Comparison of Library Preparation Methods

Nonpareil [53–55] was used with default settings to estimate diversity and coverage of virome datasets. Viral reads present in one or more datasets reflecting LASL, SISPA and MDA per sampling site were identified as follows. Redundancy was removed for each dataset by CD-hit-est clustering at 95% identity. A database was created containing all viral contigs and, using Bowtie2 [56], read cluster per library preparation method and sampling site were mapped to the database. Mapped clusters were extracted, counted and overlapping information were generated using SAMtools [57]. Viral contigs were compared between sites by an all-versus-all clustering approach (95% identity) with CD-hit-est-2D [44].

Venn diagrams were computed in R [58] using the package "venneuler" (https://cran.r-project.org/web/packages/venneuler/index.html).

2.6. Viral Taxonomic Assignment

All viral contigs per sampling site, i.e., contigs identified from all virus identification tools and library preparation methods, were combined (resulting in three datasets) and redundancies were removed by clustering with CD-hit-est at 95% nt identity. Open reading frames (ORFs) were translated from these contigs using prodigal [59] and aligned to a viral RefSeq protein database (February 2019) using DELTA-BLAST [60] with an e-value cut off of 10^{-3}. Hits were sorted by e-value and bit score and ORFs with most significant hits were aligned to the respective contigs using an in-house python script (Supplementary Information), resulting in one hit per contig. Gene sharing networks based on shared protein clusters (PCs) between viral genomes were calculated with vConTACT2 [61,62] on the iVirus platform [63] and were displayed with Cytoscape [64]. DNA contamination from cellular organisms was determined using EMIRGE [65].

2.7. Data Availability

Sequence read raw data have been made available at Sequence Read Archive accession: PRJNA530103.

3. Results

3.1. Raw Sequencing Output Statistics

The first aim of this study was to evaluate different DNA amplification techniques that may result in a sufficient amount of viral DNA for high throughput sequencing. We therefore compared three DNA amplification methods, i.e., LASL, SISPA and MDA. NASL was used as control.

MDA produced highest (quality trimmed) sequencing read numbers followed by SISPA and LASL as compared to NASL that exhibited lowest read numbers (Table S1). Significant differences (ANOVA) in quality-trimmed sequencing output were observed between NASL-MDA, NASL-SISPA and SISPA-MDA (Table 1).

Table 1. *P*-values of analysis of variance (ANOVA) of raw sequencing read and read cluster numbers between the different library preparation methods.

Number of Raw Reads				
Library Preparation	NASL	LASL	SISPA	MDA
NASL	n/a	>0.05	0.008	0.002
LASL		n/a	>0.05	0.023
SISPA			n/a	>0.05
MDA				n/a

Clusters at 90% Read Identity								
	Relative proportion				Number of clusters			
Library Preparation	NASL	LASL	SISPA	MDA	NASL	LASL	SISPA	MDA
NASL	n/a	<0.001	>0.05	>0.05	n/a	0.018	>0.05	0.008
LASL		n/a	<0.001	<0.001		n/a	>0.05	>0.05
SISPA			n/a	>0.05			n/a	>0.05
MDA				n/a				n/a

NASL: non-amplified shotgun library; LASL: linker amplification shotgun libraries; SISPA: single-primer amplification; MDA: multiple displacement amplification.

Read quality of all libraries was >97% except for libraries H51 LASL and H52 LASL for which 33.78% and 28.13% of the reads were discarded after quality trimming. However, no significant differences (ANOVA) in quality between any of the library preparation methods was observed.

PCR amplification bias may influence the evenness among sequencing reads. For example, GC-rich primers and primers with GC-stretches at their 3′-end, both present in a random primer mix, may anneal more efficiently to a target sequence than AT-rich primer oligos do. As a result, amplicons amplified from such target sequences may be favored during the amplification process what in turn leads to high numbers of identical or related DNA molecules. We therefore clustered all quality-trimmed sequencing reads with a 90% cut-off to remove this redundancy. LASL libraries produced the lowest redundancy (41.7 to 60.7% relative proportion of clusters to sequencing read numbers), with significant differences not only to non-amplified libraries (10.8 to 21.3 relative proportion of clusters to sequencing read numbers) but MDA libraries (9.1 to 17.0% relative proportion of clusters to sequencing read numbers) and SISPA libraries (5.9 to 7.4 relative proportion of clusters to sequencing read numbers) (Table 1). These data suggest an amplification bias during PCR with random primer oligomers. However, MDA libraries (together with LASL libraries) still resulted highest average numbers of read clusters (Table S1). The presence of many repetitive and homopolymeric sequencing reads (possibly sequencing artefacts) may explain the low proportion of clustered reads in NASL libraries.

3.2. Data Set Comparison

We used Nonpareil, i.e., a kmer based approach that examines the degree of overlap among individual sequence reads, to determine redundancy [53–55] among the individual reads to further assess the average coverage created from the different library data sets. NASL, SISPA and MDA libraries seem to reach a nearly full coverage while LASL libraries vary between ~20 to 80% coverage (Figure 1). However, diversity among libraries increased from NASL to SISPA, MDA and LASL, the latter being the most diverse libraries (Figure 1). These results strongly indicate the target discrimination of SISPA and MDA during PCR that results in uneven coverage of the viral metagenomes and in addition, may fail to target low abundant sequences. NASL sequencing reads dominantly consisted of repetitive and homopolymeric sequences (see also below), with most likely too low an input of DNA explaining the observed Nonpareil curve for these libraries.

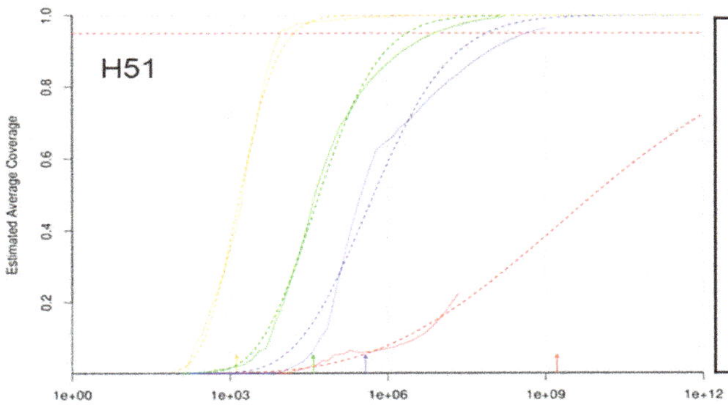

	coverage	Actual effort	Required effort	diversity
NASL	98.66%	35.07 Mbp	19.71 Kbp	7.352
LASL	29.16%	44.06 Mbp	21.87 Gbp	18.95
SISPA	98.71%	216.Mbp	10.81 Mbp	14.87
MDA	97.28%	1.298 Gbp	156.8 Mbp	16.92

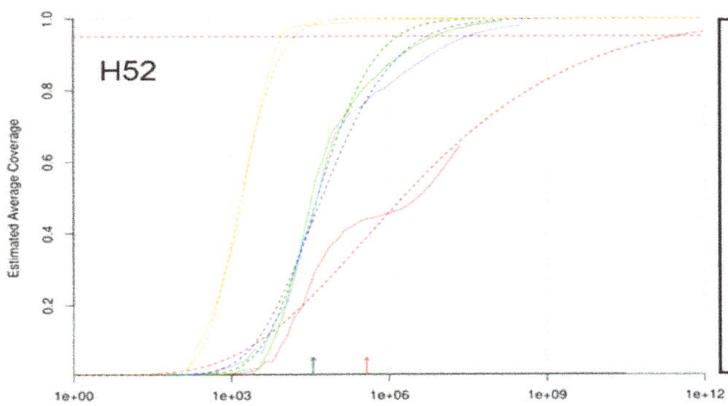

	coverage	Actual effort	Required effort	diversity
NASL	99.91%	35.34 Mbp	18.82 Mbp	7.316
LASL	69.48%	46.11Mbp	1.909 Gbp	15.77
SISPA	97.90%	204.9 Mbp	9.335 Mbp	14.8
MDA	98.58%	432.2 Mbp	33.68 Mbp	15.63

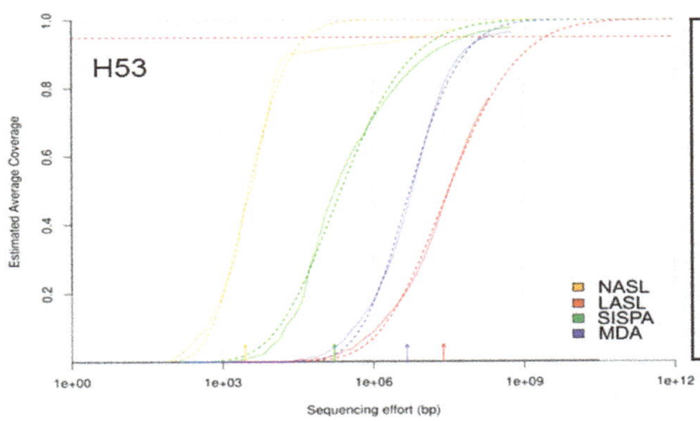

	coverage	Actual effort	Required effort	diversity
NASL	96.70%	56.78 Mbp	50.47 kbp	8.064
LASL	81.05%	266.7 Mbp	844.3 Mbp	17.97
SISPA	98.37%	763.5 Mbp	45.64 Mbp	16.18
MDA	97.72%	751.9 Mbp	189.4 Mbp	17.37

Figure 1. Comparison of Hainich groundwater viromes diversity and coverage as function of sequencing effort using Nonpareil curves [53–55]. Estimated coverage is shown as dotted lines, true coverage as solid lines. Estimated diversity is shown with arrows on the *x*-axis. Horizontal dotted line shows 95% coverage. Viral metagenome coverage, actual sequencing effort, required sequencing effort and kmer-based diversity for each library are shown in the right panel.

We were further interested in both, the number of viral reads that were exclusively detected by one of the library preparation methods and those reads that were identified from more than one library preparation method. For this, redundancy removed reads (i.e., reads that clustered at 90% identity) of LASL, SISPA and MDA libraries were independently mapped to viral contigs per individual sampling site (i.e., all viral contigs that were identified by the three virus identification tools and cross-assemblies) and counted. MDA libraries produced most reads (average: 350 k) followed by LASL (average: 143 k). Least reads were identified from SISPA libraries (average: 64 k). Overlapping information (reads found in more than one library) was rather low with 0.27 to 4.64% of reads present in all three libraries while 0.07 to 14.77% of reads were identified by two libraries (Figure 2). These data indicate target sequence discrimination between each of the library preparation methods.

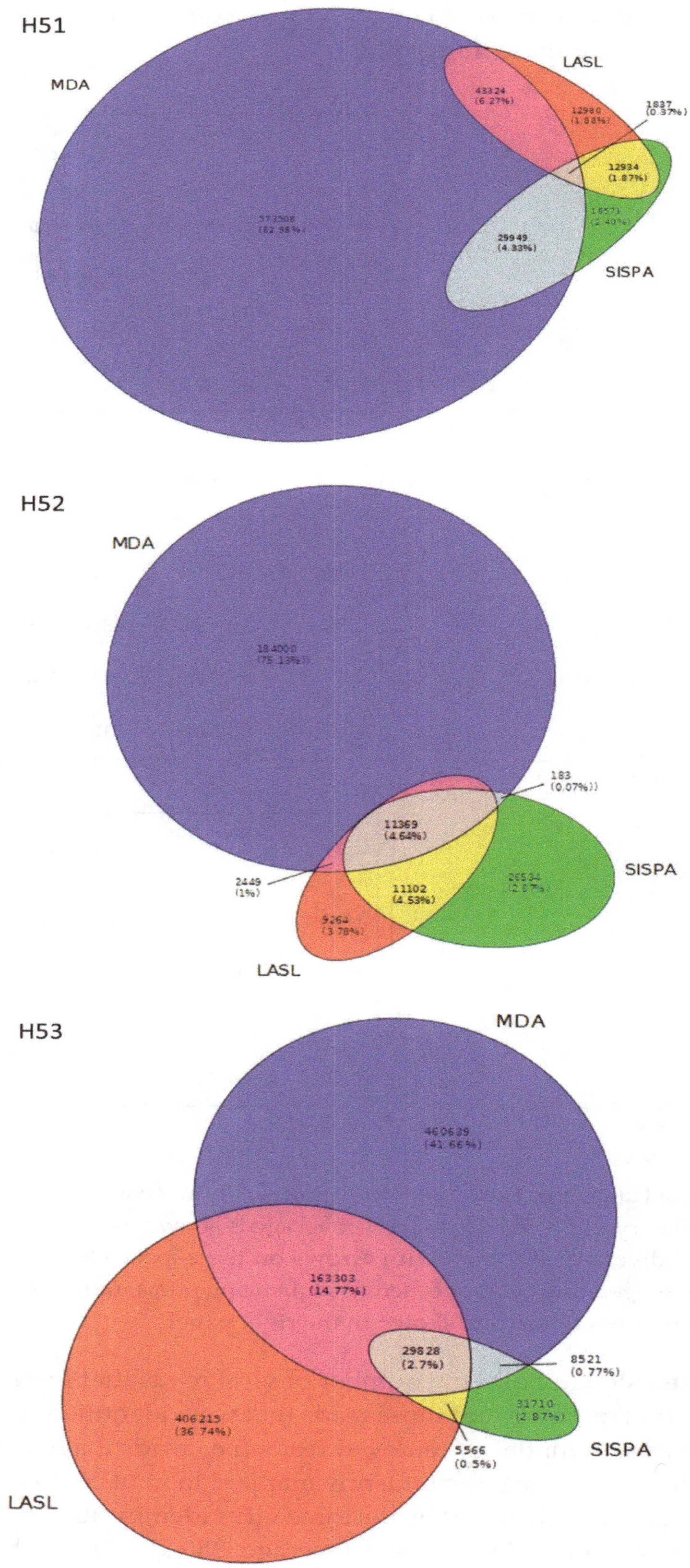

Figure 2. Overlap of sequencing read cluster (90% identity) information identified by library preparation methods, independently shown for each sampling site. Non amplified sequencing libraries (NASL) were not included in the analysis due to the homopolymeric and repetitive nature of sequences obtained from these libraries.

3.3. *Assembly Statistics and Evaluation of Viral Identification Tools*

Using both, SPAdes and SOAPdenovo-Trans, assemblies from non-amplified libraries completely failed due to repetitive and homopolymeric sequences. We therefore excluded these datasets from further analysis. Contig numbers tend to be higher for LASL and MDA libraries than for SISPA libraries (LASL-SISPA: $p = 0.062$, SISPA-MDA: $p = 0.087$; statistical test: one-way ANOVA) when assembled with SPAdes. Similar results were observed for SOAPdenovo-Trans assemblies (LASL-SISPA: $p = 0.052$, SISPA-MDA: $p = 0.028$; statistical test: one-way ANOVA). In addition, MDA library assemblies produced longer contigs (N50) when compared to LASL and SISPA ($p = 0.001$ (SPAdes), $p < 0.001$ (SOAPdenovo-Trans); statistical test: one-way ANOVA) (Figure 3, Table S2). A comparison (student's t-test) of the two assembly tools showed that SOAPdenovo-Trans may tend to produce longer contigs ($p = 0.084$), while there is no significant difference in the average contig size (N50) ($p = 0.2972$).

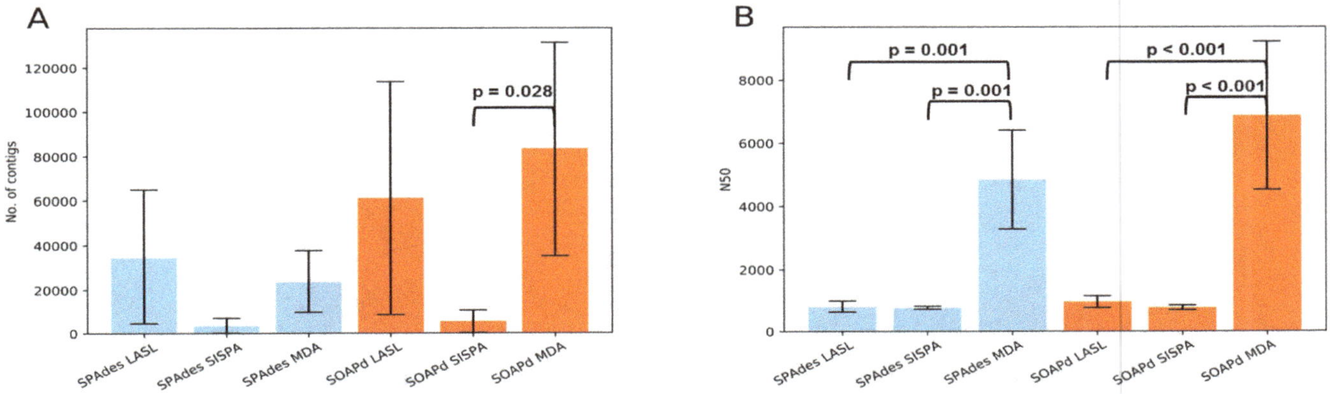

Figure 3. Number of contigs (**A**) and N50 (**B**) produced by sequence library preparation methods and assembly tools. Differences between library preparation methods were tested using analysis of variance (ANOVA). SOAPd: SOAPdenovo-Trans, NASL: non-amplified shotgun library, LASL: linker amplification shotgun libraries, SISPA: single-primer amplification, MDA: multiple displacement amplification.

Viral contigs (as identified by VirSorter, VirFinder and VrAP obtained from cross-assemblies) were clustered at 95% identity to determine a core set of sequences among the sampling sites. Only 37 contigs (0.5%) were shared by the three viromes indicating there is at least a minor common core set in the groundwater aquifers. The amount of shared contigs increased from 0.85% (H51 and H52) and 1.04% (H52 and H53) to 2.85% (H51 and H53) when two viromes were compared. However, the majority of viral contigs is exclusive for the respective virome (Figure 4). The overall viral contig number from H52 is rather low compared to H51 and H53, most likely due to the lower amount of DNA extracted from this sample. This might explain the lower contig overlap of H52 with H51 and H53, respectively, than the overlap of H51 and H53.

We used three different viral sequence identification tools that are based on the detection of viral hallmark genes (VirSorter), kmer distribution (VirFinder) and orf density (VrAP) (see more detailed description in the Materials and Methods section). VirFinder and VrAP significantly identified a higher number of viral contigs than VirSorter (One-way ANOVA $p < 0.001$). The size of viral contigs obtained by VirSorter and VirFinder were in contrast significantly higher than for VrAP (one-way ANOVA $p < 0.05$). However, each tool identified viral contigs that were not recognized by the other two revealing an advantage in the use of several identification tools for the recovery of viral sequences.

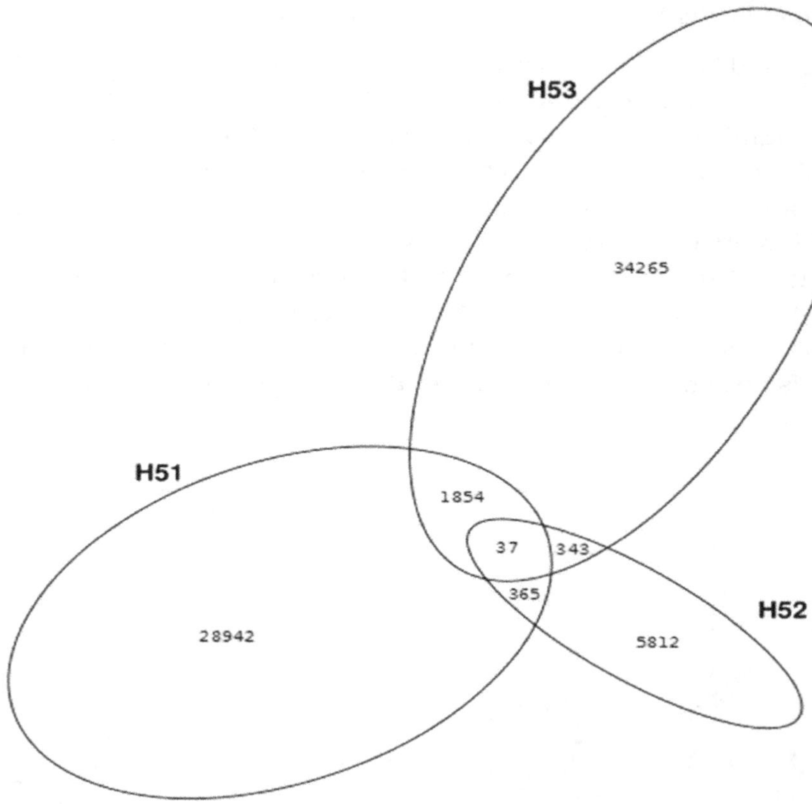

Figure 4. The venn diagram presents numbers of unique and shared viral contigs among the different viromes. Cross-assembled viral contigs (as identified by VirSorter, VirFinder and VrAP) were compared between sites by an all-versus-all clustering approach (95% identity) with CD-hit-est-2D [44].

3.4. First Insights into Viral Taxonomic Composition of Hainich Groundwater

Using cross-assembled contigs (assemblies including sequencing reads from LASL, SISPA and MDA per sampling site) and a set of three viral sequence recovery tools, we identified 27,173 (H51), 5,886 (H52) and 32,613 (H53) viral contigs from the Hainich groundwater samples (Figure 3; Table S3). These contigs were assembled from 31.19% (H51), 52.08% (H52) and 28.41% (H53) of the quality trimmed sequence reads. Among these reads we identified 19 Small subunit ribosomal ribonucleic acid sequences (8 bacterial 16S, 11 unclassified) demonstrating a low contamination with DNA from cellular organisms. Only 14.81% (H51), 18,65% (H52) and 11.92% (H53) of the viral populations could be assigned to taxonomy using delta-blast (Figure 5A). Most of them were assigned to dsDNA viruses dominated by the order *Caudovirales* (H51: 69.12%, H52: 58.20%, H53: 62.40%). Within the *Caudovirales*, members of the *Myoviridae* (40 to 46.5%) and *Siphoviridae* (40.7 to 41.9%) families were most abundant (Figure 5B, Figure S1). These findings are not surprising since *Caudovirales* have previously been presented as the most abundant group of viruses in environmental ecosystems [8,66,67]. Other identified dsDNA virus sequences belonged, for example, to the amoeba infecting giant virus families *Marseilleviridae* and *Mimiviridae*, to the algae infecting *Phycodnaviridae* family whose hosts has been shown to be present in groundwater [68], and invertebrate-infecting viruses such as *Iridoviridae* and *Poxviridae* (Figure 5B). Surprisingly, we identified only a small number of circular ssDNA viruses (Figure 5B). These viruses have been revealed as an abundant group in other environments [69,70]. We used Phi29 polymerase in MDA that preferentially amplifies circular ssDNA [35] and one could expect a bias towards overrepresentation of circular ssDNA genomes. Although this study is only a first snapshot into the Hainich groundwater virome we speculate that circular ssDNA viruses are rare in this environment. A small fraction of these DNA viromes was assigned to RNA viruses, most likely due to PCR errors and incomplete/erroneous virus reference databases.

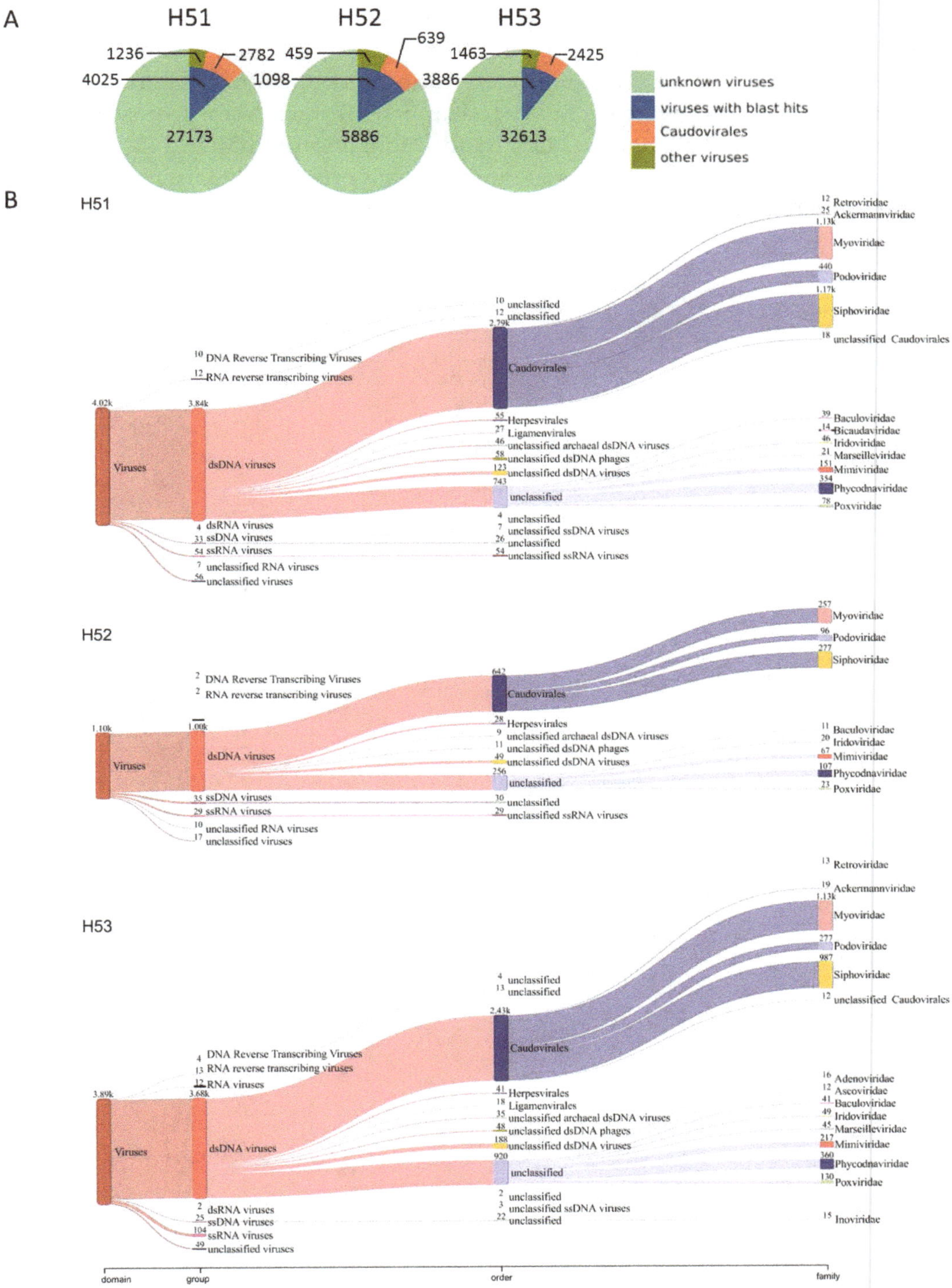

Figure 5. Taxonomic assignment of viral contigs identified from cross-assemblies. (**A**) pie charts present relative and absolute abundance of viral contigs after blastp analysis. (**B**) Taxonomic profile of viral contigs as classified by blastp (viral contigs with blast hits in figure **A**). Data were visualized with Pavian [71].

However, a high number of blast-based taxonomy assigned contigs could not be affiliated to deeper taxonomic levels but have similarity to unclassified viruses present in the viral RefSeq database (Figure 5B). These findings, together with the huge number of unknown viral contigs (without any blast hit) reveal substantial genomic and taxonomic diversity in Hainich groundwater viromes, as observed also in other environments [66,72]. To further investigate the similarity of Hainich groundwater viromes to viral RefSeq database, we used a genome-based network analysis of their shared protein content (Figure 6) [61,62]. This analysis groups viral contigs at the approximately genus level into viral clusters [61,62,73]. In total, 539 viral clusters were identified. Of those, Hainich viral contigs were found in 191 clusters, 183 of them were exclusive to Hainich viromes, among those 95 clusters exclusive to H51, 8 clusters to H52 and 23 clusters to H53. In addition, approximately 34% (H51), 64% (H52) and 63% (H53) of viral protein clusters were present in at least one other Hainich groundwater sample, suggesting some sequence conservation across these samples.

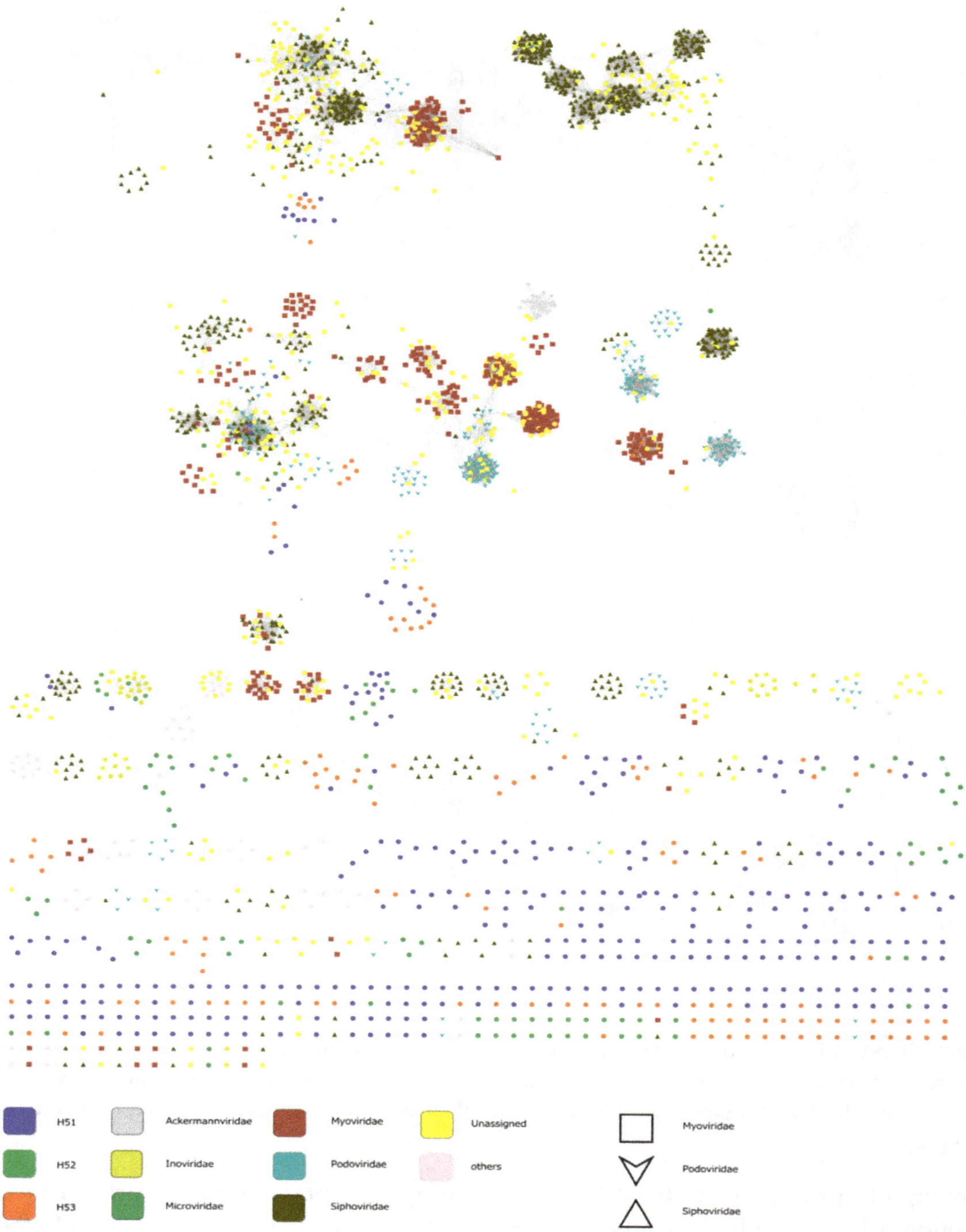

Figure 6. A network analysis of shared predicted protein content between viral RefSeq database and Hainich viral populations. Nodes (circles) indicate contigs and shared edges (lines) indicate shared protein content. Data were analysed using vConTACT2 [61,62] and displayed with cytoscape [64].

4. Discussion

Viruses play a key role in ecosystems, with most of them infecting microbes. They directly affect their hosts by lysis and horizontal gene transfer, and hence are responsible for changes in microbial community structure and composition what in turn has consequences on biogeochemical cycles and food web structures [4–6,74]. Viral metagenomics has been increasingly used to unravel viral community composition and interactions with their hosts from different ecosystems, such as marine environments and soil [66,67,72]. The terrestrial subsurface including groundwater ecosystems is at present yet underexplored [7,8,10,11,75]. A common problem is the relatively low biomass present in these difficult to obtain samples, which in return, results in only low amounts of DNA not sufficient for standard preparation of metagenome sequencing libraries [9,13,14]. Efforts have been undertaken to overcome this problem, including DNA enrichment using different DNA amplification techniques [17–19,22]. Each of these methods has its own advantages and limitations making it difficult to provide a standard protocol. Benchmark tests should therefore be performed when investigating new sample types.

Sampling procedure, virus particle isolation and nucleic acid extraction protocols are potential sources of bias [17] that have to be considered prior to sampling. Here, we focused on non-enveloped DNA viruses that passed a pore size of 200 nm after filtration and performed a benchmark study to find a method of choice to enrich viral DNA that is sufficient for sequencing. We furthermore intended to get a first snapshot of the viruses present in Hainich groundwater aquifers.

We used three DNA amplification methods, i.e., LASL, SISPA and MDA to compare one another and with NASL, using three groundwater samples. Although NASL resulted in some sequencing output none of the reads could be used for further analysis (assembly, virus sequence identification) due to their repetitive and homopolymeric nature; demonstrating that direct sequencing of NASL is not feasible with low DNA amounts. According to the Nonpareil curves, LASL was the method with the lowest amplification bias since the curves were located rightward in the plots indicating a higher diversity than for SISPA and MDA (Figure 1). Nonpareil curves for SISPA and MDA simulate a nearly full sequence coverage that emerge from redundant sequence information (Table 1, Table S1). False sequence coverage interpretation could be a result when data analysis exclusively rely on these library preparation methods. In addition, LASL resulted in the highest number of unique sequencing reads as compared to SISPA and MDA. MDA on the other hand outperformed LASL and SISPA in terms of viral contig numbers and their average contig size. In addition, MDA performed (at least in two samples) much better for taxonomic assignments in the case of *Caudovirales* families, which were dominant among the viral contigs with taxonomic affiliation (Figure S1). Considering the amount of unique viral reads per method and their low overlap (Figure 2), together with the results from cross-assemblies, it became apparent that none of the here tested DNA enrichments methods could completely detect viral sequences from pristine groundwater. However, SISPA even underperformed in terms of sequencing output, diversity and assembly statistics. Metagenomic benchmark studies using both, microbial mock communities and marine samples demonstrated the use of Mondrian and Illumina Nextera XT technologies produced high quality metagenomes from even femtogram-input DNA libraries [36,37]. These library preparation methods are comparable with the LASL protocol used in this study because all these methods use linker ligation on fragmented or tagmented DNA prior to amplification for generation of sequencing libraries. The low bias introduced by LASL on virus enriched groundwater samples from our work is consistent with these previous studies on prokaryotic metagenomes. In addition, other studies on viromes from marine and human samples showed substantial differences with respect to diversity, assembly output, types and ratio of viral sequences between LASL and MDA [18] and an outperformance of MDA over SISPA [17]. However, these studies observed an overrepresentation of circular sequences in MDA libraries as compared to LASL and SISPA. In contrast, our data identified only a few contigs that belong to circular ssDNA viruses (see also discussion below). We therefore suggest the combined use of LASL and MDA for future analysis of viral communities from pristine groundwater aquifers.

SOAPdenovo-Trans produced more contigs than SPAdes. However, average contig size was similar (Figure 3, Table S2). A combination of the assembly output seems to produce most comprehensive results but might also introduce unnecessary redundancy. Assembly for metagenomic data is already difficult, but appear to be more complex for viruses with their possibly more uneven genome coverage. Specialized tools are needed for the (de novo) assembly of viral sequences from metagenomic data [76]. The lower number of contigs for H52 could be a result of the lower amount of DNA extracted from this sample as compared to H51 and H53. Future studies will reveal whether there is a correlation between input DNA amount and contig numbers, including replicates and different yields of DNA input.

There is a high number of virus identification tools available, with all of them having their limitations [77]. We decided to use VirSorter [49], VirFinder [50] and VrAP. The latter two do not rely on database matches, increasing the chance to detect novel viruses not related to those present in public databases. Using our dataset, each tool exclusively identified some viral contigs demonstrating a combination of different virus identification tools increases the number of recovered viral contigs as also suggested previously [59,78]. However, the number of viral contigs was lower than the total number of contigs (compare Table S2 and Table S3). The experimental procedure included several steps to enrich virus particles, i.e., size filtration, chloroform treatment to remove most small-sized bacteria and digestion of free DNA that is not protected by a protein shell. Although some non-viral sequences might still be present after such methodology, one could assume the majority of the dataset consists of viral sequences and consequently includes a high number of viral contigs not recovered by one of the detection tools. Efforts should be undertaken, e.g., using machine learning, to overcome these likely limitations [78,79].

Like in many environmental studies, the taxonomy of most viral contigs remained unknown as demonstrated by blast and network analysis (Figure 5, Figure 6) [8,10,66]. Members of the order *Caudovirales* were dominating among viral contigs with taxonomic assignment. This group of tailed viruses infects a wide variety of bacteria and has been shown as one major group present in environmental ecosystems [8,66,72,80,81]. Another group of commonly highly abundant viruses, i.e., circular ssDNA viruses of the families *Microviridae* and *Circoviridae* [69,70,82], were almost entirely absent in our dataset. This is in contrast to previous results from groundwater aquifers where these viruses even dominated over dsDNA viruses among the classified sequences [10]. A technical bias seems to be unlikely since MDA is known for preferential amplification of these target sequences [35]. Future analyses including spatial and temporal variation will reveal whether these viruses are rare in pristine groundwater. We further identified viruses infecting algae, invertebrates and microeukaryotes, among the latter, contigs similar to giant viruses from the *Mimiviridae* family. These viruses should, by default, not be detected after 200 nm pore size filtration. A possible explanation could be sequence similarity of conserved mimivirus orfs, such as polymerases, to yet unknown viruses [83,84].

We show viral metagenome libraries can be produced from pristine aquifer groundwaters and suggest a combination of LASL and MDA to enrich viral DNA from these samples and to diminish an amplification bias that may occur during enrichment. We further identified new viral sequences that will help to understand the role of viruses in pristine groundwaters.

Supplementary Materials:
Figure S1: Bubble plot shows families of the Caudovirales order present in each virome., Table S1: Overview of raw sequencing read pair numbers, sequence quality and sequencing read clusters for each prepared sequencing library., Table S2: Assembly statistics for LASL, SISPA and MDA libraries per sampling site., Table S3: Overview of identified viral contigs as per virus identification tool, assembly software, sequencing library and sampling site., Supplementary information: Python script to assign orfs to contigs.

Author Contributions: Designed the experiments: R.K., M.H., M.M. and A.C.; performed the experiments: R.K. and M.H.; data analysis: R.K., M.H., R.B.T., U.N.d.R. and J.A.; prepared the manuscript: R.K., M.H., U.N.d.R., M.M. and A.C. All authors read and approved the final manuscript.

Acknowledgments: We are very grateful to Nicole Steinbach Anett Heidtmann for excellent technical assistance, to Robert Lehmann, Friedrich-Schiller-Universität Jena, who helped with sampling and to Ben Bolduc, Ohio State University, for his help in using vConTACT2. We further thank Ivonne Görlich and Marco Groth from the Core

Facility DNA sequencing of the Leibniz Institute on Aging - Fritz Lipmann Institute in Jena for their help with DNA sequencing. MH appreciates the support of the Joachim Herz Foundation by the add-on fellowship for interdisciplinary life science.

References

1. Danielopol, D.L.; Pospisil, P.; Rouch, R. Biodiversity in groundwater: A large-scale view. *Trends Ecol. Evol.* **2000**, *15*, 223–224. [CrossRef]
2. Griebler, C.; Avramov, M. Groundwater ecosystem services: A review. *Freshw. Sci.* **2015**, *34*, 355–367. [CrossRef]
3. Griebler, C.; Lueders, T. Microbial biodiversity in groundwater ecosystems. *Freshw. Biol.* **2009**, *54*, 649–677. [CrossRef]
4. Suttle, C.A. Viruses in the sea. *Nature* **2005**, *437*, 356. [CrossRef] [PubMed]
5. Suttle, C.A. Marine viruses — major players in the global ecosystem. *Nat. Rev. Microbiol.* **2007**, *5*, 801. [CrossRef] [PubMed]
6. Breitbart, M. Marine Viruses: Truth or Dare. *Annu. Rev. Mar. Sci.* **2011**, *4*, 425–448. [CrossRef]
7. Daly, R.A.; Borton, M.A.; Wilkins, M.J.; Hoyt, D.W.; Kountz, D.J.; Wolfe, R.A.; Welch, S.A.; Marcus, D.N.; Trexler, R.V.; MacRae, J.D.; et al. Microbial metabolisms in a 2.5-km-deep ecosystem created by hydraulic fracturing in shales. *Nat. Microbiol.* **2016**, *1*, 16146. [CrossRef]
8. Daly, R.A.; Roux, S.; Borton, M.A.; Morgan, D.M.; Johnston, M.D.; Booker, A.E.; Hoyt, D.W.; Meulia, T.; Wolfe, R.A.; Hanson, A.J.; et al. Viruses control dominant bacteria colonizing the terrestrial deep biosphere after hydraulic fracturing. *Nat. Microbiol.* **2019**, *4*, 352–361. [CrossRef]
9. Kyle, J.E.; Eydal, H.S.C.; Ferris, F.G.; Pedersen, K. Viruses in granitic groundwater from 69 to 450 m depth of the Äspö hard rock laboratory, Sweden. *ISME J.* **2008**, *2*, 571. [CrossRef]
10. Smith, R.J.; Jeffries, T.C.; Roudnew, B.; Seymour, J.R.; Fitch, A.J.; Simons, K.L.; Speck, P.G.; Newton, K.; Brown, M.H.; Mitchell, J.G. Confined aquifers as viral reservoirs. *Environ. Microbiol. Rep.* **2013**, *5*, 725–730. [CrossRef]
11. Pan, D.; Watson, R.; Wang, D.; Tan, Z.H.; Snow, D.D.; Weber, K.A. Correlation between viral production and carbon mineralization under nitrate-reducing conditions in aquifer sediment. *ISME J.* **2014**, *8*, 1691–1703. [CrossRef]
12. Wooley, J.C.; Ye, Y. Metagenomics: Facts and Artifacts, and Computational Challenges. *J. Comput. Sci. Technol.* **2009**, *25*, 71–81. [CrossRef] [PubMed]
13. Wilhartitz, I.C.; Kirschner, A.K.T.; Brussaard, C.P.D.; Fischer, U.R.; Wieltschnig, C.; Stadler, H.; Farnleitner, A.H. Dynamics of natural prokaryotes, viruses, and heterotrophic nanoflagellates in alpine karstic groundwater. *Microbiol. Open* **2013**, *2*, 633–643. [CrossRef] [PubMed]
14. Roudnew, B.; Lavery, T.J.; Seymour, J.R.; Smith, R.J.; Mitchell, J.G. Spatially varying complexity of bacterial and virus-like particle communities within an aquifer system. *Aquat. Microb. Ecol.* **2013**, *68*, 259–266. [CrossRef]
15. Ruby, J.G.; Bellare, P.; Derisi, J.L. PRICE: Software for the targeted assembly of components of (Meta) genomic sequence data. *G3 Bethesda Md* **2013**, *3*, 865–880. [CrossRef] [PubMed]
16. Rose, R.; Constantinides, B.; Tapinos, A.; Robertson, D.L.; Prosperi, M. Challenges in the analysis of viral metagenomes. *Virus Evol.* **2016**, *2*, vew022. [CrossRef] [PubMed]
17. Parras-Moltó, M.; Rodríguez-Galet, A.; Suárez-Rodríguez, P.; López-Bueno, A. Evaluation of bias induced by viral enrichment and random amplification protocols in metagenomic surveys of saliva DNA viruses. *Microbiome* **2018**, *6*, 119. [CrossRef] [PubMed]
18. Kim, K.-H.; Bae, J.-W. Amplification methods bias metagenomic libraries of uncultured single-stranded and double-stranded DNA viruses. *Appl. Environ. Microbiol.* **2011**, *77*, 7663–7668. [CrossRef] [PubMed]
19. Breitbart, M.; Salamon, P.; Andresen, B.; Mahaffy, J.M.; Segall, A.M.; Mead, D.; Azam, F.; Rohwer, F. Genomic analysis of uncultured marine viral communities. *Proc. Natl. Acad. Sci. USA* **2002**, *99*, 14250–14255. [CrossRef] [PubMed]
20. Thurber, R.V.; Haynes, M.; Breitbart, M.; Wegley, L.; Rohwer, F. Laboratory procedures to generate viral metagenomes. *Nat. Protoc.* **2009**, *4*, 470–483. [CrossRef]
21. Froussard, P. A random-PCR method (rPCR) to construct whole cDNA library from low amounts of RNA. *Nucleic Acids Res.* **1992**, *20*, 2900. [CrossRef] [PubMed]

22. Djikeng, A.; Halpin, R.; Kuzmickas, R.; Depasse, J.; Feldblyum, J.; Sengamalay, N.; Afonso, C.; Zhang, X.; Anderson, N.G.; Ghedin, E.; et al. Viral genome sequencing by random priming methods. *BMC Genom.* **2008**, *9*, 5. [CrossRef] [PubMed]
23. Dean, F.B.; Nelson, J.R.; Giesler, T.L.; Lasken, R.S. Rapid amplification of plasmid and phage DNA using Phi 29 DNA polymerase and multiply-primed rolling circle amplification. *Genome Res.* **2001**, *11*, 1095–1099. [CrossRef] [PubMed]
24. Angly, F.E.; Felts, B.; Breitbart, M.; Salamon, P.; Edwards, R.A.; Carlson, C.; Chan, A.M.; Haynes, M.; Kelley, S.; Liu, H.; et al. The marine viromes of four oceanic regions. *PLoS Biol.* **2006**, *4*, e368. [CrossRef] [PubMed]
25. Henn, M.R.; Sullivan, M.B.; Stange-Thomann, N.; Osburne, M.S.; Berlin, A.M.; Kelly, L.; Yandava, C.; Kodira, C.; Zeng, Q.; Weiand, M.; et al. Analysis of high-throughput sequencing and annotation strategies for phage genomes. *PloS ONE* **2010**, *5*, e9083. [CrossRef] [PubMed]
26. Solonenko, S.A.; Ignacio-Espinoza, J.C.; Alberti, A.; Cruaud, C.; Hallam, S.; Konstantinidis, K.; Tyson, G.; Wincker, P.; Sullivan, M.B. Sequencing platform and library preparation choices impact viral metagenomes. *BMC Genom.* **2013**, *14*, 320. [CrossRef] [PubMed]
27. Duhaime, M.B.; Deng, L.; Poulos, B.T.; Sullivan, M.B. Towards quantitative metagenomics of wild viruses and other ultra-low concentration DNA samples: A rigorous assessment and optimization of the linker amplification method. *Environ. Microbiol.* **2012**, *14*, 2526–2537. [CrossRef] [PubMed]
28. Székely, A.J.; Breitbart, M. Single-stranded DNA phages: From early molecular biology tools to recent revolutions in environmental microbiology. *FEMS Microbiol. Lett.* **2016**, *363*. [CrossRef]
29. Roux, S.; Solonenko, N.E.; Dang, V.T.; Poulos, B.T.; Schwenck, S.M.; Goldsmith, D.B.; Coleman, M.L.; Breitbart, M.; Sullivan, M.B. Towards quantitative viromics for both double-stranded and single-stranded DNA viruses. *PeerJ* **2016**, *4*, e2777. [CrossRef]
30. Drexler, J.F.; Corman, V.M.; Müller, M.A.; Maganga, G.D.; Vallo, P.; Binger, T.; Gloza-Rausch, F.; Cottontail, V.M.; Rasche, A.; Yordanov, S.; et al. Bats host major mammalian paramyxoviruses. *Nat. Commun.* **2012**, *3*, 796. [CrossRef]
31. Karlsson, O.E.; Belák, S.; Granberg, F. The Effect of Preprocessing by Sequence-Independent, Single-Primer Amplification (SISPA) on Metagenomic Detection of Viruses. *Biosecurity Bioterrorism Biodefense Strategy Pract. Sci.* **2013**, *11*, S227–S234. [CrossRef] [PubMed]
32. Blanco, L.; Bernad, A.; Lázaro, J.M.; Martín, G.; Garmendia, C.; Salas, M. Highly efficient DNA synthesis by the phage phi 29 DNA polymerase. Symmetrical mode of DNA replication. *J. Biol. Chem.* **1989**, *264*, 8935–8940. [PubMed]
33. Lasken, R.S.; Stockwell, T.B. Mechanism of chimera formation during the Multiple Displacement Amplification reaction. *BMC Biotechnol.* **2007**, *7*, 19. [CrossRef] [PubMed]
34. Zhang, K.; Martiny, A.C.; Reppas, N.B.; Barry, K.W.; Malek, J.; Chisholm, S.W.; Church, G.M. Sequencing genomes from single cells by polymerase cloning. *Nat. Biotechnol.* **2006**, *24*, 680. [CrossRef] [PubMed]
35. Kim, K.-H.; Chang, H.-W.; Nam, Y.-D.; Roh, S.W.; Kim, M.-S.; Sung, Y.; Jeon, C.O.; Oh, H.-M.; Bae, J.-W. Amplification of uncultured single-stranded DNA viruses from rice paddy soil. *Appl. Environ. Microbiol.* **2008**, *74*, 5975–5985. [CrossRef] [PubMed]
36. Rinke, C.; Low, S.; Woodcroft, B.J.; Raina, J.-B.; Skarshewski, A.; Le, X.H.; Butler, M.K.; Stocker, R.; Seymour, J.; Tyson, G.W.; et al. Validation of picogram- and femtogram-input DNA libraries for microscale metagenomics. *PeerJ* **2016**, *4*, e2486. [CrossRef]
37. Bowers, R.M.; Clum, A.; Tice, H.; Lim, J.; Singh, K.; Ciobanu, D.; Ngan, C.Y.; Cheng, J.-F.; Tringe, S.G.; Woyke, T. Impact of library preparation protocols and template quantity on the metagenomic reconstruction of a mock microbial community. *BMC Genom.* **2015**, *16*, 856. [CrossRef]
38. Küsel, K.; Totsche, K.U.; Trumbore, S.E.; Lehmann, R.; Steinhäuser, C.; Herrmann, M. How Deep Can Surface Signals Be Traced in the Critical Zone? Merging Biodiversity with Biogeochemistry Research in a Central German Muschelkalk Landscape. *Front. Earth Sci.* **2016**, *4*, 32. [CrossRef]
39. Kohlhepp, B.; Lehmann, R.; Seeber, P.; Küsel, K.; Trumbore, S.E.; Totsche, K.U. Aquifer configuration and geostructural links control the groundwater quality in thin-bedded carbonate–siliciclastic alternations of the Hainich CZE, central Germany. *Hydrol. Earth Syst. Sci.* **2017**, *21*, 6091–6116. [CrossRef]
40. Kumar, S.; Herrmann, M.; Thamdrup, B.; Schwab, V.F.; Geesink, P.; Trumbore, S.E.; Totsche, K.-U.; Küsel, K. Nitrogen Loss from Pristine Carbonate-Rock Aquifers of the Hainich Critical Zone Exploratory (Germany) Is Primarily Driven by Chemolithoautotrophic Anammox Processes. *Front. Microbiol.* **2017**, *8*, 1951. [CrossRef]

41. Palacios, G.; Quan, P.-L.; Jabado, O.J.; Conlan, S.; Hirschberg, D.L.; Liu, Y.; Zhai, J.; Renwick, N.; Hui, J.; Hegyi, H. Panmicrobial oligonucleotide array for diagnosis of infectious diseases. *Emerg. Infect. Dis.* **2007**, *13*, 73. [CrossRef] [PubMed]
42. Bolger, A.M.; Lohse, M.; Usadel, B. Trimmomatic: A flexible trimmer for Illumina sequence data. *Bioinformatics* **2014**, *30*, 2114–2120. [CrossRef] [PubMed]
43. Li, W.; Godzik, A. Cd-hit: A fast program for clustering and comparing large sets of protein or nucleotide sequences. *Bioinformatics* **2006**, *22*, 1658–1659. [CrossRef] [PubMed]
44. Fu, L.; Niu, B.; Zhu, Z.; Wu, S.; Li, W. CD-HIT: Accelerated for clustering the next-generation sequencing data. *Bioinformatics* **2012**, *28*, 3150–3152. [CrossRef] [PubMed]
45. Bankevich, A.; Nurk, S.; Antipov, D.; Gurevich, A.A.; Dvorkin, M.; Kulikov, A.S.; Lesin, V.M.; Nikolenko, S.I.; Pham, S.; Prjibelski, A.D. SPAdes: A new genome assembly algorithm and its applications to single-cell sequencing. *J. Comput. Biol.* **2012**, *19*, 455–477. [CrossRef]
46. Nurk, S.; Meleshko, D.; Korobeynikov, A.; Pevzner, P.A. metaSPAdes: A new versatile metagenomic assembler. *Genome Res.* **2017**, *27*, 824–834. [CrossRef]
47. Xie, Y.; Wu, G.; Tang, J.; Luo, R.; Patterson, J.; Liu, S.; Huang, W.; He, G.; Gu, S.; Li, S. SOAPdenovo-Trans: De novo transcriptome assembly with short RNA-Seq reads. *Bioinformatics* **2014**, *30*, 1660–1666. [CrossRef]
48. Hölzer, M.; Marz, M. De novo transcriptome assembly: A comprehensive cross-species comparison of short-read RNA-Seq assemblers. *GigaScience* **2019**, *8*, giz039. [CrossRef]
49. Roux, S.; Enault, F.; Hurwitz, B.L.; Sullivan, M.B. VirSorter: Mining viral signal from microbial genomic data. *PeerJ* **2015**, *3*, e985. [CrossRef]
50. Ren, J.; Ahlgren, N.A.; Lu, Y.Y.; Fuhrman, J.A.; Sun, F. VirFinder: A novel k-mer based tool for identifying viral sequences from assembled metagenomic data. *Microbiome* **2017**, *5*, 69. [CrossRef]
51. Song, L.; Florea, L.; Langmead, B. Lighter: Fast and memory-efficient sequencing error correction without counting. *Genome Biol.* **2014**, *15*, 509. [CrossRef] [PubMed]
52. Magoč, T.; Salzberg, S.L. FLASH: Fast length adjustment of short reads to improve genome assemblies. *Bioinformatics* **2011**, *27*, 2957–2963. [CrossRef] [PubMed]
53. Rodriguez-r, L.M.; Konstantinidis, K.T. Nonpareil: A redundancy-based approach to assess the level of coverage in metagenomic datasets. *Bioinformatics* **2013**, *30*, 629–635. [CrossRef] [PubMed]
54. Rodriguez, L.M.; Konstantinidis, K.T. Estimating coverage in metagenomic data sets and why it matters. *ISME J.* **2014**, *8*, 2349. [CrossRef] [PubMed]
55. Rodriguez-R, L.M.; Gunturu, S.; Tiedje, J.M.; Cole, J.R.; Konstantinidis, K.T. Nonpareil 3: Fast Estimation of Metagenomic Coverage and Sequence Diversity. *MSystems* **2018**, *3*, e00039-18.
56. Langmead, B.; Salzberg, S.L. Fast gapped-read alignment with Bowtie 2. *Nat. Methods* **2012**, *9*, 357–359. [CrossRef]
57. Li, H.; Handsaker, B.; Wysoker, A.; Fennell, T.; Ruan, J.; Homer, N.; Marth, G.; Abecasis, G.; Durbin, R. The sequence alignment/map format and SAMtools. *Bioinformatics* **2009**, *25*, 2078–2079. [CrossRef] [PubMed]
58. RStudio Team. RStudio: Integrated Development for R. Available online: http://www.rstudio.com (accessed on 27 May 2019).
59. Hyatt, D.; Chen, G.-L.; LoCascio, P.F.; Land, M.L.; Larimer, F.W.; Hauser, L.J. Prodigal: Prokaryotic gene recognition and translation initiation site identification. *BMC Bioinformatics* **2010**, *11*, 119. [CrossRef]
60. Boratyn, G.M.; Schäffer, A.A.; Agarwala, R.; Altschul, S.F.; Lipman, D.J.; Madden, T.L. Domain enhanced lookup time accelerated BLAST. *Biol. Direct* **2012**, *7*, 12. [CrossRef]
61. Bolduc, B.; Jang, H.B.; Doulcier, G.; You, Z.-Q.; Roux, S.; Sullivan, M.B. vConTACT: An iVirus tool to classify double-stranded DNA viruses that infect Archaea and Bacteria. *PeerJ* **2017**, *5*, e3243. [CrossRef]
62. Bin Jang, H.; Bolduc, B.; Zablocki, O.; Kuhn, J.H.; Roux, S.; Adriaenssens, E.M.; Brister, J.R.; Kropinski, A.M.; Krupovic, M.; Lavigne, R.; et al. Taxonomic assignment of uncultivated prokaryotic virus genomes is enabled by gene-sharing networks. *Nat. Biotechnol.* **2019**. [CrossRef] [PubMed]
63. Bolduc, B.; Youens-Clark, K.; Roux, S.; Hurwitz, B.L.; Sullivan, M.B. iVirus: Facilitating new insights in viral ecology with software and community data sets imbedded in a cyberinfrastructure. *Isme J.* **2016**, *11*, 7. [CrossRef] [PubMed]
64. Shannon, P.; Markiel, A.; Ozier, O.; Baliga, N.S.; Wang, J.T.; Ramage, D.; Amin, N.; Schwikowski, B.; Ideker, T. Cytoscape: A software environment for integrated models of biomolecular interaction networks. *Genome Res.* **2003**, *13*, 2498–2504. [CrossRef] [PubMed]

65. Miller, C.S.; Baker, B.J.; Thomas, B.C.; Singer, S.W.; Banfield, J.F. EMIRGE: Reconstruction of full-length ribosomal genes from microbial community short read sequencing data. *Genome Biol.* **2011**, *12*, R44. [CrossRef]
66. Emerson, J.B.; Roux, S.; Brum, J.R.; Bolduc, B.; Woodcroft, B.J.; Jang, H.B.; Singleton, C.M.; Solden, L.M.; Naas, A.E.; Boyd, J.A.; et al. Host-linked soil viral ecology along a permafrost thaw gradient. *Nat. Microbiol.* **2018**, *3*, 870–880. [CrossRef] [PubMed]
67. Roux, S.; Brum, J.R.; Dutilh, B.E.; Sunagawa, S.; Duhaime, M.B.; Loy, A.; Poulos, B.T.; Solonenko, N.; Lara, E.; Poulain, J.; et al. Ecogenomics and potential biogeochemical impacts of globally abundant ocean viruses. *Nature* **2016**, *537*, 689. [CrossRef] [PubMed]
68. Reisser, W. The Hidden Life of Algae Underground. In *Algae and Cyanobacteria in Extreme Environments*; Seckbach, J., Ed.; Springer Netherlands: Dordrecht, The Netherlands, 2007; pp. 47–58, ISBN 978-1-4020-6112-7.
69. Rosario, K.; Duffy, S.; Breitbart, M. Diverse circovirus-like genome architectures revealed by environmental metagenomics. *J. Gen. Virol.* **2009**, *90*, 2418–2424. [CrossRef]
70. Tucker, K.P.; Parsons, R.; Symonds, E.M.; Breitbart, M. Diversity and distribution of single-stranded DNA phages in the North Atlantic Ocean. *Isme J.* **2010**, *5*, 822. [CrossRef]
71. Breitwieser, F.P.; Salzberg, S.L. Pavian: Interactive analysis of metagenomics data for microbiomics and pathogen identification. *bioRxiv* **2016**. [CrossRef]
72. Paez-Espino, D.; Eloe-Fadrosh, E.A.; Pavlopoulos, G.A.; Thomas, A.D.; Huntemann, M.; Mikhailova, N.; Rubin, E.; Ivanova, N.N.; Kyrpides, N.C. Uncovering Earth's virome. *Nature* **2016**, *536*, 425. [CrossRef]
73. Roux, S.; Hallam, S.J.; Woyke, T.; Sullivan, M.B. Viral dark matter and virus–host interactions resolved from publicly available microbial genomes. *eLife* **2015**, *4*, e08490. [CrossRef] [PubMed]
74. Wegner, C.-E.; Gaspar, M.; Geesink, P.; Herrmann, M.; Marz, M.; Küsel, K. Biogeochemical Regimes in Shallow Aquifers Reflect the Metabolic Coupling of the Elements Nitrogen, Sulfur, and Carbon. *Appl. Environ. Microbiol.* **2019**, *85*, e02346-18. [CrossRef] [PubMed]
75. Anderson, R.E.; Brazelton, W.J.; Baross, J.A. Is the genetic landscape of the deep subsurface biosphere affected by viruses? *Front. Microbiol.* **2011**, *2*, 219. [CrossRef] [PubMed]
76. Hölzer, M.; Marz, M. Chapter Nine—Software Dedicated to Virus Sequence Analysis "Bioinformatics Goes Viral." In *Advances in Virus Research*; Beer, M., Höper, D., Eds.; Academic Press: Cambridge, MA, USA, 2017; Volume 99, pp. 233–257, ISBN 0065-3527.
77. Nooij, S.; Schmitz, D.; Vennema, H.; Kroneman, A.; Koopmans, M.P.G. Overview of Virus Metagenomic Classification Methods and Their Biological Applications. *Front. Microbiol.* **2018**, *9*, 749. [CrossRef] [PubMed]
78. Hurwitz, B.L.; Ponsero, A.; Thornton, J.; U'Ren, J.M. Phage hunters: Computational strategies for finding phages in large-scale 'omics datasets. *Virus Res.* **2018**, *244*, 110–115. [CrossRef] [PubMed]
79. Bzhalava, Z.; Tampuu, A.; Bała, P.; Vicente, R.; Dillner, J. Machine Learning for detection of viral sequences in human metagenomic datasets. *BMC Bioinform.* **2018**, *19*, 336. [CrossRef] [PubMed]
80. Wommack, K.E.; Colwell, R.R. Virioplankton: Viruses in Aquatic Ecosystems. *Microbiol. Mol. Biol. Rev.* **2000**, *64*, 69. [CrossRef] [PubMed]
81. Hurwitz, B.L.; Sullivan, M.B. The Pacific Ocean Virome (POV): A Marine Viral Metagenomic Dataset and Associated Protein Clusters for Quantitative Viral Ecology. *PLoS ONE* **2013**, *8*, e57355. [CrossRef]
82. Roux, S.; Krupovic, M.; Poulet, A.; Debroas, D.; Enault, F. Evolution and diversity of the Microviridae viral family through a collection of 81 new complete genomes assembled from virome reads. *PLoS ONE* **2012**, *7*, e40418. [CrossRef]
83. Earl, P.L.; Jones, E.V.; Moss, B. Homology between DNA polymerases of poxviruses, herpesviruses, and adenoviruses: Nucleotide sequence of the vaccinia virus DNA polymerase gene. *Proc. Natl. Acad. Sci. USA* **1986**, *83*, 3659–3663. [CrossRef]
84. Villarreal, L.P.; DeFilippis, V.R. A Hypothesis for DNA Viruses as the Origin of Eukaryotic Replication Proteins. *J. Virol.* **2000**, *74*, 7079. [CrossRef] [PubMed]

13

Recent Developments and Prospects of M13-Bacteriophage based Piezoelectric Energy Harvesting Devices

In Woo Park, Kyung Won Kim, Yunhwa Hong, Hyun Ji Yoon, Yonghun Lee, Dham Gwak and Kwang Heo *

Department of Nanotechnology and Advanced Materials Engineering, Sejong University, Seoul 05006, Korea; lovepiw@naver.com (I.W.P.); saheon7@gmail.com (K.W.K.); monkey1434@gmail.com (Y.H.); hyunji9193@naver.com (H.J.Y.); alwayscoast@gmail.com (Y.L.); whalephant13@gmail.com (D.G.)

* Correspondence: kheo@sejong.ac.kr

Abstract: Recently, biocompatible energy harvesting devices have received a great deal of attention for biomedical applications. Among various biomaterials, viruses are expected to be very promising biomaterials for the fabrication of functional devices due to their unique characteristics. While other natural biomaterials have limitations in mass-production, low piezoelectric properties, and surface modification, M13 bacteriophages (phages), which is one type of virus, are likely to overcome these issues with their mass-amplification, self-assembled structure, and genetic modification. Based on these advantages, many researchers have started to develop virus-based energy harvesting devices exhibiting superior properties to previous biomaterial-based devices. To enhance the power of these devices, researchers have tried to modify the surface properties of M13 phages, form biomimetic hierarchical structures, control the dipole alignments, and more. These methods for fabricating virus-based energy harvesting devices can form a powerful strategy to develop high-performance biocompatible energy devices for a wide range of practical applications in the future. In this review, we discuss all these issues in detail.

Keywords: virus; M13 bacteriophage; energy generator; piezoelectric; self-assembly; genetic engineering

---

1. Introduction

Recently, energy harvesting from biomechanical movement has attracted a great deal of interest in wearable, sustainable, and biomedical technologies [1–3]. Especially, piezoelectric devices have been actively studied for bio-implantable application, because they can easily generate energy by using simple motions and vibrations without any other external source [4–6]. General piezoelectric materials have the ability to generate electrical charges from applied mechanical stress. Although the energy output from piezoelectric generators may not be as large as that from other alternative energy sources, these also have their own advantages, such as simple device structures and various material groups.

After Jacques and Pierre Curie discovered piezoelectricity from quartz in 1880, various piezoelectric materials were developed from ceramics to natural biomaterials. Among them, ceramic materials such as lithium niobate ($LiNbO_3$) [7], potassium niobate ($KNbO_3$) [7], lithium tantalate ($LiTaO_3$) [8], barium titanate ($BaTiO_3$) [9], lead zirconate titanate ($Pb[Zr_xTi_{1-x}]O_3$) [10], and etc., were most investigated due to their superior piezoelectric properties. Lead zirconate titanate (PZT) has become the most common piezoelectric material in practical application today. However, as the issue of toxicity in lead-containing devices is starting to appear, extensive study has been conducted to replace it with lead-free piezoelectric materials [11]. For this purpose, several ceramics (bismuth ferrite ($BiFeO_3$), sodium niobate ($NaNbO_3$), barium titanate ($BaTiO_3$), bismuth titanate ($Bi_4Ti_3O_{12}$), quartz, etc.) [12–14] and organic materials (polyvinylidene fluoride (PVDF), polyvinylidene chloride (PVDC), etc.) [15,16] were found. Particularly, single-crystal zinc oxide (ZnO) nanostructures with wurtzite structure exhibit larger piezoelectric constants than those of bulk ZnO. However, the poor biocompatibility and brittle characteristics of these piezoelectric materials limit their applications in wearable and biomedical applications [17].

Otherwise, biomaterials have been regarded as promising alternative materials due to their good biocompatibility, non-toxicity, and environmental friendliness. After the first discovery of the piezoelectric effect of bone, the piezoelectric properties of diverse natural biomaterials such as wood, bone, hair, dentin, tendon, and collage were investigated [18–23]. Since these natural materials exhibit very weak piezoelectricity and are difficult to mass-produce, there is a limit to use for practical applications. Recently, there has been a growing interest in the biopiezoelectric materials to overcome these limitations. Polysaccharide materials [24–27], viruses [28–31], and self-assembled biomaterials [32,33] have been identified as good candidates, because their piezoelectric constants are higher than those of previous natural biomaterials, and they are possible to mass-produce. Especially, it has been found their physical and chemical properties can be modulated by their morphology, surface charges, and phases, and the piezoelectric response is directly related to these. Furthermore, the discovery of piezoelectricity in bone [34–36], which has been the most frequently studied tissue, aroused great interest because it seemed to provide an important key to understanding bone physiology. Researchers hypothesized that bone's piezoelectric signal by physical stimulation could regulate bone growth, repair, wound healing, and tissue regeneration [37–39]. In addition, piezoelectric biomaterials also have several advantages for use in sensors [40], energy storage [41], energy harvesting, and other areas [42]. Despite these advantages of biopiezoelectric materials and their potential applications, a comprehensive review of virus-based piezoelectric energy harvesting devices have not been reported.

In this short review, we provide an overview of M13 bacteriophages (phages) as superior biopiezoelectric materials for piezoelectric energy harvesting applications. In addition, we discuss in detail the piezoelectric properties of M13 phages and the fabrication of M13 phage-based piezoelectric energy harvesting devices. It is expected that this review will inspire the design of novel biomaterials and the development of functional devices for energy harvesting, sensing, biomedical applications, and other applications.

2. Biological Building Block for Piezoelectric Energy Harvesting Devices: M13 Bacteriophages

Due to their unique structural, biological, and physical properties, the M13 phage is the most attractive candidate in biomaterials for mimicking natural structures and developing novel piezoelectric energy harvesting devices. Especially, from the engineering point of view, M13 phages have several advantages, such as (1) structural similarity with collagens; (2) mass-producibility by bacteria infection and mass-amplification; (3) surface tunability through genetic engineering; (4) possibility of forming a highly-ordered structure via self-assembly; and (5) superior piezoelectric properties.

The M13 phage is a filamentous bacteriophage composed of circular single-stranded deoxyribonucleic acid (ssDNA) and capsid proteins. ssDNA is encapsulated in approximately 2700 copies of the helically arranged major coat protein pVIII, and five to seven copies of two different minor coat proteins (pIX, pVI, pIII, pVII) on the ends (Figure 1a). The diameter and length of M13 phages are about 6.6 nm and 880 nm, respectively [43]. Since the structural characteristic of phages is very similar to the structure of the human collagen, the M13 phages are quite capable of mimicking nature's hierarchical structures based on collagen [44]. These M13 phages, which are perfectly identical copies, can be mass-produced using the living characteristic of viruses. M13 phages are any group of viruses which carry out a lysogenic infection in which the phage inserts its genome into the bacterial genome. The minor coat protein pIII attaches to the receptor of the host bacteria and infects the bacteria. A huge amount of phages can be produced in an infected bacteria using the metabolic reactions of the host cell (Figure 1b). The infected cells are not involved in the cell lysis, but a decrease in the rate of cell growth [45].

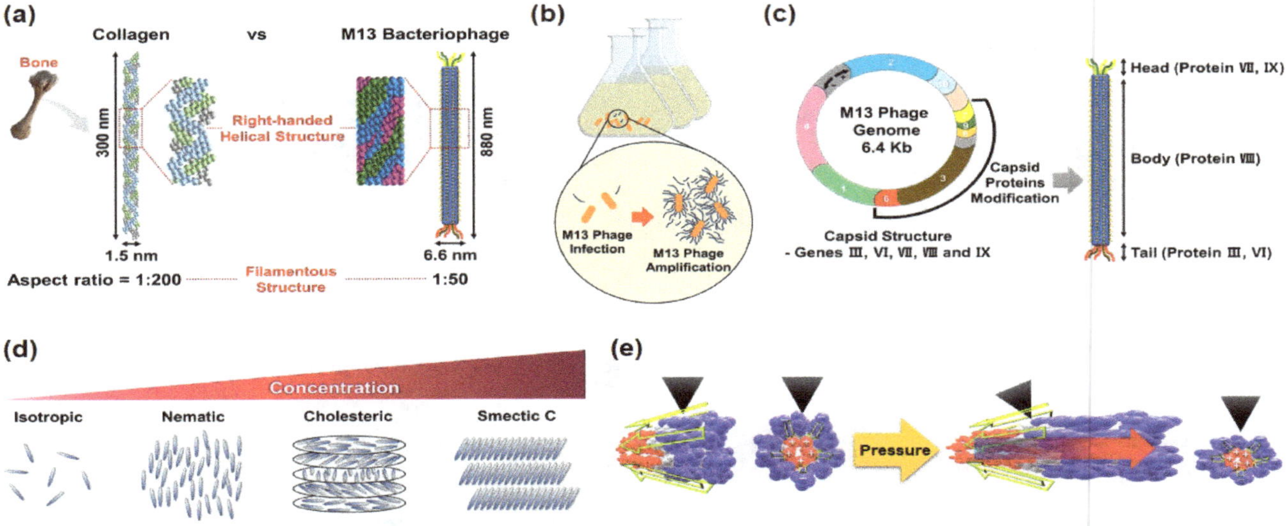

Figure 1. Schematic diagram showing the main characteristics of M13 bacteriophages (phages). (**a**) Structural similarity between natural collagen and M13 phages. Both biomaterials are filamentous structures with high aspect ratio and have right-handed helical structures. (**b**) Infection and mass-amplification of M13 phages. A huge amount of phages can be produced in infected bacteria by using the metabolic reactions of the host cell. (**c**) M13mp phage vector and surface modification through genetic engineering. We can design the molecular structures of the phage's outer surfaces in accordance with the desired properties and display the related peptide motif on the coat proteins of M13 phages. (**d**) Liquid crystalline phase transition of M13 phages. M13 phages exhibit a lyotropic liquid crystalline phase transition due to their helical structure, nanofibrous shape, monodispersity, and functional motifs. (**e**) Piezoelectric properties of M13 phages, which enable us to make piezoelectric energy harvesting devices.

Recent advances in genetic engineering make it possible to modulate the peptide sequence of phage proteins as desired. By using the recombinant DNA technique and M13mp phage vectors, we can design the molecular structures of surfaces according to the required properties and easily display the related peptide motif on the coat proteins of M13 phages (Figure 1c). This ability of M13 phages is a unique feature that distinguishes them from other nano and biomaterials [45].

In addition, M13 phages exhibit a lyotropic liquid crystalline phase due to their helical structure, nanofibrous shape, monodispersity, and expressed functional motifs (Figure 1d). According to the concentration of the phage suspension, the resulting structures of M13 phage films change from an isotropic phase to a cholesteric phase in a controlled manner [43]. Owing to these characteristics, we can prepare highly-ordered crystalline structures in a large area, which allows us to fabricate functional devices.

Lastly, recent studies have shown that M13 phages have excellent piezoelectric properties, which are larger than other natural biomaterials [28]. This makes it possible to fabricate high-performance piezoelectric energy harvesting devices (Figure 1e).

3. Introduction to Piezoelectric Effect

Piezoelectricity is a phenomenon of coupling between the electrical and mechanical states of a material by crystal deformation. When piezoelectric materials are mechanically stressed and deformed, the positive and negative charge centers shift in the materials, which then results in an external electrical field and a current flow. The opposite can also happen. When an electrical field is applied to the materials, the piezoelectric materials are stretched or compressed (Figure 2a).

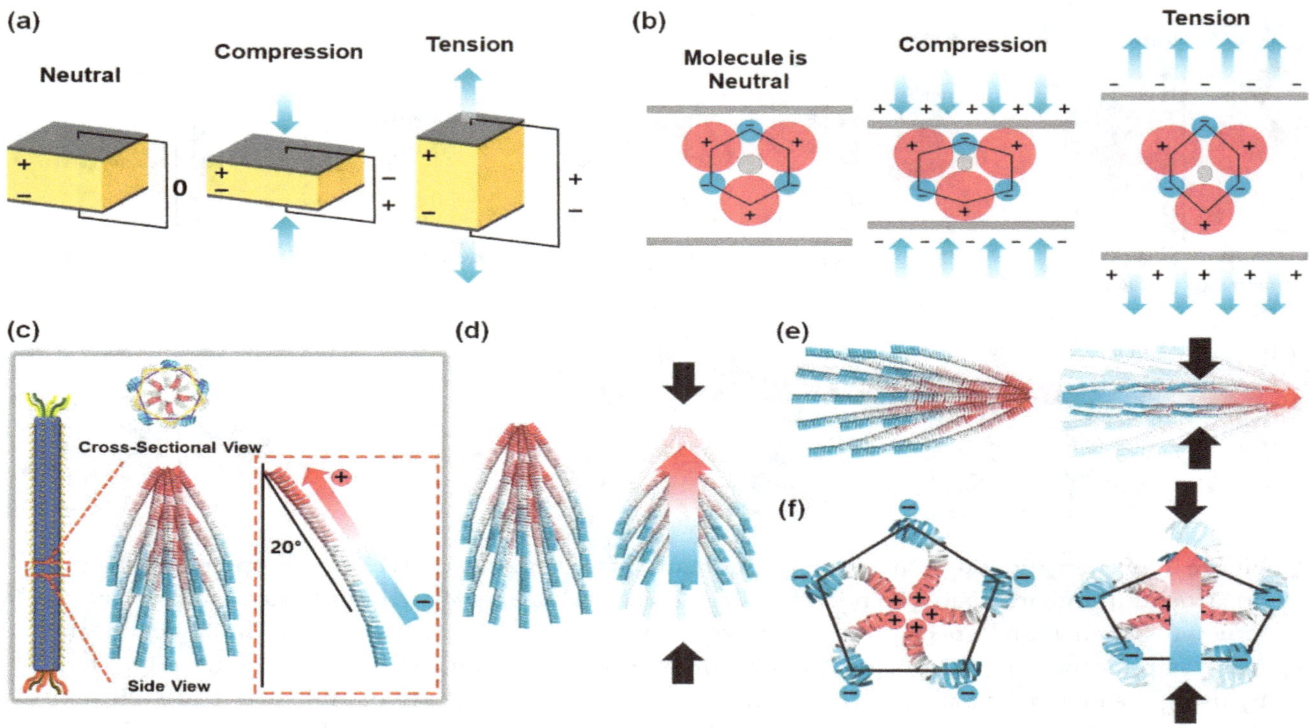

Figure 2. Schematic diagram showing the working mechanism of the piezoelectric effect. (**a**) Diagrammatic demonstration of the piezoelectric effect. (**b**) Schematic illustration showing the working mechanism of the piezoelectric effect in crystal structures. In equilibrium, the charges of the unit cell are neutral (no net dipole moment). When mechanical stresses are applied, a net dipole moment and electrical polarization arise in piezoelectric materials. (**c**) Schematic diagram showing structural characteristics of M13 bacteriophages. M13 phages have five-fold rotational symmetry, two-fold screw symmetry, and no inversion center. Schematic illustrations showing the working mechanism of the piezoelectric effect based on M13 phages when the stress is applied along the phage long axis (**d**) and phage short axis (**e**,**f**).

This direct piezoelectric effect was first discovered in 1880 by Paul-Jacques Curie, Pierre, and Marie Curie. They combined the knowledge of pyroelectricity with their understanding of crystal structures and behavior, and demonstrated the first piezoelectric effect by using crystals of quartz and Rochelle salt. Since then, many researchers have discovered and reported the piezoelectric properties of organic [15,16], inorganic [7–14,17], and biomaterials [18–33].

These piezoelectric characteristics originate from the deformation of the crystal structure and charge rearrangement within the material. In the equilibrium state, the arrangement of charges within the material lattice is neutral. However, there will be a charge redistribution within the unit cell and this induces net charges on the faces of the unit cell under the mechanical stress, which results in a net dipole moment. The sum of these dipole moments from all the unit cells leads to charge separation and generates electrical polarization in piezoelectric materials (Figure 2b). The most important thing in this system is that the materials must not have a center of symmetry, because the sum of net dipole moments is zero if the materials have a symmetry center.

Fortunately, M13 phages have structural properties that can make them suitable for piezoelectric properties. M13 phages are composed of ssDNA and capsid proteins, and each capsid protein has a dipole moment because these capsid proteins are made up of three parts—a positively charged area, a neutral charged area, and a negatively charged area. Especially, 2700 pVIII major coat proteins, which are the body-coat proteins, are assembled on the ssDNA with a 20° tilt angle with respect to the phage long axis and are arranged in right-handed helical structures. These major coat proteins form a pentagonal structure, which means that M13 phages have five-fold rotational symmetry, two-fold screw symmetry, and no inversion center (Figure 2c) [46,47]. Therefore, M13 phages can have piezoelectric properties. Interestingly, these complex structural characteristics of M13 phages allow us to use various piezoelectric properties. When stress is applied along the phage long axis, net dipole moments and electrical polarization are generated along the direction of applied stress (Figure 2d). On the other hand, when the stress is applied along the body (phage short axis), net dipole moments and electrical polarization are generated in two different directions (Figure 2e,f). Due to these characteristics, various types of piezoelectric energy harvesting devices can be developed.

4. Surface Modification of M13 Bacteriophages through Genetic Engineering

One of the great advantages of M13 phages as functional materials is the possibility of surface modification through genetic engineering. The most frequently used and well-established method to modify the genes of phages is recombinant DNA technology, which involves the insertion of foreign genes into the bacterial plasmids. Especially, M13mp phage vectors are usually used for engineering M13 phages. By incorporating foreign DNA, converting certain DNA into foreign DNA, and deleting specific DNA with enzymes, a high density of functional peptides and proteins can be simultaneously displayed on the M13 phage's coat proteins [45]. This technique enables us to design the surface molecular structures of M13 phages according to their purpose. For example, J.-W. Oh et al. developed the highly trinitrotoluene (TNT)-selective sensors based on phage colorimetric structures by expressing the AXXXWHWQXXDP (WHW) peptide sequence (which shows excellent binding affinity to TNT molecules) on pVIII major coat proteins [48]. J. Wang et al. reported that RGD phages induce osteogenesis and angiogenesis by activating the endothelialization and osteogenic differentiation of mesenchymal stem cells. In this work, RGD and RGD/PHSRN (combination of RGD and PHSRN) peptides, which interact with integrin, have a key role in adhesion with fibronectin [49,50].

These genetic modifications are also very useful in fabricating energy harvesting devices. Most energy harvesting devices (e.g., piezoelectric and triboelectric devices) have a direct correlation to surface charges and dipole moments. Therefore, the number of charges on the outer surfaces of M13 phages should be increased to improve the power of energy generators. For this purpose, B. Y. Lee et al. expressed AEGDP (1E), AEEGDP (2E), AEEEGDP (3E), and AEEEEDP (4E) peptide sequences on the outer surfaces of the pVIII major coat protein of M13 phages (Figure 3a) [28]. In the case of vertically aligned phages, the HHHHHH peptide sequence was expressed at the N-terminus of the pIII minor coat protein with a spacer GGGS as a specific binder with Ni-NTA surface (Figure 3b). The YEEE peptide was also expressed between the first and fourth residues at the N-terminal of the pVIII major coat protein for enhancing mechanical stability by Y-Y cross-linkage between phages through UV illumination [29].

Likewise, we can improve the physical and chemical properties and extend the range of applications through genetic molecular design. In the future, we expect that chemical modification (e.g., bioconjugate techniques and cross-linking) as well as genetic modification will be used for the further improvement of physical properties.

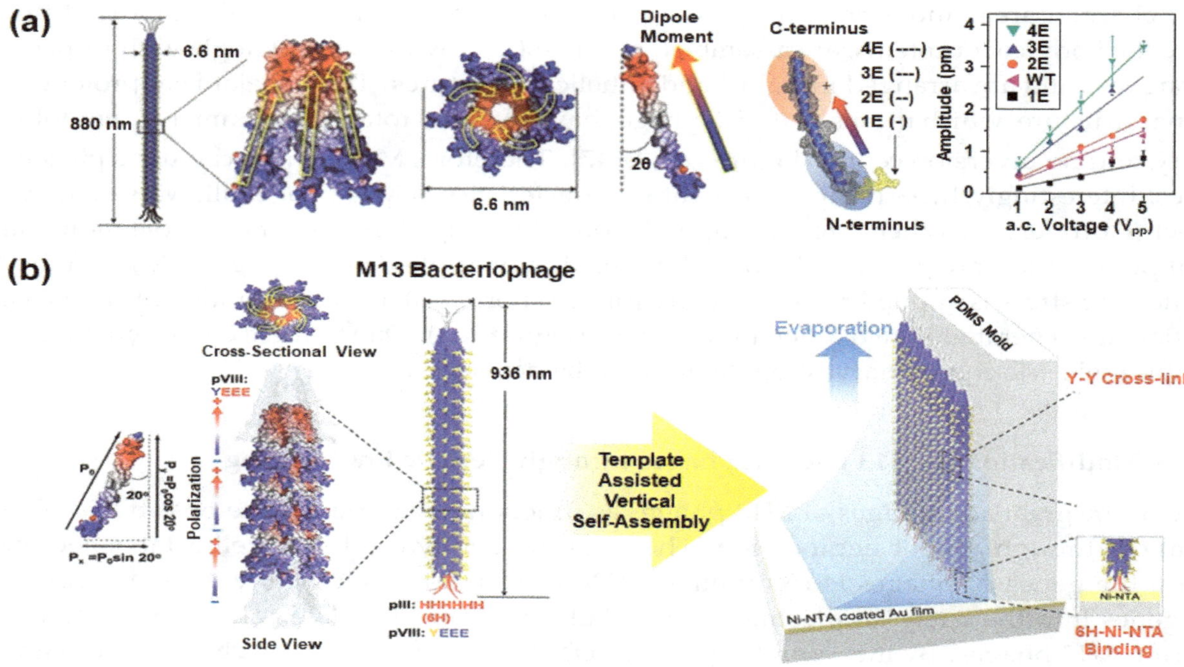

Figure 3. Surface modification of M13 bacteriophages through genetic engineering. (**a**) AEGDP (1E), AEEGDP (2E), AEEEGDP (3E), and AEEEEDP (4E) peptide sequences were expressed on the outer surfaces of the pVIII major coat protein of phages to enhance the piezoelectric properties. The M13 phage is 880 nm in length and 6.6 nm in diameter, which is covered by 2700 pVIII coat proteins and has five copies each of pIII and pIX proteins. The dipole moment is directed from the N-terminus (blue) to the C-terminus (red). Reproduced with permission from [28]. Copyright Nature Research, 2012. (**b**) HHHHHH peptide sequence expressed at the end of the N-terminus of the pIII minor coat protein for specific binding of phages on the Ni-NTA surface. The YEEE peptide motif is expressed between the first and fourth residues at the N-terminal of the pVIII major coat protein to enhance the mechanical stability by Y-Y cross-linkage. Reproduced with permission from [29]. Copyright American Chemical Society, 2019.

5. Piezoelectric Properties of M13 Bacteriophages

In general, the M13 phage is covered with 2700 pVIII coat major proteins, and the individual coat protein of phages is roughly divided into three sections: a positively charged region (C-terminus), a neutral region, and a negatively charged region (N-terminus). Owing to this adequate arrangement of charges, each coat protein has a dipole moment which is directed from the N-terminus to the C-terminus. Furthermore, the positively charged region of coat proteins is bound to central single-stranded DNA with a 20° tilt angle with respect to the phage long axis and α-helical structure when they are released from the host cell. The resulting structures of assembled pVIII coat proteins have five-fold rotational symmetry, two-fold screw symmetry, and no inversion center. Based on this fundamental study of phage structures, we can easily predict that M13 phages can present strong piezoelectric properties due to their permanent axial polarization caused by the net dipole moment in the pVIII proteins [28]. In 2012, B.Y. Lee et al. successfully observed the piezoelectric properties of M13 phages by using piezoresponse force microscopy (PFM) (Figure 4a). For this study, they prepared the phage monolayer sample by vertically pulling an octadecanethiol (ODT)/cysteamine patterned substrate from the phage suspension at a constant speed. Then, the piezoelectric properties of wild-type, 1E, 2E, 3E and 4E phages were measured by PFM. Vertical PFM measurements revealed that the effective piezoelectric coefficients (d_{eff}) of the wild-type phage was 0.30 ± 0.03 pm V^{-1}. The coefficients of 1E, 2E, 3E, and 4E phages were 0.14 ± 0.03 pm V^{-1}, 0.35 ± 0.03 pm V^{-1}, 0.55 ± 0.03 pm V^{-1}, and 0.70 ± 0.05 pm V^{-1}, respectively. This indicates that the coefficients improve as the surface charges of phages increase [28].

To further enhance the piezoelectric properties, they fabricated a multilayer phage film with 100 nm thickness. The multilayer film exhibited an increased effective piezoelectric coefficient (3.9 ± 0.05 pm V^{-1}). Although this value is lower than the d_{33} values of periodically-poled lithium niobate (PPLN) (13.2 pm V^{-1}), it is higher than collagen (1.1 pm V^{-1}) and other natural biomaterials. In addition, the effective piezoelectric coefficient of M13 phage films is further enhanced by fabricating vertically aligned phage nanostructures. These vertically assembled phages exhibited unidirectionally oriented piezoelectric polarization with an effective vertical piezoelectric coefficient of 13.2 pm V^{-1} (Figure 4b). Therefore, M13 phages are the best natural biomaterials for developing piezoelectric energy generators based on biomaterials [28].

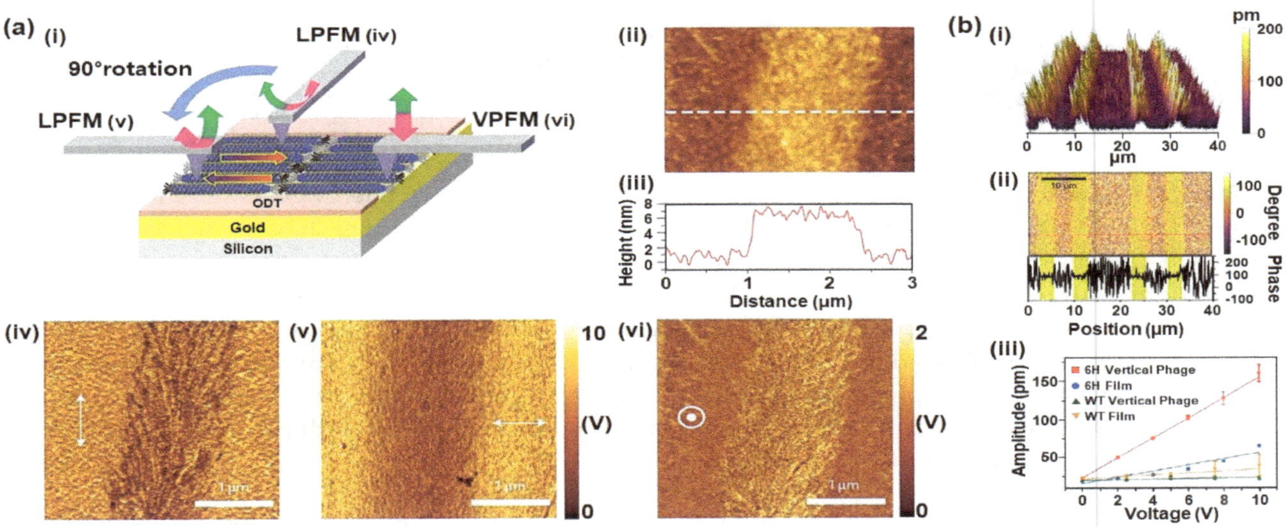

Figure 4. Piezoelectric properties of M13 bacteriophages. (**a**) Schematic of piezoresponse force microscopy (PFM) measurement (i); AFM topography (ii); height profile (iii); lateral PFM image along the phage long axis direction (iv); lateral PFM image obtained after changing the scanning direction by 90° (v); and vertical PFM image (vi) of the phage monolayer film. The resulting effective piezoelectric coefficients of 1E, 2E, 3E, and 4E phages were 0.14 ± 0.03 pm V^{-1}, 0.35 ± 0.03 pm V^{-1}, 0.55 ± 0.03 pm V^{-1}, and 0.70 ± 0.05 pm V^{-1}, respectively. Reproduced with permission from [28]. Copyright Nature Research, 2012. (**b**) PFM image (i), PFM phase image (ii) of vertically aligned phages which exhibits unidirectional polarization in the out-of-plane direction, and comparison of out-of-plane PFM amplitude versus applied voltage along the aligned direction (iii). The resulting effective piezoelectric coefficients (d_{eff}) of 6H vertical phage, 6H film, WT vertical phage, and WT film were ~13.2 pm V^{-1}, ~3.96 pm V^{-1}, ~0.35 pm V^{-1} and 1.22 pm V^{-1}, respectively. Reproduced with permission from [29]. Copyright American Chemical Society, 2019.

6. Developments and Applications of M13 Bacteriophage Based Piezoelectric Energy Harvesting Devices

Due to their excellent piezoelectric properties, the group of Prof. Lee at UC Berkeley first fabricated M13 phage-based piezoelectric energy generators in 2012 [28]. They prepared well-ordered self-assembled multilayer films based on M13 phages onto gold-coated flexible substrates by using a drop and evaporation method (Figure 5a). During the evaporation process, M13 phages were self-assembled and formed long-range ordered smectic-phase liquid-crystalline films by their chiral and monodisperse characteristics. When they overlaid a counter gold-coated flexible substrate on the film and embedded the device between two 2.5-mm-thick polydimethylsiloxane (PDMS) matrices, they could fabricate phage-based energy generators (Figure 5b). The generating power of the device can be modulated by the surface modification of M13 phages through genetic engineering, and the device produced a current of 6 nA and a voltage of 400 mV when they use 4E phages (Figure 5c). This was a sufficient energy output to turn on a liquid-crystal display [28].

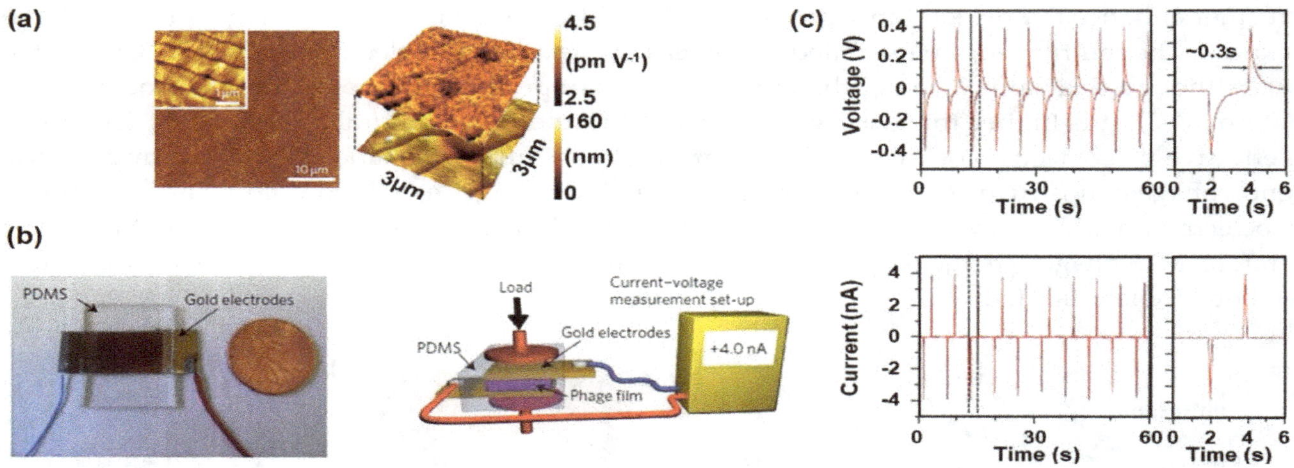

Figure 5. Characterization of an M13-bacteriophage-based piezoelectric energy generator (drop-casted film). (**a**) AFM topography image of phage film prepared by drop-casting, showing the long-range ordered smectic-phase liquid-crystalline film structure. Reproduced with permission from [28]. Copyright Nature Research, 2012. (**b**) Photograph of an M13 phage-based piezoelectric energy generator and schematic of piezoelectric energy generation measurement set-up. Reproduced with permission from [28]. Copyright Nature Research, 2012. (**c**) Open-circuit voltage and short-circuit current signal from the M13 phage-based piezoelectric energy generator. The device produced a current of 6 nA and a voltage of 400 mV when they use 4E phages. Reproduced with permission from [28]. Copyright Nature Research, 2012.

One of effective strategy for enhancing the power of phage-based energy generators is modulating the film morphology. Recently, K. Heo et al. reported a novel biomimetic assembly method for fabricating phage-based hierarchical structures with diverse surface morphologies by mimicking nature's self-assembly system [30]. They modulated the meniscus by controlling the thermodynamic and kinetic parameters (i.e., phage concentration, ionic concentration, phage surface charge, and pulling speed) and created a hierarchically organized phage film with diverse morphology in a controlled manner as the meniscus can serve as a transient scaffold to guide phage self-assembly (Figure 6a).

In this process, the shape of the meniscus can be systematically modulated due to a combination of multiple factors, such as fingering instability, Rayleigh instability, and elastocapillary instability. All of the resulting phage structures were long-range-ordered chiral phage films showing multiple levels of hierarchical organization from nano- to macro-scale (single phage–phage filaments–fiber bundles–mesoscale periodic structure–macroscale band) (Figure 6b) [30]. When they fabricated piezoelectric energy generators based on these hierarchically organized phage films, the device power was improved compared to previous drop-casted phage films. The continuous and line film patterns exhibited 6.3 and 56 nA short-circuit current and 0.36 and 0.75 V open-circuit voltage, respectively (Figure 6c). The 2D-dot patterns showed the highest piezoelectric performance, which exhibited peak values of 94 nA current and 0.95 V voltage. They claimed that these enhanced piezoelectric properties of 2D-dot phage patterns were mainly due to the enhanced crystallinity of the phage nanofilaments that were periodically organized in an active array, in contrast to other films. The enhanced power was available to display the words "VIRUS" and "LEE LAB" on a liquid-crystal display.

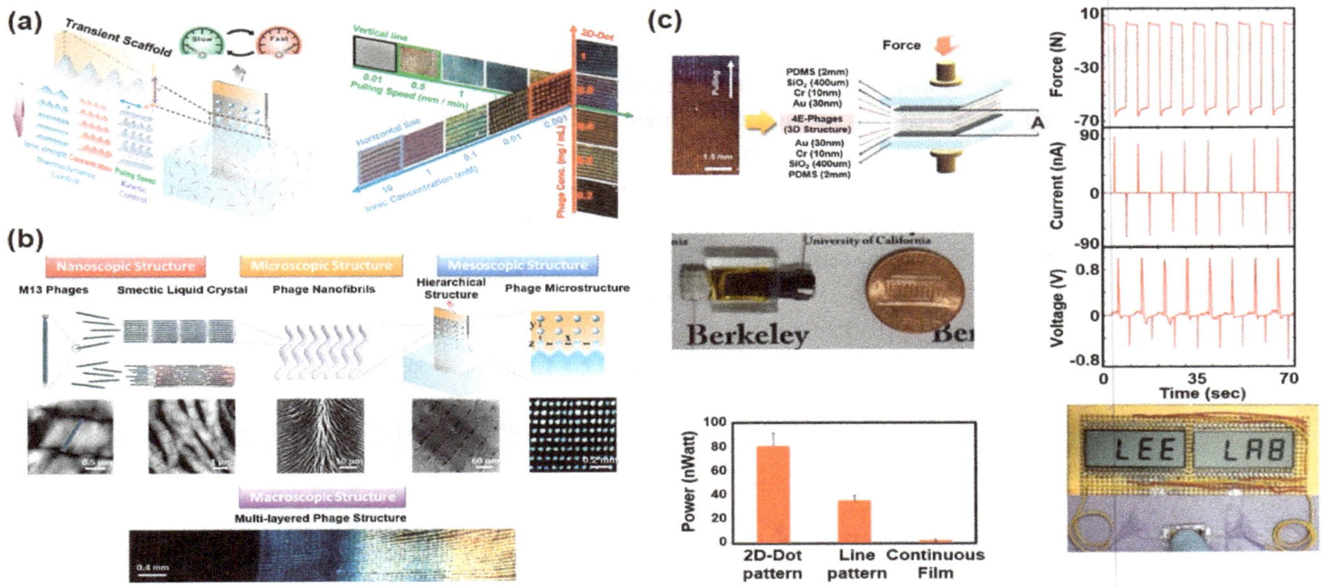

Figure 6. Characterization of an M13 bacteriophage-based piezoelectric energy generator. (biomimetic hierarchical structures). (**a**) Schematic illustration of the biomimetic hierarchical assembly process and structural transition diagram of phage films depending on phage concentration, ionic concentration, and pulling speed. Reproduced with permission from [30]. Copyright Elsevier, 2019. (**b**) Hierarchical structures of the phage-based films from nanoscale to macroscale. (single phage–phage filaments–fiber bundles–mesoscale periodic structure–macroscale band). (**c**) Piezoelectric characterization depending on phage structures. The 2D-dot patterns showed the highest piezoelectric performance, exhibiting peak values of 94 nA current and 0.95 V voltage. Reproduced with permission from [30]. Copyright Elsevier, 2019.

Another strategy for improving the piezoelectric power is to change the direction of the mechanical force applied to the phage. As mentioned in Section 3, if we consider the direction of the dipole moment in the individual M13 phages, it is predicted that the accumulated charges are maximized when the mechanical force is applied in the vertical direction rather than the lateral direction (Figure 2d). However, we could not carry out the related research because of the absence of an effective process to vertically align the phages. Recently, D.-M. Shin et al. developed a robust and facile method to prepare vertically aligned the phages for the first time (Figure 7a) [31]. They extruded phage suspension into a porous anodic aluminium oxide (AAO) template at precisely controlled speeds and repeated this process until all holes of the porous template were completely filled with phages, resulting in the formation of phage nanopillars (PNPs). During this process, negatively charged 4E phages were randomly adsorbed on the positively charged inner surface of the porous template and spontaneously accumulated inside the pores due to their liquid-crystalline characteristics. Afterwards, they deposited bottom and top Au electrodes on the AAO template including the PNPs and encapsulated the whole device using PDMS to improve their stability. The 4E PNP-based energy generator produced a 232 mV open-circuit voltage and 1.1 nA short-circuit current (Figure 7b) [31]. The relatively low power of this device compared to what was expected is presumed to be a result of the difficulty of forming well-ordered liquid crystalline phage structures, and the direction of the dipole moments of individual phages is therefore randomly oriented.

To overcome the limitation of the dipole alignment at the vertically oriented phage structures, J.H. Lee et al. developed a powerful method by combining the self-assembly of phages in a micro-fluidic channel and the surface modification of phages through genetic engineering (Figure 8a) [29]. To align the phages in the vertical direction, they tried to use a PDMS mold with micro-channels. When they dropped the phage suspension on the substrate and put on the PDMS mold, phage suspensions were infiltrated inside the micro-channels. The phages were vertically assembled on the wall of the

PDMS micro-channels and cross-linked with each other as the solvent was evaporated with UV light exposure. The morphology and filling density of vertical phage structure could be controlled by the initial phage concentration. Furthermore, they controlled the direction of the dipole moment of the phages by changing the peptide sequence of minor coat proteins at the same time. Because they inserted hexa-histidine (6H) at the N-terminal of the minor coat protein (pIII) of the phages through genetic modification, the pIII proteins of all phages were strongly specifically bound with the nickel-nitrilotriacetic acid (Ni-NTA) modified substrate, which polarized the dipole moment of the M13 phages. Finally, they fabricated piezoelectric energy harvesters using the resulting vertically aligned and unipolarized phages, and the peak voltage reached 2.8 V, with a current of 120 nA (Figure 8b). This is the largest power among phage-based energy generators. Five integrated energy generators demonstrated the operation of a liquid-crystal display reading "UC Berkeley" [29].

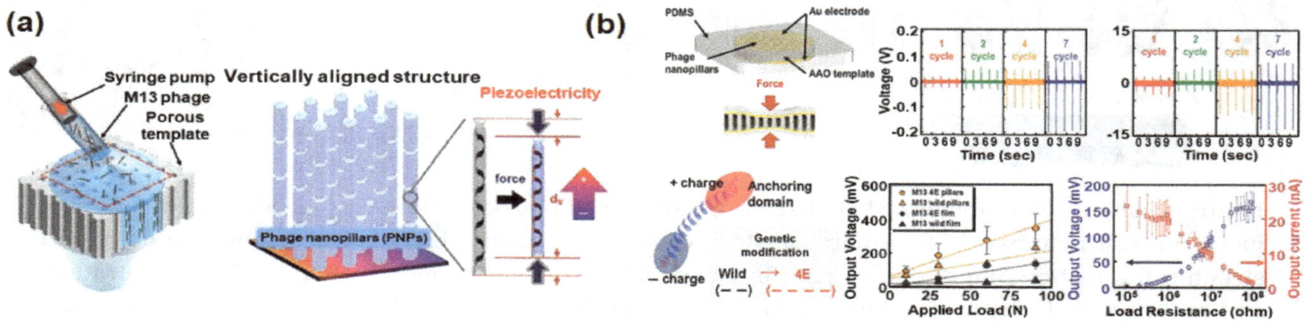

Figure 7. Characterization of an M13 bacteriophage-based piezoelectric energy generator (vertically aligned phage nanopillars). (**a**) Schematic showing the fabrication process and the resulting structures of vertically aligned M13 bacteriophage nanopillars. Reproduced with permission from [31]. Copyright The Royal Society of Chemistry, 2015. (**b**) Piezoelectric characterization of the phage nanopillar (PNP)-based energy generators. The 4E PNP-based energy generator produced 232 mV open−circuit voltage and 11.1 nA short-circuit current. Reproduced with permission from [31]. Copyright The Royal Society of Chemistry, 2015. AAO: anodic aluminium oxide.

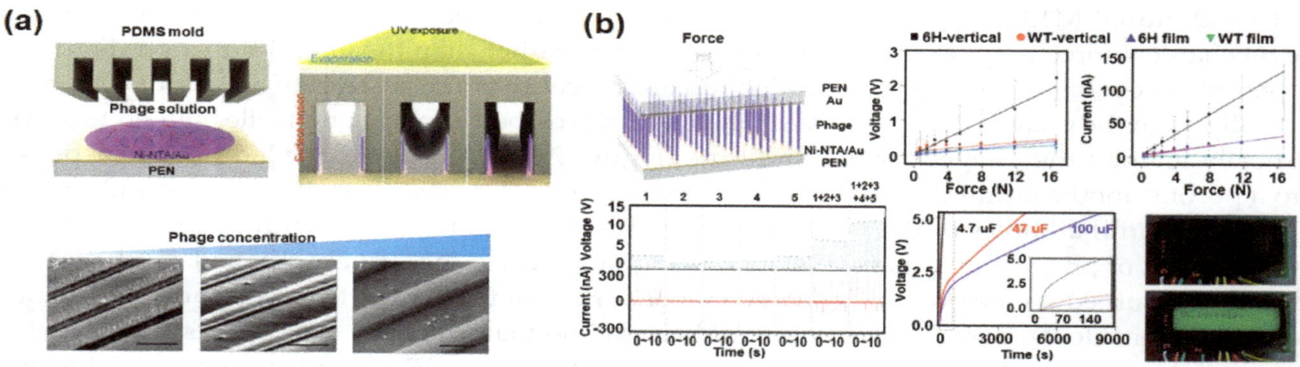

Figure 8. Characterization of an M13 bacteriophage-based piezoelectric energy generator (vertically aligned and unipolarized phage structures). (**a**) Schematic diagram of the fabrication process and scanning electron microscope (SEM) images of vertically aligned phages. Reproduced with permission from [29]. Copyright American Chemical Society, 2019. (**b**) Characterization of vertically aligned phage-based piezoelectric energy harvesters. This device exhibited a peak voltage of 2.8 V and a current of 120 nA. This is the largest power among phage-based energy generators. Reproduced with permission from [29]. Copyright American Chemical Society, 2019.

Although the power of phage-based energy generators is improving with diverse strategies, the power of the devices is still too low. For practical applications of these devices, many ways for enhancing electrical properties and developing mass-production methods should be contrived.

Because these methods are to prepare the M13 phage film by self-assembly, the performance of the devices may be reduced when they are used for a long time. Fortunately, laterally assembled structures are very stable and robust, allowing the devices to run reliably for a long time [28,30]. However, vertically aligned structures are likely to be vulnerable to long-term use. This can be solved by using chemically cross-linked M13 phages [29] and filling rubbery buffer materials in the empty space.

As this technique is at an early stage of research, studies of the devices' ideal operating conditions and toxicity issues have not been adequately carried out. However, by inferring from previous research based on M13 phages, the characteristics of these device can be predicted. Because the M13 phage is a biological material, optimized conditions for operating these devices will be room temperature (30 °C–70 °C) and low humidity [28]. However, these conditions can be modulated by surface modification. Through genetic engineering, we can modify the surface peptide motif to increase the hydrophobicity and cross-link phages to each other. Further, the M13 phage is known to be benign to humans because its host is *Escherichia coli* bacteria, not human cells [51–53]. Removing the infection motif in the pIII protein through genetic modification is also expected to be a good way to block the toxicity issues. Nevertheless, the study of M13 phages' toxicity should be conducted in the near future.

Since these technologies are still in their early stages of research, it is too early to discuss mass-production for practical applications. Most of the techniques discussed here are not suitable for mass production, because they use new process methods rather than conventional fabrication techniques. However, because these novel fabrication processes are very simple and facile, there is a strong possibility of mass-production and scale-up in the future. Although one of the main issues for scale-up is mass-production of the M13 phages, we can solve this problem using huge fermenters in the factories, like with biosimilar drug and alcohol manufacturing. Although the manufacturing cost of these devices is more expensive than existing devices, the M13 phage-based devices have several strong advantages which are very important in the biomedical fields. The M13 phage has very high piezoelectric coefficient compared to other biomaterials and their surfaces can be easily modified by genetic engineering. Further, it is also possible to mass-produce them.

7. Conclusions and Future Perspective

Even though the piezoelectric properties of biomaterials are lower than other inorganic materials, it is very important to design novel piezoelectric biomaterials and develop functional devices because of their specific applications in biomedical field. In particular, M13 bacteriophages are very attractive materials due to their unique features which distinguish them from other materials, such as their similar structures with collagens, mass-amplification, genetic modification, liquid-crystalline phase transition, and excellent piezoelectric properties. Recently, taking advantage of these characteristics, many researchers have made a great deal of efforts to fabricate M13 phage-based piezoelectric energy harvesting devices. Among these devices, vertically aligned phage films exhibited the highest performance—a peak voltage of 2.8 V and a peak current of 120 nA [29].

However, it is still a challenge to develop high-performance piezoelectric energy generators based on M13 phages owing to the limitations of surface modification, structural, and dipole alignment control. Thus, the novel design of phage structures through genetic and chemical modification may improve the performance of devices. Further, fabricating triboelectric devices based on M13 phages will also be an effective way to enhance the power of devices.

Another strategy for enhancing the power of devices is to develop composite structures composed of organic and inorganic biomaterials. Recently, novel methods for coating inorganic materials on biomaterial surfaces are attracting the attention of many researchers because of their various applications in biomedical field. For example, some researchers have reported effective methods to coat the inorganic materials on M13 bacteriophage surfaces via biomineralization [54–56], while other researchers developed the strategies to cast metals on the surface of biological materials by using protein cage systems and self-assembly [57–59]. These methods are expected to be used to produce precursors for energy-harvesting devices and maximize the power of devices.

High-performance energy harvesting devices based on biomaterials can be used in various fields, such as chemical/bio-sensors, artificial skin, bioimplantable energy devices, flexible electronics, soft robotics, and more (Figure 9). Especially, because there are many reports indicating that the surface charges and electrical signal can affect tissue regeneration, these piezoelectric biomaterials are also expected to be utilized for the development of biodegradable scaffolds for tissue engineering in the future.

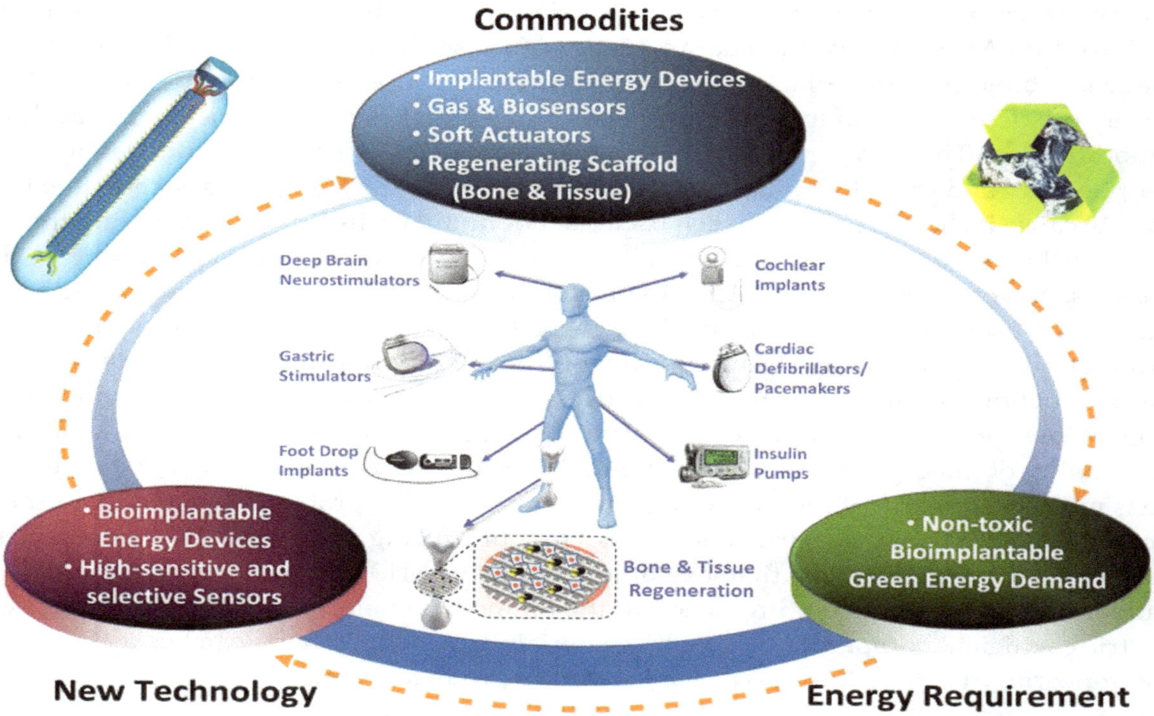

Figure 9. Schematic diagram showing the practical applications of M13 bacteriophage-based piezoelectric energy harvesting devices in the future.

Author Contributions: K.H., I.W.P., and K.W.K. developed the idea of the review article and wrote the paper. Y.H., H.J.Y., Y.L., and D.G. revised and improved the manuscript. K.H. supervised the manuscript. All authors have read and agreed to the published version of the manuscript.

References

1. Zheng, Q.; Zou, Y.; Zhang, Y.L.; Liu, Z.; Shi, B.J.; Wang, X.X.; Jin, Y.M.; Ouyang, H.; Li, Z.; Wang, Z.L. Biodegradable triboelectric nanogenerator as a life-time designed implantable power source. *Sci. Adv.* **2016**, *2*, e1501478. [CrossRef] [PubMed]
2. Parida, K.; Xiong, J.; Zhou, X.; Lee, P.S. Progress on triboelectric nanogenerator with stretchability, self-healability and bio-compatibility. *Nano Energy* **2019**, *59*, 237. [CrossRef]
3. Ali, F.; Raza, W.; Li, X.; Gul, H.; Kim, K.-H. Piezoelectric energy harvesters for biomedical applications. *Nano Energy* **2019**, *57*, 879. [CrossRef]
4. Zhou, Q.F.; Lau, S.T.; Wu, D.W.; Shung, K. Piezoelectric films for high frequency ultrasonic transducers in biomedical applications. *Prog. Mater. Sci.* **2011**, *56*, 139. [CrossRef] [PubMed]
5. Hwang, G.T.; Byun, M.; Jeong, C.K.; Lee, K.J. Flexible Piezoelectric Thin-Film Energy Harvesters and Nanosensors for Biomedical Applications. *Adv. Healthc. Mater.* **2015**, *4*, 646. [CrossRef] [PubMed]
6. Zheng, Q.; Shi, B.J.; Li, Z.; Wang, Z.L. Recent Progress on Piezoelectric and Triboelectric Energy Harvesters in Biomedical Systems. *Adv. Sci.* **2017**, *4*, 1700029. [CrossRef] [PubMed]
7. Guo, Y.P.; Kakimoto, K.; Ohsato, H. $(Na_{0.5}K_{0.5})NbO_3$-$LiTaO_3$ lead-free piezoelectric ceramics. *Mater. Lett.* **2005**, *59*, 241. [CrossRef]
8. Smith, R.T.; Welsh, F.S. Temperature Dependence of Elastic, Piezoelectric, and Dielectric Constants of Lithium Tantalate and Lithium Niobate. *J. Appl. Phys.* **1971**, *42*, 2219. [CrossRef]

9. Setter, N.; Damjanovic, D.; Eng, L.; Fox, G.; Gevorgian, S.; Hong, S.; Kingon, A.; Kohlstedt, H.; Park, N.Y.; Stephenson, G.B.; et al. Ferroelectric thin films: Review of materials, properties, and applications. *J. Appl. Phys.* **2006**, *100*, 051606. [CrossRef]
10. Haertling, G.H. Ferroelectric ceramics: History and technology. *J. Am. Ceram. Soc.* **1999**, *82*, 797. [CrossRef]
11. Saito, Y.; Takao, H.; Tani, T.; Nonoyama, T.; Takatori, K.; Homma, T.; Nagaya, T.; Nakamura, M. Lead-free piezoceramics. *Nature* **2004**, *432*, 84. [CrossRef] [PubMed]
12. Egerton, L.; Dillon, D.M. Piezoelectric and Dielectric Properties of Ceramics in the System Potassium-Sodium Niobate. *J. Am. Ceram. Soc.* **1959**, *42*, 482. [CrossRef]
13. Leontsev, S.O.; Eitel, R.E. Dielectric and Piezoelectric Properties in Mn-Modified (1-x)$BiFeO_3$-x$BaTiO_3$ Ceramics. *J. Am. Ceram. Soc.* **2009**, *92*, 2957. [CrossRef]
14. Rodel, J.; Jo, W.; Seifert, K.T.P.; Anton, E.M.; Granzow, T.; Damjanovic, D. Perspective on the Development of Lead-free Piezoceramics. *J. Am. Ceram. Soc.* **2009**, *92*, 1153. [CrossRef]
15. Nunes, J.S.; Wu, A.; Gomes, J.; Sencadas, V.; Vilarinho, P.M.; Lanceros-Mendez, S. Relationship between the microstructure and the microscopic piezoelectric response of the alpha- and beta-phases of poly(vinylidene fluoride). *Appl. Phys. A Mater. Sci. Process.* **2009**, *95*, 875.
16. Guo, H.F.; Li, Z.S.; Dong, S.W.; Chen, W.J.; Deng, L.; Wang, Y.F.; Ying, D.J. Piezoelectric PU/PVDF electrospun scaffolds for wound healing applications. *Colloids Surf. B* **2012**, *96*, 29. [CrossRef]
17. Wang, Z.L. Towards Self-Powered Nanosystems: From Nanogenerators to Nanopiezotronics. *Adv. Funct. Mater.* **2008**, *18*, 3553. [CrossRef]
18. Fukada, E. Piezoelectric Properties of Biological Polymers. *Q. Rev. Biophys.* **1983**, *16*, 59. [CrossRef]
19. Minary-Jolandan, M.; Yu, M.F. Nanoscale characterization of isolated individual type I collagen fibrils: Polarization and piezoelectricity. *Nanotechnology* **2009**, *20*, 085706. [CrossRef]
20. Vivekananthan, V.; Alluri, N.R.; Purusothaman, Y.; Chandrasekhar, A.; Selvarajan, S.; Kim, S.J. Biocompatible Collagen Nanofibrils: An Approach for Sustainable Energy Harvesting and Battery-Free Humidity Sensor Applications. *ACS Appl. Mater. Interfaces* **2018**, *10*, 18650. [CrossRef]
21. Williams, W.S.; Breger, L. Analysis of Stress Distribution and Piezoelectric Response in Cantilever Bending of Bone and Tendon. *Ann. N. Y. Acad. Sci.* **1974**, *238*, 121. [CrossRef] [PubMed]
22. Liboff, A.R.; Shamos, M.H. Piezoelectric Effect in Dentin. *J. Dent. Res.* **1971**, *50*, 516. [CrossRef] [PubMed]
23. Fukada, E. Piezoelectricity of Natural Biomaterials. *Ferroelectrics* **1984**, *60*, 285. [CrossRef]
24. Maiti, S.; Kumar Karan, S.; Lee, J.; Kumar Mishra, A.; Bhusan Khatua, B.; Kon Kim, J. Bio-waste onion skin as an innovative nature-driven piezoelectric material with high energy conversion efficiency. *Nano Energy* **2017**, *42*, 282. [CrossRef]
25. Ghosh, S.K.; Mandal, D. High-performance bio-piezoelectric nanogenerator made with fish scale. *Appl. Phys. Lett.* **2016**, *109*, 103701. [CrossRef]
26. Praveen, E.; Murugan, S.; Jayakumar, K. Investigations on the existence of piezoelectric property of a bio-polymer-Chitosan and its application in vibration sensors. *RSC Adv.* **2017**, *7*, 35490. [CrossRef]
27. Karan, S.K.; Maiti, S.; Kwon, O.; Paria, S.; Maitra, A.; Si, S.K.; Kim, Y.; Kim, J.K.; Khatua, B.B. Nature driven spider silk as high energy conversion efficient bio-piezoelectric nanogenerator. *Nano Energy* **2018**, *49*, 655. [CrossRef]
28. Lee, B.Y.; Zhang, J.; Zueger, C.; Chung, W.J.; Yoo, S.Y.; Wang, E.; Meyer, J.; Ramesh, R.; Lee, S.W. Virus-based piezoelectric energy generation. *Nat. Nanotechnol.* **2012**, *7*, 351. [CrossRef]
29. Lee, J.H.; Lee, J.H.; Xiao, J.; Desai, M.S.; Zhang, X.; Lee, S.W. Vertical Self-Assembly of Polarized Phage Nanostructure for Energy Harvesting. *Nano Lett.* **2019**, *19*, 2661. [CrossRef]
30. Heo, K.; Jin, H.-E.; Kim, H.; Lee, J.H.; Wang, E.; Lee, S.-W. Transient self-templating assembly of M13 bacteriophage for enhanced biopiezoelectric devices. *Nano Energy* **2019**, *56*, 716. [CrossRef]
31. Shin, D.-M.; Han, H.J.; Kim, W.-G.; Kim, E.; Kim, C.; Hong, S.W.; Kim, H.K.; Oh, J.-W.; Hwang, Y.-H. Bioinspired piezoelectric nanogenerators based on vertically aligned phage nanopillars. *Energy Environ. Sci.* **2015**, *8*, 3198. [CrossRef]
32. Lee, J.H.; Heo, K.; Schulz-Schonhagen, K.; Lee, J.H.; Desai, M.S.; Jin, H.E.; Lee, S.W. Diphenylalanine Peptide Nanotube Energy Harvesters. *ACS Nano* **2018**, *12*, 8138. [CrossRef] [PubMed]
33. Kholkin, A.; Amdursky, N.; Bdikin, I.; Gazit, E.; Rosenman, G. Strong piezoelectricity in bioinspired peptide nanotubes. *ACS Nano* **2010**, *4*, 610. [CrossRef] [PubMed]
34. Fukada, E.; Yasuda, I. On the Piezoelectric Effect of Bone. *J. Phys. Soc. Jpn.* **1957**, *12*, 1158. [CrossRef]

35. Andrew, C.; Bassett, L. Electrical effects in bone. *Sci. Am.* **1965**, *213*, 18.
36. Shamos, M.H.; Lavine, L.S.; Shamos, M.I. Piezoelectric Effect in Bone. *Nature* **1963**, *197*, 81. [CrossRef]
37. Tian, J.J.; Shi, R.; Liu, Z.; Ouyang, H.; Yu, M.; Zhao, C.C.; Zou, Y.; Jiang, D.J.; Zhang, J.S.; Li, Z. Self-powered implantable electrical stimulator for osteoblasts' proliferation and differentiation. *Nano Energy* **2019**, *59*, 705. [CrossRef]
38. Yu, P.; Ning, C.; Zhang, Y.; Tan, G.; Lin, Z.; Liu, S.; Wang, X.; Yang, H.; Li, K.; Yi, X.; et al. Bone-Inspired Spatially Specific Piezoelectricity Induces Bone Regeneration. *Theranostics* **2017**, *7*, 3387. [CrossRef]
39. Ren, L.; Yang, P.; Wang, Z.; Zhang, J.; Ding, C.; Shang, P. Biomechanical and biophysical environment of bone from the macroscopic to the pericellular and molecular level. *J. Mech. Behav. Biomed. Mater.* **2015**, *50*, 104. [CrossRef] [PubMed]
40. Chorsi, M.T.; Curry, E.J.; Chorsi, H.T.; Das, R.; Baroody, J.; Purohit, P.K.; Ilies, H.; Nguyen, T.D. Piezoelectric Biomaterials for Sensors and Actuators. *Adv. Mater.* **2019**, *31*, 1802084. [CrossRef] [PubMed]
41. Yuan, H.; Lei, T.; Qin, Y.; He, J.-H.; Yang, R. Design and application of piezoelectric biomaterials. *J. Phys. D* **2019**, *52*, 194002. [CrossRef]
42. Ben Amar, A.; Kouki, A.B.; Cao, H. Power Approaches for Implantable Medical Devices. *Sensors* **2015**, *15*, 28889–28914. [CrossRef] [PubMed]
43. Chung, W.J.; Oh, J.W.; Kwak, K.; Lee, B.Y.; Meyer, J.; Wang, E.; Hexemer, A.; Lee, S.W. Biomimetic self-templating supramolecular structures. *Nature* **2011**, *478*, 364. [CrossRef] [PubMed]
44. Jin, H.E.; Jang, J.; Chung, J.; Lee, H.J.; Wang, E.; Lee, S.W.; Chung, W.J. Biomimetic Self-Templated Hierarchical Structures of Collagen-Like Peptide Amphiphiles. *Nano Lett.* **2015**, *15*, 7138. [CrossRef] [PubMed]
45. Rakonjac, J.; Bennett, N.J.; Spagnuolo, J.; Gagic, D.; Russel, M. Filamentous Bacteriophage: Biology, Phage Display and Nanotechnology Applications. *Curr. Issues Mol. Biol.* **2011**, *13*, 51. [PubMed]
46. Goldbourt, A. Structural characterization of bacteriophage viruses by NMR. *Prog. Nucl. Magn. Reson. Spectrosc.* **2019**, *114–115*, 192. [CrossRef]
47. Xu, J.; Dayan, N.; Goldbourt, A.; Xiang, Y. Cryo-electron microscopy structure of the filamentous bacteriophage IKe. *Proc. Natl. Acad. Sci. USA* **2019**, *116*, 5493. [CrossRef]
48. Oh, J.W.; Chung, W.J.; Heo, K.; Jin, H.E.; Lee, B.Y.; Wang, E.; Zueger, C.; Wong, W.; Meyer, J.; Kim, C.; et al. Biomimetic virus-based colourimetric sensors. *Nat. Commun.* **2014**, *5*, 3043. [CrossRef]
49. Wang, J.L.; Wang, L.; Yang, M.Y.; Zhu, Y.; Tomsia, A.; Mao, C.B. Untangling the Effects of Peptide Sequences and Nanotopographies in a Biomimetic Niche for Directed Differentiation of iPSCs by Assemblies of Genetically Engineered Viral Nanofibers. *Nano Lett.* **2014**, *14*, 6850. [CrossRef]
50. Wang, J.L.; Yang, M.Y.; Zhu, Y.; Wang, L.; Tomsia, A.P.; Mao, C.B. Phage Nanofibers Induce Vascularized Osteogenesis in 3D Printed Bone Scaffolds. *Adv. Mater.* **2014**, *26*, 4961. [CrossRef]
51. Frenkel, D.; Solomon, B. Filamentous phage as vector-mediated antibody delivery to the brain. *Proc. Natl. Acad. Sci. USA* **2002**, *99*, 5675. [CrossRef]
52. Yoo, S.Y.; Kobayashi, M.; Lee, P.P.; Lee, S.-W. Early osteogenic differentiation of mouse preosteoblasts induced by collagen-derived DGEA-peptide on nanofibrous phage tissue matrices. *Biomacromolecules* **2011**, *12*, 987. [CrossRef] [PubMed]
53. Zhu, H.; Cao, B.; Zhen, Z.; Laxmi, A.A.; Li, D.; Liu, S.; Mao, C. Controlled growth and differentiation of MSCs on grooved films assembled from monodisperse biological nanofibers with genetically tunable surface chemistries. *Biomaterials* **2011**, *32*, 4744. [CrossRef] [PubMed]
54. Tom, S.; Jin, H.E.; Heo, K.; Lee, S.W. Engineered phage films as scaffolds for $CaCO_3$ biomineralization. *Nanoscale* **2016**, *8*, 15696. [CrossRef] [PubMed]
55. Kilper, S.; Jahnke, T.; Aulich, M.; Burghard, Z.; Rothenstein, D.; Bill, J. Genetically Induced In Situ-Poling for Piezo-Active Biohybrid Nanowires. *Adv. Mater.* **2019**, *31*, e1805597. [CrossRef] [PubMed]
56. Jeong, C.K.; Kim, I.; Park, K.I.; Oh, M.H.; Paik, H.; Hwang, G.T.; No, K.; Nam, Y.S.; Lee, K.J. Virus-Directed Design of a Flexible $BaTiO_3$ Nanogenerator. *ACS Nano* **2013**, *7*, 11016. [CrossRef]
57. Slocik, J.M.; Crouse, C.A.; Spowart, J.E.; Naik, R.R. Biologically tunable reactivity of energetic nanomaterials using protein cages. *Nano Lett.* **2013**, *13*, 2535. [CrossRef]
58. Perriman, A.W.; Colfen, H.; Hughes, R.W.; Barrie, C.L.; Mann, S. Solvent-free protein liquids and liquid crystals. *Angew. Chem. Int. Ed.* **2009**, *48*, 6242. [CrossRef]
59. Ryadnov, M.G. A self-assembling peptide polynanoreactor. *Angew. Chem. Int. Ed.* **2007**, *46*, 969. [CrossRef]

14

Protein Structure-Guided Hidden Markov Models (HMMs) as a Powerful Method in the Detection of Ancestral Endogenous Viral Elements

Heleri Kirsip [1,*] and Aare Abroi [2,*]

1 Department of Bioinformatics, University of Tartu, Tartu, 51010, Riia 23, Estonia
2 Institute of Technology, University of Tartu, Tartu, 50411, Nooruse 1, Estonia
* Correspondence: heleri16@ut.ee (H.K.); abroi@ut.ee (A.A.)

Abstract: It has been believed for a long time that the transfer and fixation of genetic material from RNA viruses to eukaryote genomes is very unlikely. However, during the last decade, there have been several cases in which "virus-to-host" gene transfer from various viral families into various eukaryotic phyla have been described. These transfers have been identified by sequence similarity, which may disappear very quickly, especially in the case of RNA viruses. However, compared to sequences, protein structure is known to be more conserved. Applying protein structure-guided protein domain-specific Hidden Markov Models, we detected homologues of the Virgaviridae capsid protein in Schizophora flies. Further data analysis supported "virus-to-host" transfer into Schizophora ancestors as a single transfer event. This transfer was not identifiable by BLAST or by other methods we applied. Our data show that structure-guided Hidden Markov Models should be used to detect ancestral virus-to-host transfers.

Keywords: endogenous viral elements; bioinformatics; horizontal gene transfer; virus-to-host gene transfer; HMM; tobacco mosaic virus; *Drosophila*; capsid protein

---

1. Introduction

Viruses are one of the most abundant and prevalent biological entities on Earth and thus are an important and integral part of the biosphere (for viral importance in environments, see [1–4]). The importance of viruses for humans and the biosphere is illustrated by the fact that, in our bodies, there are 10 times more bacteria than human cells (~10^{13} cells), and the number of viruses is even higher than the number of bacteria by an order of magnitude [5]. The same prevalence also applies to viruses in other biotopes; for example, there are approximately 10^8 viruses in one litre of water [6] and approximately 10^6–10^7 viruses per m^3 in the atmosphere [7]. In most biotopes, the number of viruses exceeds the number of prokaryotes by an order of magnitude [4]. The overall virus–host interaction is an important aspect driving viral evolution and the role of viruses in evolution. These roles are not restricted to the host–parasite arms race. It has been known for a long time that bacteriophages mediate transduction in bacteria. It is also known that retroviruses integrate into their hosts' genomes and thus can affect the genomic organisation of the vertebrate's cells [8]. However, Retroviridae is one viral family among approximately one hundred viral families that infect eukaryotes, and vertebrates constitute only a small fraction of eukaryotes. What is less known is that other, non-retroviral viruses are able to do the same, i.e., integrate viral genes into the chromosomes of infected cells, both in vertebrates and in other eukaryotic cells, of course not in such direct way as retroviruses. [9,10]. The viral genome elements that have integrated into eukaryotic genomes and have become fixed are called endogenous viral elements (EVEs) [9]. It has been shown that both cytoplasm- and nucleus-replicating non-retroviral viruses with different replication strategies ((+)ssRNA, (−)ssRNA, dsRNA, ssDNA and dsDNA viruses) can integrate into animal [11,12], plant [13] and fungal [14] genomes (for a thorough review, see [15]). Thus, the integration of viral genes into eukaryotic genomes is quite widespread. In contrast, the function and expression of retroviral EVEs have only been thoroughly researched in model organisms [16], and for non-retroviral EVEs, additional analyses of EVE function and expression have been performed only for human and squirrel endogenous bornavirus-like nucleoprotein (EBLN) elements [17–19] and tobacco plant geminivirus-related DNA sequences (GRDs) [9,20]. The exact pathway by which viral genes integrate and become fixed in eukaryotic organisms is not clear. However, to be effectively spread over the population, EVEs must also benefit the organism (although spreading via genetic drift, founder effects and population bottlenecks cannot be excluded).

The main strategy for detecting EVEs in organisms was formalised by Kondo et al. [21]. The detection protocol for EVEs consists of three main stages: (a) EVE detection with bioinformatics tools, mostly with BLAST searches in different databases; (b) EVE detection and confirmation in experiments, mostly with PCR and DNA sequencing to confirm integration; and (c) EVE phylogeny construction combined with viral sequences to identify EVEs closest relatives and to evaluate the integration time.

Following this protocol, scientists have discovered a large amount of viral-like sequences in different eukaryotic genomes. However, there is a problem with detecting distant homologues, which was noted by Brenner et al.; BLAST does not work well with sequences under 30% of identity [22]. Park et al. [23] showed that Intermediate Sequence Search (ISS) method, Hidden Markov Models (HMMs) and PSI-BLAST work considerably more effectively than pairwise sequence comparison methods (e.g., BLAST), especially when the sequences have changed greatly over time. Profile-HMMs are probabilistic models that can be used to describe the patterns shared by sets of protein/domain sequences (HMMs are generated from constructed multiple sequence alignments (MSAs), and the generated model covers both the diversity and conservation within the input MSA). It is possible to search for a sequence that is the most similar to the model, which is especially important when dealing with viral sequences because the virus mutation speed is much higher, especially in RNA viruses (approximately 10^{-2} to 10^{-7} substitutions per position per year compared to about 10^{-9} substitutions per position per year in mammals [24–26]), and confident pairwise similarity between homologous sequences may disappear. Thus, profile-HMMs should detect distant homologous sequences more efficiently.

Advantages of other methods in comparison with BLAST in the case of viruses have been shown in practise. Kuchibhatla et al. started from the assumptions above and used sequence-profile comparisons (such as HMMER3) and profile–profile comparisons (HHpred) in addition to BLAST analyses [27]. They showed that sequences that were previously classified as "ORFans" (more accurately "taxonomically restricted to only one taxon") according to BLAST analyses, had distant homologues in various viruses and that this approach could also be applied to organisms. Thus, considering the fast evolution of viruses, profile-based algorithms that can detect distant homologues should be more suitable for detecting (distant) homologues of viral proteins [28,29].

It is also known that protein structure is more conserved than sequence [30,31]. Therefore, the protein structure is also important to consider when detecting distant homologous sequences. When dealing with the identification of similarities between evolutionarily distant, but homologous sequences, taking protein structure-based information into account may improve the results. Challis and Schmidler showed that, when including structural information, the phylogenetic inference for distant relationships improves [32]. Additionally, Herman et al. showed that structure contains more information that can be obtained from sequences; hence, it vastly reduces the alignment and phylogenetic tree topology uncertainty [33]. This finding is especially important in the case of viruses, which lose sequence similarity within a relatively short time, especially compared to organisms. Thus, structural information should be included when studying deep evolutionary relationships.

In this study, we tested whether a complex method, such as profile-HMM together with protein structure-guided information, is a suitable and effective first step in EVE detection and whether it should be implemented in the EVE-detection protocol. To complete this aim, a comparison between BLAST and profile-HMM search results was made with the objective of finding whether profile-HMMs could detect more distant EVEs and how many more EVEs are identified when implementing a structure-guided profile-HMM search. This work shows that using profile-HMMs with additional structure-guided information enables the detection of more distant homologous sequences; in this case, endogenous *Tobamovirus* (family Virgaviridae; +ssRNA viruses) coat protein-like (eTCPL) elements in different fly genomes that have not been detected previously by using pairwise sequence comparison method BLAST.

2. Materials and Methods

2.1. EVE Detection Using BLAST Analysis

To detect eukaryotic endogenous *Tobamovirus* coat protein-like (eTCPL) elements, the BLAST method was used (Figure 1). Different *Tobamovirus* (family Virgaviridae, +ssRNA viruses), *Hordeivirus* (family Virgaviridae, +ssRNA viruses), *Tobravirus* (family Virgaviridae, +ssRNA viruses) and *Pecluvirus* (family Virgaviridae, +ssRNA viruses) coat proteins (CPs) were used as queries (the sequences used are shown in Table S1) in NCBI BLASTp and tBLASTn (protein sequence against the nucleotide database translated into 6 reading frames) (BLAST+ 2.2.30+, updated 6.10.2014; using the NCBI BLAST server http://blast.ncbi.nlm.nih.gov/ [34]) searches in eukaryotes (NCBI Taxonomy ID 2759). The search was performed against nucleotide collection (nt/nr), Transcriptome Shotgun Assembly (TSA), expressed sequence tag (Est) and whole-genome shotgun contig (WGS) databases using default parameters with an additional implemented *E*-value threshold of 1×10^{-5} (May 2017) (results in Table S2). additionally, BLASTp (protein sequence against protein database) was run using the same query sequences and search parameters (including an *E*-value threshold of 1×10^{-5}) against a non-redundant protein database ("nr", which also includes most of the UniProtKB sequences). Searches and databases used are visualised in Figure 1.

2.2. EVE Detection Using the Profile-HMM Method

As an alternative method, searches were also performed with the HMMER3 package (Figure 1) (version 1.8; using the web server http://www.ebi.ac.uk/Tools/hmmer/ [35]). The searches were performed using the same *Tobamovirus*, *Hordeivirus*, *Tobravirus* and *Pecluvirus* CP sequences as described above for BLAST analyses with taxonomic restriction to Eukaryotes (NCBI Taxonomy ID 2759) using the default parameters with an additional *E*-value threshold of 1×10^{-7}.

In the "phmmer" search type (protein sequence against protein sequence database), the databases used were UniProtKB (version v.2017_05), Swiss-Prot (version v.2017_05), Ensemble All (version v.88) and Ensemble Genomes (version v.35) because no other databases were available during the analyses (May 2017) (results in Table S3).

In "hmmscan" (protein sequence against the profile-HMM database), viral sequences were used to search against protein families in the Pfam-A [36] and SUPERFAMILY [37] databases (April 2017). Both resources contain protein domain-specific HMM profiles and assign these profiles to available protein sequences in different databases. The "hmmscan" search results give the protein domain-specific HMM profile ID. Further, it is also possible to analyse these HMM profiles and the sequences that are assigned to the profile. Eukaryotic sequences assigned to these profiles can be analysed as possible eTCPLs. The *E*-value threshold for belonging to the respective domain family was set to 1×10^{-7} (results in Table S4).

For the "hmmsearch" search (protein alignment/profile-HMM against protein sequence database), the *Tobamovirus* CP MSA and the *Tobamovirus-Pecluvirus-Tobravirus-Hordeivirus* CP MSA were constructed using the MUSCLE algorithm (with default parameters) within MEGA (version 7.0.20; [38]).

The search was performed in the UniProtKB and Swiss-Prot databases (April 2017) (results in Table S5). Next, "hmmsearch" was performed using previous "hmmscan" results (predefined protein family models from the Pfam-A/SUPERFAMILY databases) as queries (May 2017) (results in Table S6).

2.3. Testing for False Positive Hits with Alternative Methods

As stated by Pearson, "When a scientifically unexpected alignment appears to be statistically significant, investigators should consider alternate strategies for estimating statistical significance" [39]; we applied alternative methods to confirm the homology of high scoring hits (Figure 2). First, to test for annotation artefacts, database annotations of sequence's scaffold/contig/chromosome (their origin), overall length, sequence location and surrounding area were scanned. Data confirmation was needed to exclude the possibility of sample contamination that could have occurred during cloning or sequence assembly and/or a misannotation of the protein itself (for example, nucleic acid is isolated from tissues infected with a virus, but all sequences are still annotated as host sequences).

Second, to exclude false positive hits (sequences that are more similar to other sequences or domains), a reciprocal sequence similarity search was performed with default parameters using tBLASTn for protein sequences (against the NCBI nr/nt database) and BLASTn (nucleotide sequence against the nucleotide database) for nucleotide sequences (against the NCBI nr/nt database). When the reciprocal search gives the most significant hit to primary query sequences or their close relatives (in viruses in the same viral genus or family), it is noted as such in all of the tables and in other cases, the best hit is described; additionally, "no hits" are marked with "−". Other viral hits that were not closely related to Virgaviridae viruses were classified as false positives for eTCPL but will remain possible positive hits for other EVEs and will be analysed in the future. Additionally, a "hmmscan" (protein sequence against the profile-HMM database) search was performed on the HMMER webpage to test whether the sequence belongs to another protein domain family with much higher confidence.

Third, to determine possible false positive results with alternative methods, protein structure prediction was performed with the LOMETS [40] meta-server for all remaining positive hits from both the BLAST and HMMER3 analyses. Additionally, this step was important for the positive hits (both eukaryotic and viral hits) obtained from the SUPERFAMILY and Pfam-A databases to confirm that profile-HMM models have not falsely assigned sequences to the models. The LOMETS meta-server simultaneously uses many different algorithms to predict possible protein structures from sequences using the protein threading method. If the sequence is a real homologue, the server should recognise the tobamoviral coat protein or its structural relatives in SCOP, i.e., the protein domains with "SCOP concise classification string" (SCCS) (also named "SCOP superfamily identifier") staring with "a.24.5" as a template for high-confidence models. Additionally, the confidence score, the alignment length and the template coverage should be high, and at least one algorithm not based on HMMs should give a high confidence score. In addition, the Z-score was considered, for which a higher score is considered to be better than a lower score. All possible eTCPL hits that were not classified as false positives or non-determined (ND) were used in the phylogenetic analyses.

In general, all the sequences that reciprocal BLAST classified as similar to tobamoviruses or other Virgaviridae but only had RNA data available were classified as ND because we cannot distinguish between actual viral sequences and new EVEs.

2.4. Phylogenetic Analysis for eTCPL

For the phylogenetic analysis, a joint list of viral coat protein sequences and eukaryotic EVE sequences were used. Viral sequences were taken from the NCBI full genomes database, belonging to tobamoviruses, pecluviruses, hordeiviruses, tobraviruses and goraviruses (Table S1).

The eukaryotic sequences used in this study are from two datasets. The first dataset consists of eukaryotic sequences from the SUPERFAMILY database that have been assigned to eTCPL elements (Table S7). Sequences that were included in the phylogeny were those that remained after the false positives were excluded. For these sequences, only the eTCPL region that was assigned to the SUPERFAMILY HMM model was used. The second dataset (search results) consists of the HMMER analysis results (Table S5) that were obtained when using Virgaviridae viral MSAs as queries. One additional sequence (*Lasioglossum albipes*; NCBI accession ID: ANOB01025386.1), which was the one positive hit from the tBLASTn analyses using the viral sequence as a query, was added to the second dataset. All phylogenetic analyses were performed using the protein domain part of the sequences (*Drosophila melanogaster*, FlyBase gene ID FBgn0029799, UniProt ID Q9W483, protein region 122–259). To determine the domain part of the sequences (where the domain part was not previously designated), the two datasets were merged and aligned within MEGA using ClustalW and were edited using the first eukaryotic dataset as a baseline to cut out the regions that were outside of the eTCPL protein domain. Additionally, Jalview [41] was used to remove identical sequences (remove redundancy threshold of 100%).

All the results were additionally filtered to ensure data quality because most of the partial domains in proteins are alignment and/or annotation artefacts [42]. Protein sequences shorter than 130 AA were removed and not used in the MSA. All sequences were aligned using the Mafft alignment algorithm (version 7) with the default parameters [43]. To select the best model for phylogeny construction, the ProtTest server was used [44]. The best model was selected using the Akaike information criterion framework: LG + I + G + F. The phylogenetic tree was constructed using the MEGA program package with maximum likelihood methods using LG with frequencies (+F) model along with the gamma distribution with invariant sites (G + I). The number of discrete gamma categories was 5. Gaps and missing data were dealt with via partial deletion using a site coverage cut-off of 95%. The initial tree was performed automatically (using the NJ/BioNJ method). For the phylogeny test, the bootstrap method (500 replications) was used.

2.5. Additional BLAST Analysis to Broaden the Search to Databases that Are not Accessible in HMMER3

For the final step in detecting EVEs, an additional BLAST analysis was performed to identify potential EVEs not yet annotated as proteins (Figure 1). This analysis was done because the HMMER3 web server has few directly connected databases, none of which are nucleotide databases. Hence, performing an additional BLAST analysis enables us to broaden the phylogenomic coverage to incomplete genomes and/or nonannotated proteins. Eukaryotic sequences (Table S8), obtained through protein domain databases (SUPERFAMILY), were used as queries to search for eukaryotes (NCBI Taxonomy ID 2759) using tBLASTn with the same parameters described previously and with *E*-value 1×10^{-7}.

2.6. eTCPL Synteny in Complete Fly Genomes

To determine the possible integration events in further detail, completely sequenced and annotated genomes that harbour potential eTCPLs were explored in more detail. From all the results, only *Drosophila* species were fully sequenced; thus, they were used in the synteny analysis. All the work was performed using the FlyBase database [45] and its genome browser (June 2017). eTCPL genes in corresponding fly genomes were determined, and the gene regions surrounding it on both sides were analysed. Additionally, the copy number of the eTCPLs, the localisation of the surrounding genes and the mobile element existing near the eTCPL site were included in the analysis.

3. Results

3.1. Searching for EVEs Using "Sequence versus Sequence" Search Algorithms (BLAST and "Phmmer")

First, to identify potential homologues of *Tobamovirus* coat proteins (tobamo-CPs) from eukaryotic genomes, a BLAST analysis [34], as a standard protocol for detecting EVEs, was used. In general, the viral protein sequences were used as queries in BLAST analyses (a list of viral sequences used can be found in Table S1), but the analyses gave no informative *Tobamovirus*-related eukaryotic results (Figure 1 and Table S2). In BLASTp (protein vs. protein search), one result (UniProt ID: P93362_TOBAC) had a significant E-value and bit-score; however, closer examination of the annotation showed that its viral sequences have been annotated as a part of the *Nicotiana tabacum* plant genome.

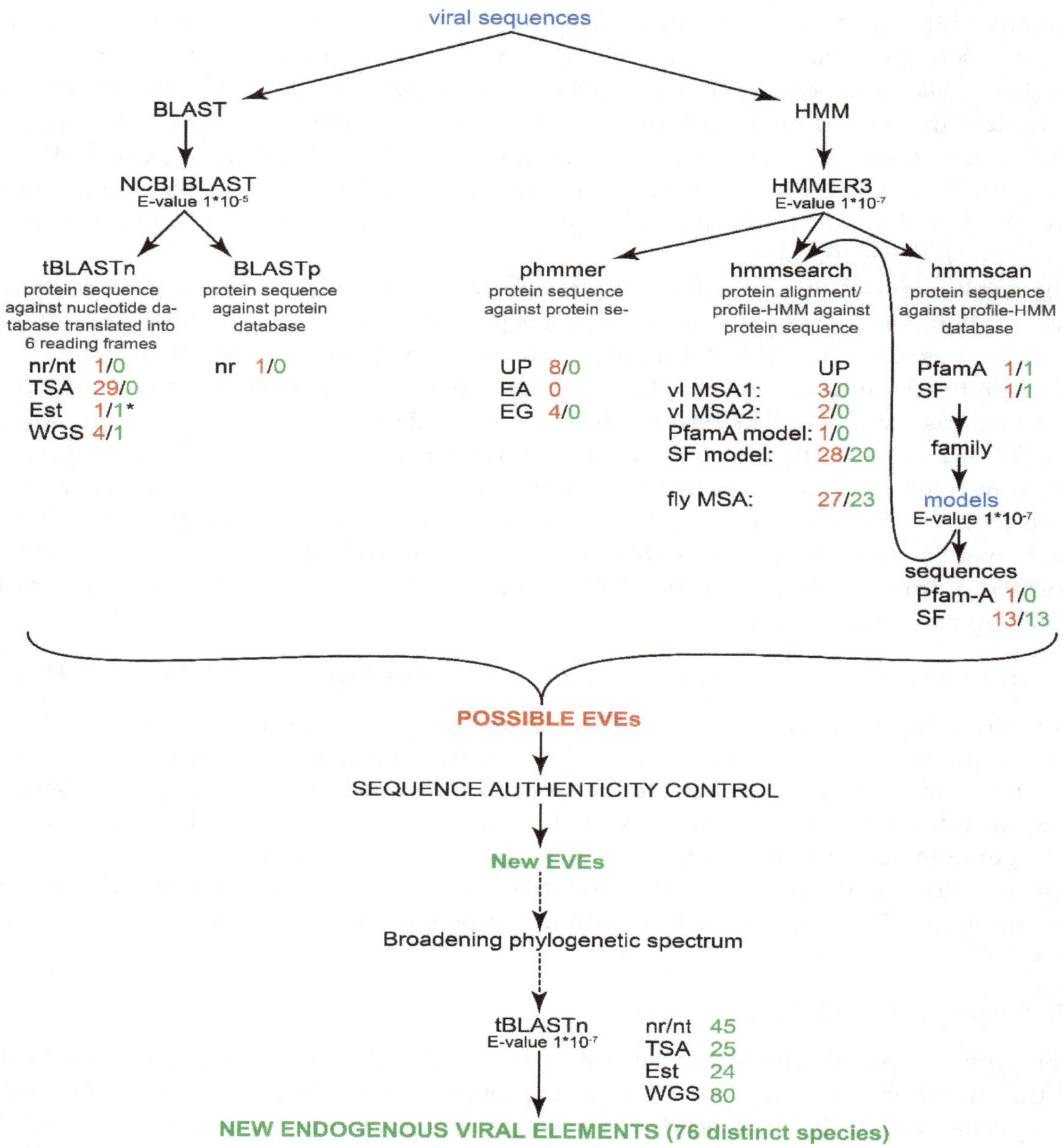

Figure 1. Workflow of the identification of new EVEs. The first step is the search against eukaryotic sequences from different databases (the Swiss-Prot database results are left out because this database gave no hits) using viral sequences as a query. All the search inputs (sequences and models) are indicated in blue script. All the hits are then examined to determine false positives and mis-annotations. All the numbers after the databases (written in red) indicate the distinct hit IDs that were collected from the databases. The second number (written in green) indicates (after all the noted sequence authenticity controls) how many possible EVEs were identified from the databases. The "*" denotes one recombinant sequence that was found in the analyses. "vl MSA1" denotes *Tobamovirus*-specific MSA, "vl MSA2" denotes *Tobamovirus-Hordeivirus-Pecluvirus-Tobravirus* specific MSA and "fly MSA" denotes an MSA that was constructed using 13 *Drosophila* fly sequences from the SUPERFAMILY database. Abbreviations: UniProtKB, UP; Ensembl All, EA; Ensembl Genomes, EG; SUPERFAMILY, SF. Note that "nr" database in BLASTp search include also UniProtKB. The sequence authenticity control workflow is described in detail in Figure 2.

We extended our search to tBLASTn (protein sequence as a query against nucleotide database translated into 6 reading frames) to find EVEs not yet annotated or not coding a protein. Using tobamo-CP as a query in tBLASTn against the nr/nt database, no true positives were found. However, against WGS (Whole Genome Sequence) database, four significant hits were obtained (Figure 1 and Table S2). In reciprocal BLAST analyses, one sequence was highly similar to Polydnaviridae, and two of the sequences seemed to be relatives of the tobamo-CP. However, since the contig length of the DNA was similar or shorter than the known *Tobamovirus* genome length, these sequences were classified as ND (non-determined because of a lack of information to confidently classify this sequence as part of the eukaryotic genome). One of these, *Lasioglossum albipes* (bee) sequence ANOB01025384.1 (NCBI accession ID) in the WGS database, however, did have a DNA scaffold over 40 Mbp (40,000 bp) in length. Protein structure prediction of the region that harbours the potential EVE indicated that the tobamoviral coat protein was the best match. However, the database annotations also state that DNA was extracted from the whole body; hence, it is currently not possible to determine whether the scaffold is part of the actual bee genome or is from some other organism found in the bee's holobiont.

We also performed a similar tBLASTn search against the TSA (Transcription Shotgun Assembly) and Est (Expressed Sequenced Tags) databases. No true positive hits were obtained because, while sequencing the whole-body RNA, no evidence was given that the RNA was transcribed from the eukaryotic genome (Figure 1 and Table S2).

To determine whether using a more complex method (while still using the protein sequence as a query) can give us more information than a pairwise comparison method, additional analyses were performed. Hidden Markov Models have been shown to identify distant homologous sequences more successfully than BLAST. HMMER3 [35] was used for this purpose because it has been shown to be more sensitive while not losing computational speed. When using the same *Tobamovirus* coat protein sequences as queries in HMMER3 "phmmer" search against the protein sequence databases, no new significant results were found (Table S3), but this result may be due to the small number of databases that can be searched against (UniProt, SwissProt and Ensemble). None of the 18 primary hits from this analysis met the other criteria described in the Materials and Methods Section.

Thus, sequence versus sequence searches gave one possible true hit (*Lasioglossum albipes*; NCBI accession ID: ANOB01025384.1). Despite expanding the search queries to include *Hordeivirus* (family Virgaviridae; +ssRNA viruses), *Pecluvirus* (family Virgaviridae; +ssRNA viruses) and *Tobravirus* (family Virgaviridae; +ssRNA viruses) capsid proteins (sequences listed in Table S1), no additional EVEs were found.

3.2. *Searching for EVEs Using "MSA versus Sequence" search Algorithms ("hmmsearch")*

As noted above, it has been known for a long time that profile-based methods work better than sequence-based methods, especially in distant homologous sequence detection. We used *Tobamovirus* and *Tobamovirus-Hordeivirus-Pecluvirus-Tobravirus* capsid proteins (sequences listed in Table S1) to generate MSAs and the corresponding profiles, and then we searched the sequence databases using these profiles. All these analyses gave hits (Table S5) that had been previously classified as false positives, whether by viral contamination or by belonging to other protein domains. Hence, the "MSA versus sequence" search did not result in any new EVEs (Figure 1).

3.3. *Identification of EVEs Using Profile-HMMs from the Pfam-A Database*

Additionally, when searching single sequences against databases that use protein domain profile-HMMs (e.g., Pfam-A, TIGRFAM, and SUPERFAMILY), HMMER3 "hmmscan" gave highly reliable results when using the tobamo-CP as a query (Table S4). The Pfam-A database gave the result of protein family "Virus coat protein (TMV like)" (Pfam-A code PF00721.20). This Pfam-A family consists of sequences of *Tobamovirus*, *Hordeivirus*, *Pecluvirus*, *Tobravirus* (Virgaviridae, +ssRNA virus), *Furovirus* (Virgaviridae, +ssRNA virus), *Pomovirus* (Virgaviridae, +ssRNA virus) and *Benyvirus* (Benyviridae, +ssRNA virus) coat proteins. However, this Pfam-A family does not contain any eukaryotic sequences (neither in Pfam-A version 28 nor in versions 29, 30 and 31), except for one *Nicotiana tabacum* protein (UniProt accession ID: P93362_TOBAC) that has previously identified as a viral sequence. Pfam-A version 31 is based on UniProt release 2016_10 and does not include sequences deposited after October 2016. To overcome this limitation, we used Pfam-A's own profile-HMM model as a search query against the UniProt database ("hmmsearch") on the HMMER webpage (https://www.ebi.ac.uk/Tools/hmmer/), which uses a monthly updated version of the UniProt. However, no new results were found (Table S6).

3.4. Searching for EVEs Using Structure Guided Profile-HMMs from the SUPERFAMILY Database

SUPERFAMILY is the other profile-HMM database, where the protein domain family is presented as a profile-HMM [37]. It identified "TMV-like viral coat proteins" (SUPERFAMILY accession ID: 47195 or SF_47195) as a positive hit (Table S4). SUPERFAMILY is a resource of genomic assignment of protein structural domains. This database constructs profile-HMMs for protein structural domains at the SCOP [46] superfamily level (throughout this article, we used term "superfamily" as it is defined in SCOP) and then uses completely sequenced and annotated genomes to identify distant homologous sequences that belong to the same structural superfamily. According to the SCOP authors, a superfamily level groups protein domains with a common ancestor.

According to the SCOP classification, the superfamily "TMV-like viral coat protein" (SCOP concise classification string "sccs" a.24.5) is composed of four *Tobamovirus* coat protein structures (PDB ID: 1EI7, 1VTM, 1RMV, 1CGM). Obviously, the *Tobamovirus* capsid sequences (22 genomes) are assigned to HMM models of this superfamily. Additionally, the capsid sequences of tobraviruses (three genomes), hordeiviruses (one genome) and pecluviruses (two genomes) are assigned to the "TMV-like viral coat protein" superfamily. Additionally, three sequences from Bymoviruses (Potyviridae, +ssRNA virus), each from different genomes, gave highly significant hits. However, these sequences are not capsid proteins but are from a RNA2-encoded polyprotein (P2). Recall that SUPERFAMILY resources use non-redundant, complete viral genomes from NCBI as a source for viral sequences.

In addition to viruses, thirteen *Drosophila* fly species are also assigned to "TMV-like viral coat protein" superfamily (Table S7). All thirteen *Drosophila* species were assigned with very high confidence (an E-value less than 1×10^{-33}) and with an almost full alignment length. Thus, these sequences are most likely homologous to tobamo-CPs. Although this database uses completely sequenced and annotated genomes and the profile-HMMs are trained to be highly specific, false positives classifications should not be overlooked; hence, protein structure prediction was performed to confirm the SUPERFAMILY database results. Additionally, when using HMM models of the SF_47195 SCOP classification from the SUPERFAMILY database as queries in the "hmmsearch" on the HMMER webpage with up-to-date databases, several hits were found (Table S6). Most of these species belonged to different *Drosophila* species, and others belonged to different fly species from the Schizophora section (*Bactrocera* species, *Ceratitis capitata*, *Lucilia cuprina*, *Musca domestica*, *Stomoxys calcitrans* and *Glossina* species). Similar results were also observed when using the SUPERFAMILY *Drosophila* sequence (protein domain part)-based MSA as the "hmmsearch" query (Table S5).

3.5. Confirming SUPERFAMILY Hits as True Homologues to TMV-CP

To exclude false positives, the LOMETS metaserver [40] was chosen for protein structure prediction (Figure 2). This program uses protein threading, which attempts to fit the given sequence to previously identified protein structures, aligns each amino acid accordingly, and then it evaluates how well the target fits the template. The results of the LOMETS metaserver for these *Drosophila* sequences show that the best modelling templates, according to several different prediction algorithms (including those that do not use HMMs), belong to tobamo-CP structures. This result confirms that *Drosophila* fly sequences are highly likely to have structures very similar to tobamo-CPs and thus have a high likelihood of being real homologues.

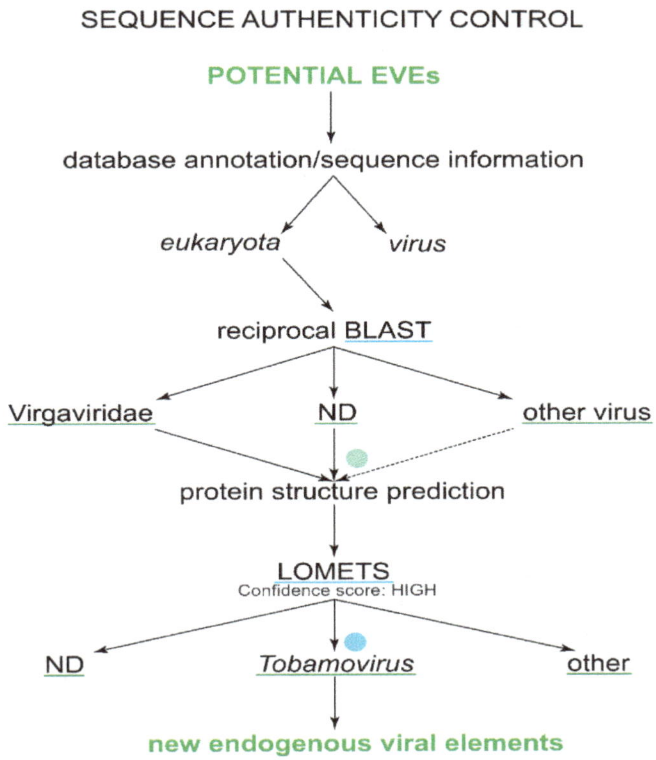

Figure 2. A detailed sequence authenticity control workflow. Each step describes what kind of results can be obtained from each step and what results can be used in further analyses. The programs used in the analyses are underlined with blue line and the results of the analyses are underlined with green line. The green dot shows that we cannot distinguish between viruses and EVEs when only RNA-seq data are available. Similarly, the blue dot shows that, even when protein structure prediction may indicate that the sequence is homologous to a certain virus, this sequence may have a close relative whose structure is not yet solved. However, the protein structure prediction still indicates that the potential EVE sequence most likely belongs to the respective structural superfamily (in terms of the SCOP).

When looking at the surrounding region of the eTCPL in *Drosophila* (and in other true positive hits for SUPERFAMILY HMMs) in more detail (i.e., when analysing contig length and the presence of other viral domains), no other indication of a misannotated viral origin was found. Thus, the *Drosophila* eTCPL is coded by the real part of the genome and not by viral sequences accidently annotated as a part of the genome. The eTCPL is found in several *Drosophila* species, which significantly reduces the probability of sequencing artefacts. In conclusion, the SUPERFAMILY HMMs found several true positives among Diptera.

3.6. Phylogenetic Analysis of Virgaviridae CP and Diptera eTCPLs

The above analysis shows that *Drosophila* and other flies contain the eTCPL domain. This result by itself does not address transfer direction. To evaluate the phylogenetic relationship of eTCPLs and exogenous viruses, we performed a comprehensive phylogenetic analysis using CP sequences of corresponding Virgaviridae viruses (tobamoviruses, tobraviruses, pecluviruses, hordeiviruses and goraviruses) (Table S1); and the integrated eTCPLs in eukaryotes (only high-confidence eTCPLs were used).

As shown in the phylogenetic tree in Figure 3, all viral genera (tobamoviruses, hordeiviruses, pecluviruses and tobraviruses) have been clearly separated according to their known phylogeny. The overall distribution of viral genera is very similar to the known divergence of these viruses, with hordeiviruses and pecluviruses being very closely related, as determined based on both replication proteins and capsid protein phylogenies of Virgaviridae [47]. These data show that the phylogenetic analysis is biologically significant and meaningful.

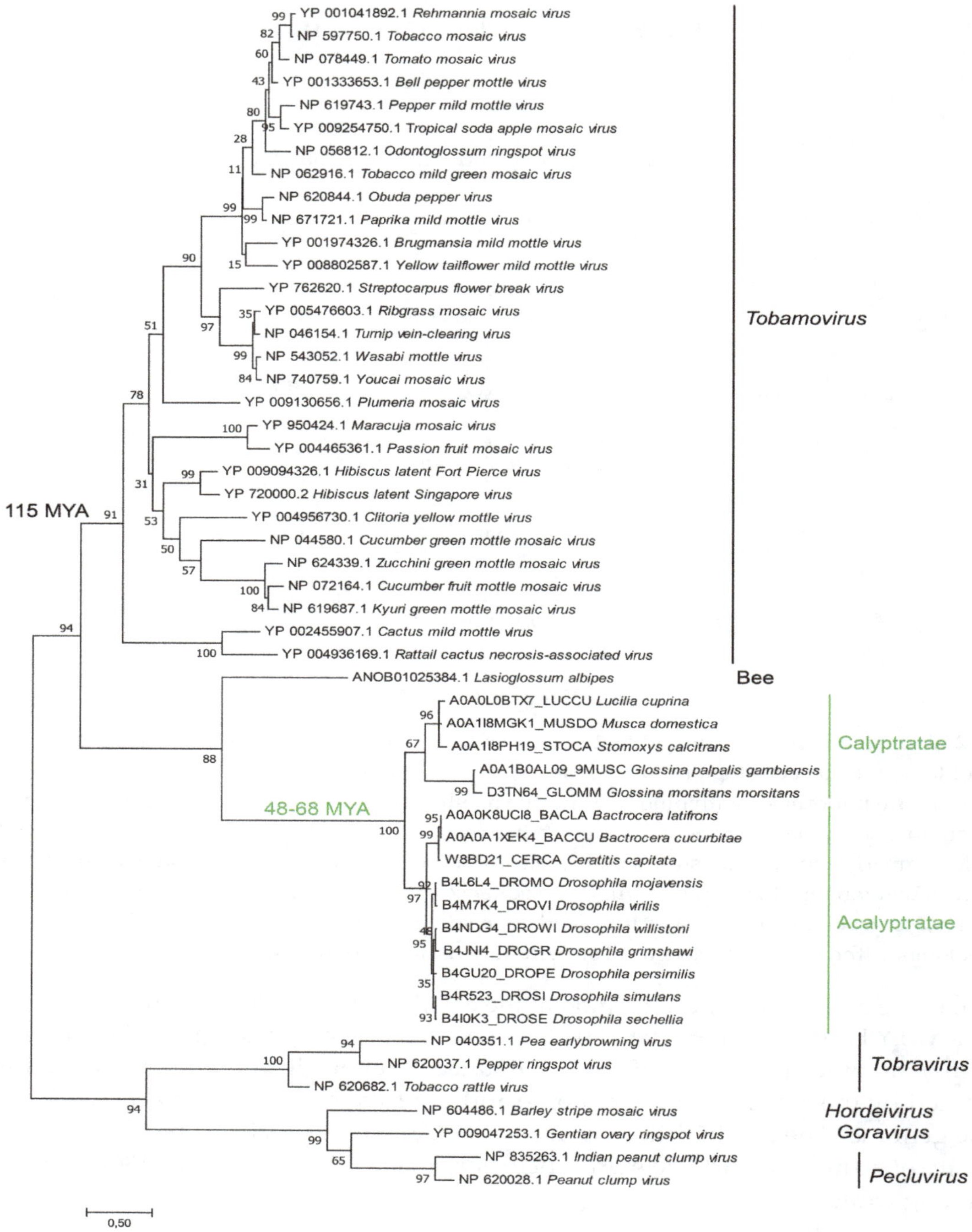

Figure 3. Phylogeny of eTCPLs and the corresponding viral sequences. The tree was constructed using the maximum likelihood method (LG + I + G + F model) using the MEGA program package [48]. All viral genera are monophyletically grouped according to their phylogeny. All the eukaryotic sequences are also grouped, and the pattern follows known fly phylogeny. The Acalyptratae-Calyptratae divergence from other flies [49–52] and *Tobamovirus* divergence from other Virgaviridae [53,54] have been noted.

All flies are distributed to two main groups based on the two subsections of Schizophora flies: Acalyptratae and Calyptratae. All the eTCPLs from different flies cluster together with very high confidence and form a sister clade with extant viruses. The phylogenetic pattern inside the eTCPL clade is consistent with the divergence of flies in the Calyptratae and Acalyptratae subsections and resembles the phylogeny of the flies itself [55].

It has been shown that the Acalyptratae-Calyptratae split occurred approximately 63–68 MYA according to three independent studies [49–51]. A more focused and through study showed that the split occurred approximately 48 MYA [52]. On the other hand, the ancestral split of *Tobamovirus* from other Virgaviridae has been shown to have occurred approximately 115 MYA [53,54]. Overall, when the phylogenetic tree is supplemented with available divergence dates, it provides strong support for the virus-to-host transfer.

3.7. eTCPL Synteny in Completely Sequenced Fly Genomes

In this work, eTCPLs have been found in many different fly genomes. Of these, only a few genomes have been fully sequenced, assembled and deposited into the FlyBase database [45]. All of them are from *Drosophila* species (Table S7). Using their assembled genomes, the eTCPLs and the surrounding region were analysed. All the eTCPLs are present in a single copy, and the protein domain itself is coded by a single exon (in *Drosophila willistoni* and *Drosophila ananassae*, the respective genes have two exons, but the eTCPL domain is coded by a single exon). The genes surrounding *Drosophila melanogaster* eTCPL are IntS6, CG4078, CG15771 and lin-52 (the eTCPL gene overlaps with the last gene but not in the coding sequence). The same trend is apparent in all of the *Drosophila* species (Figure 4). However, some changes have occurred in either *Drosophila* or the *Sophophora* subgenus. For *Drosophila virilis* and *Drosophila grimshawi*, the genes IntS6 and CG4078 are located further away and are located in different scaffolds for *Drosophila mojavensis*. Additionally, *Drosophila sechellia* acquired a new gene (GM19534) that has no known orthologues. Henceforth, a possible viral gene could have integrated in this locus before the *Drosophila* fly split.

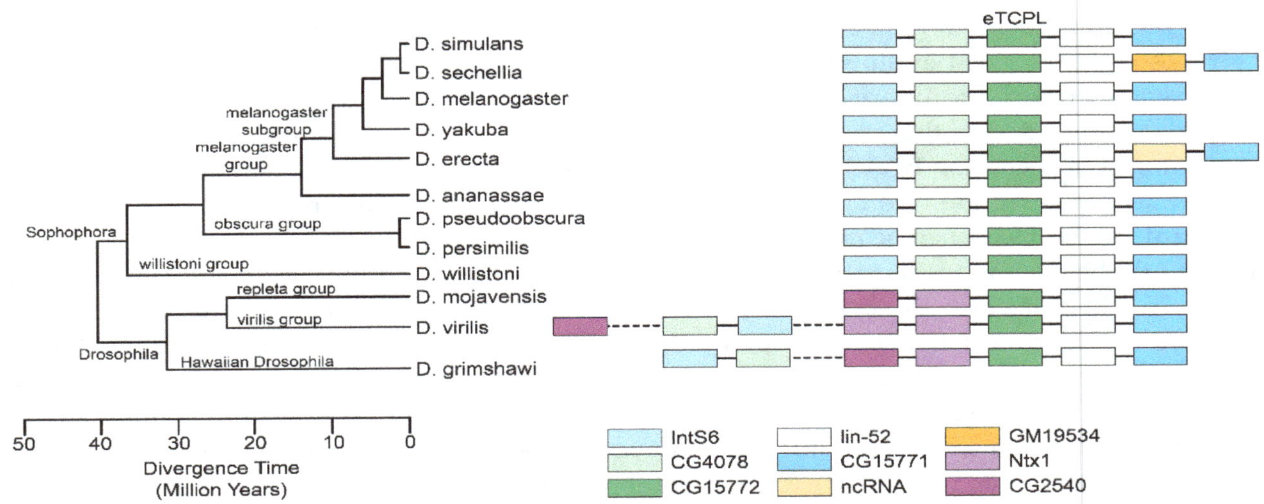

Figure 4. eTCPL gene synteny comparisons on fully sequenced genomes from *Drosophila* flies. The eTCPL is noted in dark green and was taken as an observation centre. *Drosophila melanogaster*, the most well-researched fly, was taken as a baseline and two to three genes surrounding the eTCPL were also considered. All data were collected from the FlyBase database [45], including the tree picture. In some flies whose genes surrounding the eTCPL were different, previously noted genes (IntS6 and CH4078) were also located for this analysis.

3.8. EVE Detection from Incompletely Sequenced Genomes

True positive eukaryotic hits of the SF_47195 from the SUPERFAMILY database (which uses the complete genomes and UniProt) indicated flies are hosts to eTCPLs. The next step in the analysis was an additional BLAST analysis to evaluate their phylogenomic distribution in uncomplete genomes, whose proteins may not yet be annotated.

In our approach, HMMs can recognise deeper evolutionary relationships, and BLAST will give a better phylogenomic coverage since the HMM search was limited to a few protein databases and

BLAST has a much wider possible database usage range and allows searching against nucleotide databases (including subsets that do not have annotated translations). Using the fly eTCPLs as queries, 76 different species (including 13 species from SUPERFAMILY) were identified as containing true positive eTCPLs in their genome (Table S8). The results mostly consist of both Acalyptratae and Calyptratae species (65 species belonging to the Schizophora section) covering both of their phylogenies as best as possible with the data available.

Most of the species (in the Schizophora group) also have the same genes surrounding the eTCPLs as in *Drosophila melanogaster* (genes IntS6, CG15771 and CG4078 located on the same contig as the eTCPL) with the exception of a few species (data not shown), most likely because those species are not sequenced as fully as others. This also indicates that eTCPLs' synteny shown for *Drosophilidae* (Figure 4) is also extended to Calyptratae and thus for all Schizophora.

We can use protein structure-guided profile-HMM analyses with an additional tBLASTn search to cover the un-annotated coding sequences. With this approach, it is possible to detect eTCPL sequences in a large number of eukaryotic genomes, which was not possible when using only a BLAST search.

4. Discussion

During the last decade, many EVEs have been detected in eukaryotes. According to the published literature, the number of viral families that have donated EVEs to eukaryotes is at least 34 for identified nucleic acid transfers, including at least 21 viral families where the coding potential of integrated viral sequences have remained (data collected from the literature by H. Kirsip). The recipient eukaryotic organisms have been fungi, plants, vertebrates, arthropods and others. Even when using sequence-based EVE search strategies, the impact of non-retroviral viruses in cellular genomes/proteomes is (and has been) significant. Currently applied methods do not take into account that viruses have very fast evolutionary rates, and thus, sequence similarity may not be detectable using pairwise sequence comparison methods.

The data presented in this study show that, as expected from the literature, there exist methods that are more sensitive than BLAST to detect "virus-to-host" transfers of genetic material. It should be noted that the same method could also be applied to the "host-to-virus" approach. According to the literature, HMM-based searches should be more sensitive than BLAST, and structure-guided HMMs should be even more sensitive than sequence-based HMM searches in the detection of remote homologues. In our case, BLAST found one possible homologue in eukaryotes for *Tobamovirus* capsid proteins. Additionally, using different HMMER3 search capabilities ("phmmer" and the viral MSA-based "hmmsearch", both based on HMMs) did not give any new reliable results. At the same time, structure-guided HMMs from SUPERFAMILY found distant homologues of tobamoviral proteins in eukaryotes. Our data show that, as a "proof of principle", structure-guided HMMs can be used to detect V2H transfers that are not detectable by other methods.

In modern high throughput times, the annotations of the data in different databases vary, particularly from the point of view that relevant aspects of annotations vary for different researchers (and for their scientific questions). In our case, the authenticity of sequences (their origin) must be carefully tested because very often the taxonomic annotations of the origin of a nucleic acid correspond to major species in holobionts and not to their real origin. This phenomenon has also been used by scientists in a positive way, extracting, for example, the viruses (real viruses, not EVEs) infecting insects from the data with taxonomic annotation to insects [56]. If the sequence data are not linked to publications (or if the methods of how the probe was isolated and prepared have not been properly described), it is very hard to evaluate the real origin of a sequenced nucleic acid. Some types of data, by definition, do not allow us to distinguish between viruses and EVEs (for example, RNA-seq data). Curated datasets such as complete proteomes or complete genomes seem to have fewer un-authentic sequences compared to WGS. However, using WGS and similar databases are very useful for acquiring taxonomic coverage that is as wide as possible. Of course, every method may have false positive hits;

therefore, we analysed all the search hits with multiple alternative methods to obtain more confidence results. Unfortunately, these methods are also not as "high throughput".

Structure-guided HMMs help to detect remote homology; however, detecting homology is not enough to classify the sequence as an EVE. In addition to vertical inheritance from primordial worlds, there are two processes that may lead to homologous proteins in viruses and organisms: "virus-to-host" transfer and "host-to-virus" transfer. To detect EVEs, the transfer direction of homologues found in viruses and cells must be determined. Determination of the transfer direction for more ancestral transfers is not as trivial as in the case of more recent events. In the latest case, it is possible to use the highly confident outgroup sequences to root the phylogenetic tree and polarised it on time. At the same time, when dealing with more older events, determining a good outgroup is a challenge. Studying the evolutionary history of a single viral protein (or a protein domain) selection of outgroups is complicated by rampant horizontal gene transfer between viruses. Even such an unlikely event as gene transfer between ssDNA viruses and RNA viruses has been reported [10,57,58]. Known divergence times of cellular and viral taxa would also help to polarise the phylogenetic tree. Unfortunately, very few virus group divergence times have been determined. Therefore, to determine the transfer direction, different independent approaches should be used.

To determine the transfer direction of proteins under study, we combined several approaches. First, we tested the synteny of eTCPLs in well-annotated species in the FlyBase database. The eTCPL and its surrounding area in fully sequenced genomes were analysed. The data in Figure 4 show the synteny of the eTCPL in these species. In different *Drosophila* fly genomes, the eTCPL is in the same locus, is coded by a single exon, and the surrounding genes are similar. The exception is between *Drosophila* and the *Sophophora* subgenus, where some gene region reorganisation has occurred in one of them. Respective species (where the eTCPL has the same location according to FlyBase) diverged approximately 48 MYA (the known divergence time for *Drosophila* species) [52], and at least for these species, the eTCPL shows a clear monophyletic origin with a single transfer event according to the proposed phylogeny of the *Drosophila*/*Sophophora* common ancestor. We also reconstitute the phylogenetic tree of viral tobamo-CP and cellular eTCPL sequences together. This tree (and several other trees with different alignment and tree building algorithms [59]) shows very strong bootstrap confidence intervals for viral genera as well as for the Calyptratae/Acalyptratae split (from an alignment with a length of ~130 AA), indicating the biological significance of the tree. Seeing as both Schizophora subsections (Acalyptratae and Calyptratae) have eTCPL in their genomes and the eTCPL phylogeny is consistent with known fly phylogeny [55], it can be concluded that the integration of viral elements occurred before these two groups diverged. One group of flies, Aschiza, diverged before the Acalyptratae/Calyptratae divergence and are closest relatives of Schizophora. In this group of flies, there are three organisms with fully sequenced genomes (according to the NCBI Genomes database): *Megaselia abdita*, *Megaselia scalaris* and *Eristalis dimidiata*. None of the organisms could be identified as carrying the eTCPL gene; however, the neighbouring genes of eTCPL were detected using BLAST [59]. Therefore, it could be concluded that the main integration of the tobamo-CP gene was into the ancestor of the Schizophora flies. Supplementing the phylogenomic tree with known divergence times of tobamoviruses, the Calyptratae/Acalyptratae split, and synteny analysis, tobamoviruses are much older than eTCPLs in Schizophora. Therefore, taking all this information into account, the most likely single transfer event took place between 115 and 48 MYA with a direction from viruses to insects.

The presence of plant virus genes in insect genomes may seem very unlikely, but we must take into account that insects are common plant virus vectors and otherwise have very intimate contact with plants. According to that, the described transfer is no longer unexpected. Additionally, single V2H transfer events in one organism must be somehow beneficial for hosts to be fixed and to spread in populations and maintained over millions of years. The detected EVE seems to be biologically active, as the respective mRNA has been observed in several fly species (Table S8). In *Drosophila melanogaster*, the expression is observed in different developmental stages and different tissues (our in silico analysis of preexisting data deposited in different databanks, data not shown). In addition, the peptide corresponding to amino acids 229–246 in UniProt Q9W483 (a product of gene CD15772, i.e., The *D. melanogaster* eTCPL) have been observed in adult *Drosophila melanogaster* heads according to a search in the PeptideAtlas database (www.peptideatlas.org).

However, more detailed experiments need to be performed to determine the full function of the eTCPLs in flies.

5. Conclusions

This work showed that fast evolving viral protein homologues can be identified in eukaryotic genomes using structure-guided HMM searches, even when the integration event itself is very ancient.

Author Contributions: Conceptualisation, A.A.; methodology, A.A.; software, H.K.; validation, H.K., A.A.; formal analysis, H.K. and A.A.; investigation, A.A. and H.K.; data curation, H.K.; writing—original draft preparation, H.K.; writing—review and editing, A.A. and H.K.; visualisation, H.K.; and supervision, A.A.

Acknowledgments: The authors are grateful to Lauri Saag for their support with phylogenetic trees.

References

1. Thingstad, T.F.; Lignell, R. Theoretical models for the control of bacterial growth rate, abundance, diversity and carbon demand. *Aquat Microb. Ecol.* **1997**, *13*, 19–27. [CrossRef]
2. Middelboe, M.; Lyck, P.G. Regeneration of dissolved organic matter by viral lysis in marine microbial communities. *Aquat Microb. Ecol.* **2002**, *27*, 187–194. [CrossRef]
3. Suttle, C.A. Marine viruses–major players in the global ecosystem. *Nat. Rev. Microbiol.* **2007**, *5*, 801–812. [CrossRef]
4. Cobián Güemes, A.G.; Youle, M.; Cantú, V.A.; Felts, B.; Nulton, J.; Rohwer, F. Viruses as Winners in the Game of Life. *Annu. Rev. Virol.* **2016**, *3*, 197–214. [CrossRef] [PubMed]
5. Mokili, J.L.; Rohwer, F.; Dutilh, B.E. Metagenomics and future perspectives in virus discovery. *Curr. Opin. Virol.* **2012**, 63–77. [CrossRef]
6. Bergh, Ø.; Børsheim, K.Y.; Bratbak, G.; Heldal, M. High abundance of viruses found in aquatic environments. *Nature* **1989**, *340*, 467–468. [CrossRef] [PubMed]
7. Whon, T.W.; Kim, M.-S.; Roh, S.W.; Shin, N.-R.; Lee, H.-W.; Bae, J.-W. Metagenomic Characterization of Airborne Viral DNA Diversity in the Near-Surface Atmosphere. *J. Virol.* **2012**, *86*, 8221–8231. [CrossRef]
8. Lander, E.S.; Linton, L.M.; Birren, B.; Nusbaum, C.; Zody, M.C.; Baldwin, J.; Devon, K.; Dewar, K.; Doyle, M.; Fitzhugh, W.; et al. Initial sequencing and analysis of the human genome. *Nature* **2001**, *409*, 860–921. [CrossRef]
9. Bejarano, E.R.; Khashoggi, A.; Witty, M.; Lichtenstein, C. Integration of multiple repeats of geminiviral DNA into the nuclear genome of tobacco during evolution. *Proc. Natl. Acad. Sci. USA* **1996**, *93*, 759–764. [CrossRef]
10. Stedman, K.M. Deep Recombination: RNA and ssDNA Virus Genes in DNA Virus and Host Genomes. *Annu. Rev. Virol.* **2015**, *2*, 203–217. [CrossRef] [PubMed]
11. Katzourakis, A. Endogenous Viral Elements in Animal Genomes. *PLoS Genet.* **2010**. [CrossRef] [PubMed]
12. Horie, M.; Tomonaga, K. Non-retroviral fossils in vertebrate genomes. *Viruses* **2011**, *3*, 1836–1848. [CrossRef] [PubMed]
13. Chu, H.; Jo, Y.; Cho, W.K. Evolution of endogenous non-retroviral genes integrated into plant genomes. *Curr. Plant Biol.* **2014**, 55–59. [CrossRef]
14. Frank, A.C.; Wolfe, K.H. Evolutionary capture of viral and plasmid DNA by yeast nuclear Chromosomes. *Eukaryot. Cell* **2009**. [CrossRef]
15. Aiewsakun, P.; Katzourakis, A. Endogenous viruses: Connecting recent and ancient viral evolution. *Virology* **2015**. [CrossRef] [PubMed]
16. Frank, J.A.; Feschotte, C. Co-option of endogenous viral sequences for host cell function. *Curr. Opin. Virol.* **2017**, 81–89. [CrossRef] [PubMed]
17. Myers, K.N.; Barone, G.; Ganesh, A.; Staples, C.J.; Howard, A.E.; Beveridge, R.D.; Maslen, S.; Skehel, J.M.; Collis, S.J. The bornavirus-derived human protein EBLN1 promotes efficient cell cycle transit, microtubule organisation and genome stability. *Sci. Rep.* **2016**, *6*. [CrossRef]
18. He, P.; Sun, L.; Zhu, D.; Zhang, H.; Zhang, L.; Guo, Y.; Liu, S.; Zhou, J.; Xu, X.; Xie, P. Knock-down of endogenous bornavirus-like nucleoprotein 1 inhibits cell growth and induces apoptosis in human oligodendroglia cells. *Int. J. Mol. Sci.* **2016**, *17*, 435. [CrossRef]

19. Fujino, K.; Horie, M.; Honda, T.; Merriman, D.K.; Tomonaga, K. Inhibition of Borna disease virus replication by an endogenous bornavirus-like element in the ground squirrel genome. *Proc. Natl. Acad. Sci. USA* **2014**, *111*, 13175–13180. [CrossRef]
20. Murad, L.; Bielawski, J.P.; Matyasek, R.; Kovarík, A.; Nichols, R.A.; Leitch, A.R.; Lichtenstein, C.P. The origin and evolution of geminivirus-related DNA sequences in Nicotiana. *Heredity* **2004**, *92*, 352–358. [CrossRef]
21. Kondo, H.; Chiba, S.; Suzuki, N. Detection and analysis of non-retroviral RNA virus-like elements in plant, fungal, and insect genomes. *Methods Mol. Biol.* **2015**, *1236*, 73–88. [CrossRef]
22. Brenner, S.E.; Chothia, C.; Hubbard, T.J. Assessing sequence comparison methods with reliable structurally identified distant evolutionary relationships. *Proc. Natl. Acad. Sci. USA.* **1998**, *95*, 6073–6078. [CrossRef]
23. Park, J.; Karplus, K.; Barrett, C.; Hughey, R.; Haussler, D.; Hubbard, T.; Chothia, C. Sequence comparisons using multiple sequences detect three times as many remote homologues as pairwise methods. *J. Mol. Biol.* **1998**, *284*, 1201–1210. [CrossRef] [PubMed]
24. Abroi, A.; Gough, J. Are viruses a source of new protein folds for organisms?—Virosphere structure space and evolution. *BioEssays* **2011**, *33*, 626–635. [CrossRef] [PubMed]
25. Aiewsakun, P.; Katzourakis, A. Time-Dependent Rate Phenomenon in Viruses. *J. Virol.* **2016**, *90*, 7184–7195. [CrossRef] [PubMed]
26. Duffy, S.; Shackelton, L.A.; Holmes, E.C. Rates of evolutionary change in viruses: Patterns and determinants. *Nat. Rev. Genet.* **2008**, *9*, 267–276. [CrossRef] [PubMed]
27. Kuchibhatla, D.B.; Sherman, W.A.; Chung, B.Y.W.; Cook, S.; Schneider, G.; Eisenhaber, B.; Karlin, D.G. Powerful Sequence Similarity Search Methods and In-Depth Manual Analyses Can Identify Remote Homologs in Many Apparently "Orphan" Viral Proteins. *J. Virol.* **2014**, *88*, 10–20. [CrossRef]
28. Dunbrack, R.L. Sequence comparison and protein structure prediction. *Curr. Opin. Struct. Biol.* **2006**, 374–384. [CrossRef] [PubMed]
29. Karlin, D.; Belshaw, R. Detecting remote sequence homology in disordered proteins: Discovery of conserved motifs in the N-termini of mononegavirales phosphoproteins. *PLoS ONE* **2012**, *7*. [CrossRef] [PubMed]
30. Illergård, K.; Ardell, D.H.; Elofsson, A. Structure is three to ten times more conserved than sequence—A study of structural response in protein cores. *Proteins Struct. Funct. Bioinform.* **2009**, *77*, 499–508. [CrossRef]
31. Chothia, C.; Lesk, M. The relation between the divergence of sequence and structure in proteins. *EMBO J.* **1986**, *5*, 823–826. [CrossRef] [PubMed]
32. Challis, C.J.; Schmidler, S.C. A stochastic evolutionary model for protein structure alignment and phylogeny. *Mol. Biol. Evol.* **2012**, *29*, 3575–3587. [CrossRef] [PubMed]
33. Herman, J.L.; Challis, C.J.; Novák, A.; Hein, J.; Schmidler, S.C. Simultaneous Bayesian Estimation of Alignment and Phylogeny under a Joint Model of Protein Sequence and Structure. *Mol. Biol. Evol.* **2014**, *31*, 2251–2266. [CrossRef] [PubMed]
34. Altschul, S.F.; Madden, T.L.; Schäffer, A.A.; Zhang, J.; Zhang, Z.; Miller, W.; Lipman, D.J. Gapped BLAST and PSI-BLAST: A new generation of protein database search programs. *Nucleic Acids Res.* **1997**, *25*, 3389–3402. [CrossRef]
35. Finn, R.D.; Clements, J.; Eddy, S.R. HMMER web server: Interactive sequence similarity searching. *Nucleic Acids Res.* **2011**, *39*, 29–37. [CrossRef]
36. Finn, R.D.; Coggill, P.; Eberhardt, R.Y.; Eddy, S.R.; Mistry, J.; Mitchell, A.L.; Potter, S.C.; Punta, M.; Qureshi, M.; Sangrador-Vegas, A.; et al. The Pfam protein families database: Towards a more sustainable future. *Nucleic Acids Res.* **2016**, *44*, D279–D285. [CrossRef] [PubMed]
37. Gough, J.; Karplus, K.; Hughey, R.; Chothia, C. Assignment of homology to genome sequences using a library of hidden Markov models that represent all proteins of known structure. *J. Mol. Biol.* **2001**, *313*, 903–919. [CrossRef] [PubMed]
38. Kumar, S.; Stecher, G.; Tamura, K. MEGA7: Molecular Evolutionary Genetics Analysis Version 7.0 for Bigger Datasets. *Mol. Biol. Evol.* **2016**, *33*, 1870–1874. [CrossRef] [PubMed]
39. Pearson, W.R. An introduction to sequence similarity ("homology") searching. *Curr. Protoc. Bioinform.* **2013**. [CrossRef]
40. Wu, S.; Zhang, Y. LOMETS: A local meta-threading-server for protein structure prediction. *Nucleic Acids Res.* **2007**, *35*, 3375–3382. [CrossRef]
41. Waterhouse, A.M.; Procter, J.B.; Martin, D.M.; Clamp, M.; Barton, G.J. Jalview Version 2-A multiple sequence alignment editor and analysis workbench. *Bioinformatics* **2009**, *25*, 1189–1191. [CrossRef] [PubMed]
42. Triant, D.A.; Pearson, W.R. Most partial domains in proteins are alignment and annotation artifacts. *Genome Biol.* **2015**, *16*, 1–12. [CrossRef] [PubMed]

43. Katoh, K.; Rozewicki, J.; Yamada, K.D. MAFFT online service: Multiple sequence alignment, interactive sequence choice and visualization. *Brief Bioinform.* **2017**. [CrossRef] [PubMed]
44. Abascal, F.; Zardoya, R.; Posada, D. ProtTest: Selection of best-fit models of protein evolution What can I use ProtTest for?—Introduction The program: Using ProtTest. *Bioinformatics* **2005**, *21*, 1–17. [CrossRef]
45. Attrill, H.; Falls, K.; Goodman, J.L.; Millburn, G.H.; Antonazzo, G.; Rey, A.J.; Marygold, S.J. FlyBase Consortium. Flybase: Establishing a gene group resource for Drosophila melanogaster. *Nucleic Acids Res.* **2016**, *44*, D786–D792. [CrossRef] [PubMed]
46. Hubbard, T.J.P.; Ailey, B.; Brenner, S.E.; Murzin, A.G.; Chothia, C. SCOP: A structural classification of proteins database. *Nucleic Acids Res.* **1999**, 254–256. [CrossRef]
47. Adams, M.J.; Antoniw, J.F.; Kreuze, J. Virgaviridae: A new family of rod-shaped plant viruses. *Arch Virol.* **2009**, *154*, 1967–1972. [CrossRef] [PubMed]
48. Tamura, K.; Stecher, G.; Peterson, D.; Filipski, A.; Kumar, S. MEGA6: Molecular Evolutionary Genetics Analysis Version 6.0. *Mol. Biol. Evol.* **2013**, *30*, 2725–2729. [CrossRef] [PubMed]
49. Junqueira, A.C.M.; Azeredo-Espin, A.M.L.; Paulo, D.F.; Marinho, M.A.T.; Tomsho, L.P.; Drautz-Moses, D.I.; Purbojati, R.W.; Ratan, A.; Schuster, S.C. Large-scale mitogenomics enables insights into Schizophora (Diptera) radiation and population diversity. *Sci. Rep.* **2016**, *6*. [CrossRef]
50. Thomas, J.A.; Trueman, J.W.H.; Rambaut, A.; Welch, J.J. Relaxed phylogenetics and the palaeoptera problem: Resolving deep ancestral splits in the insect phylogeny. *Syst. Biol.* **2013**, *62*, 285–297. [CrossRef]
51. Ding, S.; Li, X.; Wang, N.; Cameron, S.L.; Mao, M.; Wang, Y.; Xi, Y.; Yang, D. The phylogeny and evolutionary timescale of muscoidea (Diptera: Brachycera: Calyptratae) inferred from mitochondrial genomes. *PLoS ONE* **2015**, *10*, e0134170. [CrossRef] [PubMed]
52. Wiegmann, B.M.; Yeates, D.K.; Thorne, J.L.; Kishino, H. Time Flies, a New Molecular Time-Scale for Brachyceran Fly Evolution Without a Clock. *Syst. Biol.* **2003**, *52*, 745–756. [CrossRef]
53. Gibbs, A.J.; Wood, J.; Garcia-Arenal, F.; Ohshima, K.; Armstrong, J.S. Tobamoviruses have probably co-diverged with their eudicotyledonous hosts for at least 110 million years. *Virus Evol.* **2015**, *1*, vev019. [CrossRef] [PubMed]
54. Stobbe, A.H.; Melcher, U.; Palmer, M.W.; Roossinck, M.J.; Shen, G. Co-divergence and host-switching in the evolution of tobamoviruses. *J. Gen. Virol.* **2012**, *93*. [CrossRef] [PubMed]
55. Wiegmann, B.M.; Trautwein, M.D.; Winkler, I.S.; Barr, N.B.; Kim, J.-W.; Lambkin, C.; Bertone, M.A.; Cassel, B.K.; Bayless, K.M.; Heimberg, A.M.; et al. Episodic radiations in the fly tree of life. *Proc. Natl. Acad. Sci. USA* **2011**, *108*, 5690–5695. [CrossRef] [PubMed]
56. Kondo, H.; Chiba, S.; Maruyama, K.; Andika, I.B.; Suzuki, N. A novel insect-infecting virga/nege-like virus group and its pervasive endogenization into insect genomes. *Virus Res.* **2017**. [CrossRef] [PubMed]
57. Roux, S.; Enault, F.; Bronner, G.; Vaulot, D.; Forterre, P.; Krupovic, M. Chimeric viruses blur the borders between the major groups of eukaryotic single-stranded DNA viruses. *Nat. Commun.* **2013**, *4*. [CrossRef]
58. Diemer, G.S.; Stedman, K.M. A novel virus genome discovered in an extreme environment suggests recombination between unrelated groups of RNA and DNA viruses. *Biol. Direct.* **2012**, *7*. [CrossRef]
59. Kirsip, H.; University of Tartu, Tartu, Estonia. Phylogenetic and synteny analyses for the eTCPL in Diptera genomes, 2018.

15

Hierarchical Cluster Analysis of Medical Chemicals Detected by a Bacteriophage-Based Colorimetric Sensor Array

Chuntae Kim [1,†], Hansong Lee [2,3,†], Vasanthan Devaraj [4,†], Won-Geun Kim [1], Yujin Lee [1], Yeji Kim [1], Na-Na Jeong [5], Eun Jung Choi [4], Sang Hong Baek [6], Dong-Wook Han [7,*], Hokeun Sun [2,3,*] and Jin-Woo Oh [1,4,8,*]

1 Department of Nanofusion Technology, Pusan National University, Busan 46241, Korea; chuntae1122@gmail.com (C.K.); kim1guen@gmail.com (W.-G.K.); pinky204@hanmail.net (Y.L.); kkyeaj0153@naver.com (Y.K.)
2 Interdisciplinary Program of Genomic Data Science, Pusan National University, Busan 46241, Korea; hansong798@naver.com
3 Department of Statistics, Pusan National University, Busan 46241, Korea
4 Research Center for Energy Convergence and Technology, Pusan National University, Busan 46241, Korea; vasanth@pusan.ac.kr (V.D.); eunjung721203@gmail.com (E.J.C.)
5 BK21PLUS Program in Embodiment: Health-Society Interaction, Department of Public Health Sciences, Graduate School, Korea University, Seoul 02841, Korea; nana8931@naver.com
6 Laboratory of Cardiovascular Disease, Division of Cardiology, School of Medicine, The Catholic University of Korea, Seoul 06591, Korea; whitesh@catholic.ac.kr
7 Department of Cogno-Mechatronics Engineering, College of Nanoscience & Nanotechnology, Pusan National University, Busan 46241, Korea
8 Department of Nanoenergy Engineering, Pusan National University, Busan 46241, Korea
* Correspondence: nanohan@pusan.ac.kr (D.-W.H.); hsun@pusan.ac.kr (H.S.); ojw@pusan.ac.kr (J.-W.O.)
† These authors contributed equally to this work.

Abstract: M13 bacteriophage-based colorimetric sensors, especially multi-array sensors, have been successfully demonstrated to be a powerful platform for detecting extremely small amounts of target molecules. Colorimetric sensors can be fabricated easily using self-assembly of genetically engineered M13 bacteriophage which incorporates peptide libraries on its surface. However, the ability to discriminate many types of target molecules is still required. In this work, we introduce a statistical method to efficiently analyze a huge amount of numerical results in order to classify various types of target molecules. To enhance the selectivity of M13 bacteriophage-based colorimetric sensors, a multi-array sensor system can be an appropriate platform. On this basis, a pattern-recognizing multi-array biosensor platform was fabricated by integrating three types of sensors in which genetically engineered M13 bacteriophages (wild-, RGD-, and EEEE-type) were utilized as a primary building block. This sensor system was used to analyze a pattern of color change caused by a reaction between the sensor array and external substances, followed by separating the specific target substances by means of hierarchical cluster analysis. The biosensor platform could detect drug contaminants such as hormone drugs (estrogen) and antibiotics. We expect that the proposed biosensor system could be used for the development of a first-analysis kit, which would be inexpensive and easy to supply and could be applied in monitoring the environment and health care.

Keywords: M13 bacteriophage; multi-array sensors; hierarchical cluster analysis; high selectivity

1. Introduction

M13 bacteriophage, one kind of filamentous bacteriophages, have been utilized as receptors in biosensors [1,2]. By means of genetic engineering techniques, it is possible to modify and achieve a better binding affinity of M13 bacteriophage towards desired target molecules [3–5]. Among various types of M13 bacteriophage-based biosensors, colorimetric sensor systems have been intensively investigated due to their facile fabrication process and sensing method [6–8]. Colorimetric sensors fabricated by self-assembly of M13 bacteriophage result in nanostructures with varying size and periodicity [9]. When white light is illuminated onto the nanostructure, specific wavelengths determined by Bragg's law are scattered more dominantly from the nanostructures. After the penetration of external chemicals, the self-assembled nanostructures swell, resulting in a change of the wavelengths that are scattered [10]. The observed color change is detected by a complementary metal–oxide–semiconductor detector, and this is followed by image analysis.

The image analysis results consist of numerical values that can be used to determine the type and concentration of the external chemicals. Using genetic engineering techniques, M13 bacteriophage-based colorimetric sensors can display sensitive color changes toward desired target materials. Furthermore, by integrating various types of genetically engineered M13 bacteriophage-based colorimetric sensors on a single chip to fabricate a sensor array, a number of target molecules can be classified by analyzing the pattern of a color change. Even though M13 bacteriophage-based multi-array sensor systems have been successfully demonstrated as a powerful platform for the detection of extremely small amounts of target molecules, the discrimination of many types of target molecules is still both necessary and challenging.

However, as the types of genetically engineered M13 bacteriophage integrated on sensor arrays increase, the analysis of the sensing results becomes complicated due to the huge amount of numerical data. In this work, we introduce a statistical method, henceforth called hierarchical cluster analysis, to classify many types of medical chemicals. Antibiotics are used widely in the livestock industry on a daily basis, but the excessive use of antibiotics raises the problem of increased antibiotic resistance in animals and humans [11].

If estrogen compounds, such as estrone (E1), 17β-estradiol (E2), and estriol (E3), and oral contraceptives released from humans and animals such as endocrine disrupters (similar to 17α-ethinylestradiol (EE2)) contaminate the environment through sewage and the excrement of animals, they can adversely affect the ecology in the water [12]. Estrogenic and antibiotic substances are found in trace amounts (~ng/L) in effluent, fresh water, river water, and even in drinking water [13,14], as a consequence of their ineffective removal at wastewater treatment plants (WWTP). In general, as any particular substance is not always present by itself, an effective method to comprehensively classify unknown compounds is necessary [15]. In this regard, M13 bacteriophage-based sensor arrays and hierarchical cluster analysis are a suitable sensor system to detect and classify various types of unknown compounds. Figure 1 is a schematic illustration describing our sensor system.

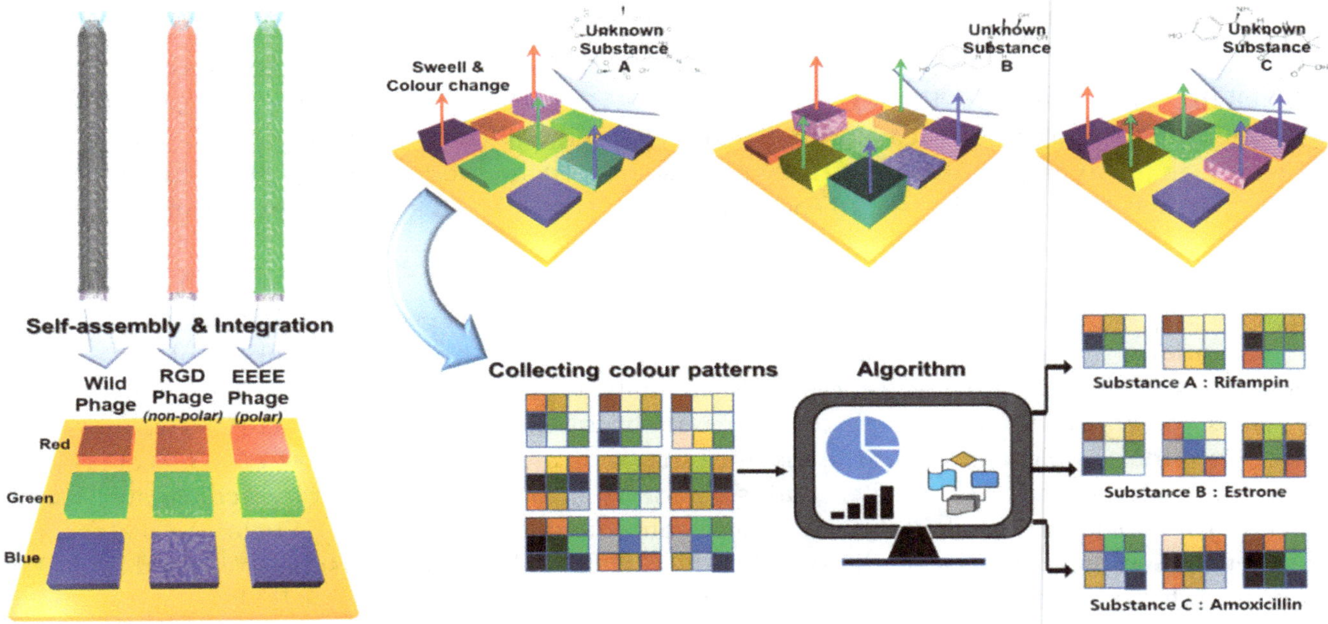

Figure 1. Schematic illustration of the M13 bacteriophage-based colorimetric sensor array. The sensor array consists of a functional M13 bacteriophage with a modified major coat protein (pVIII). When a color band is reacting with a target analyte, each type of sensor chip shows its own color change value according to the individual M13 bacteriophage's characteristics. A color pattern is formed as a unique response value, and it is possible to construct a sensor platform which can discriminate unknown textures through pattern analysis.

2. Materials and Methods

2.1. *Sensor Analysis Analytes*

Four types of estrogen drugs and four types of antibiotics were purchased from Sigma-Aldrich (Seoul, South Korea) and a local pharmacy (Medipharm, Busan, South Korea), ground, and used directly. In real-world situations, all such analytes exist in various physical conditions that cause molecules to vibrate or diffuse. To mimic these kinds of physical conditions, sensing experiments were carried out at four different temperatures (30, 60, 90, and 120 °C) for each material.

2.2. *Development of Functional Bioreporter Materials*

The wild types of filamentous phage particles were based on M13KE, which was purchased from New England Biolabs (Ipswich, MA, USA). The major coat protein (pVIII) of an M13 bacteriophage with an arginine-, glycine-, and aspartic acid-based functional peptide sequence (-RGD-) [16] and a glutamic acid-based peptide sequence (-EEEE-) were fabricated on M13KE phage [17].

2.3. *Phage Colorimetric Sensor-Based Multi-Array Chip*

A silicon wafer with a 100 nm gold film on a 5 nm platinum adhesion layer was selected as a substrate to deposit the M13 bacteriophage. Simple pulling methods [18] were utilized to prepare three kinds of colorimetric sensors consisting of wild-, RGD-, and EEEE-type M13 bacteriophages. The three types of colorimetric sensors consisted of red-, green-, and blue-colored M13 bacteriophage films that were determined by pulling speeds of 30 μm/min, 40 μm/min, and 50 μm/min, respectively. The fabricated colorimetric sensors were integrated into a single chip.

2.4. Colorimetric Signal Analysis and Data Processing

The color change was observed using a handheld digital microscope (Celestron DELUXE HANDHELD DIGITAL MICROSCOPE, Torrance, CA, USA) in the chamber of a color sensor array set made from phage chips. The colorimetric signal of the unit cell constituting each sensor chip was measured as the average value of a ~1000 pixel square. A method of setting the values of ΔR, ΔG, and ΔB was used, and the changes in each variable were used as the criteria for the amount of color change [7].

2.5. Analytical Data Statistics

Essentially, model-less clustering analyses, such as principal component analysis or hierarchical clustering analysis, use hierarchical analysis because they are unsuitable for predictive (i.e., classificatory) use. Hierarchical clustering analysis is a statistical analysis that classifies nearby objects into the same group using the distance that indicates the similarity of each object [19]. The method is called hierarchical cluster analysis because it forms a tree-like hierarchical structure by starting from objects at the closest distance and combining them. Generally, Euclidean distances are commonly used for distance calculations. On the other hand, the color distance * is used so that the color characteristics of RGB can be utilized, as the data used for the analysis are RGB data, representing a color change value that occurs when one sample reacts with another sample.

$$\sum_{i=1}^{n} \sqrt{\left(\left(R_i^a - R_i^b\right)^2 + \left(B_i^a - B_i^b\right)^2 + \left(G_i^a - G_i^b\right)^2\right)}$$

* Color distance: The color distance metric calculates the Euclidean distance in color space between each pair of clusters, ignoring their size. The distance between images a and b (using RGB) was calculated as in the above expression, where n is the number of bins.

3. Results and Discussion

Figure 1 depicts a schematic illustration of the M13 bacteriophage-based multi-array sensor chip and color-pattern analysis process. The multi-array sensor chip shows specific color patterns according to the applied materials due to the different chemical groups expressed on the M13 bacteriophage. Furthermore, microstructures of colorimetric sensors determined by the pulling speed contribute to the color change, owing to their different surface-to-volume ratios [7,8,10,18]. After collecting the color patterns, a statistical analysis process determines the types of unknown substances. Figure 2 displays the specific color patterns toward each target chemical at a temperature of 120 °C. The specific color patterns were generated by calculating pixel differentiation between initial color and color after exposure to the target chemical. As shown in Figure 2, the M13 bacteriophage-based sensor array displays specific color change according to the exposed chemicals. However, because this was a qualitative analysis, there were limitations in the exact classification of the target chemicals. The M13 bacteriophage-based sensor array, different from specialized sensors such as litmus paper or a pregnancy tester, is a non-specific sensor system, which means small amounts of color changes should be analyzed and classified exactly. Signal detection and data analysis by the sensor of the drug component are based on the inherent chemical properties of the material. The analysis of female hormone-related drugs, antibiotic components, and their clustering tendencies were analyzed through various features. Through the RGB distance values, the Euclidean distance was set to the RGB distance, and a hierarchical classification was performed using the Ward Linkage (Ward.D) method [20] (see Figure 3). The similarity between two clusters was measured on the basis of the increment of the error sum of squares (ESS) when two clusters were combined. The increase of ESS was measured as the distance between two clusters when the distance matrix was obtained. This method was less sensitive to noise or outlier data than the Single Linkage Method [20]. The clustering results showed a close relationship with each chemical composition. They confirmed that estrogen-based drugs and antibiotics-based drugs can be classified within a large frame and partially. In particular, the similar chemical structures of antibiotic substances made their distances very close to each other. Their vapor pressure values are similar to each-other at various temperatures, and these compounds react comparably with sensor array chips. The similar composition of the pharmaceutical forms containing each compound also had an impact.

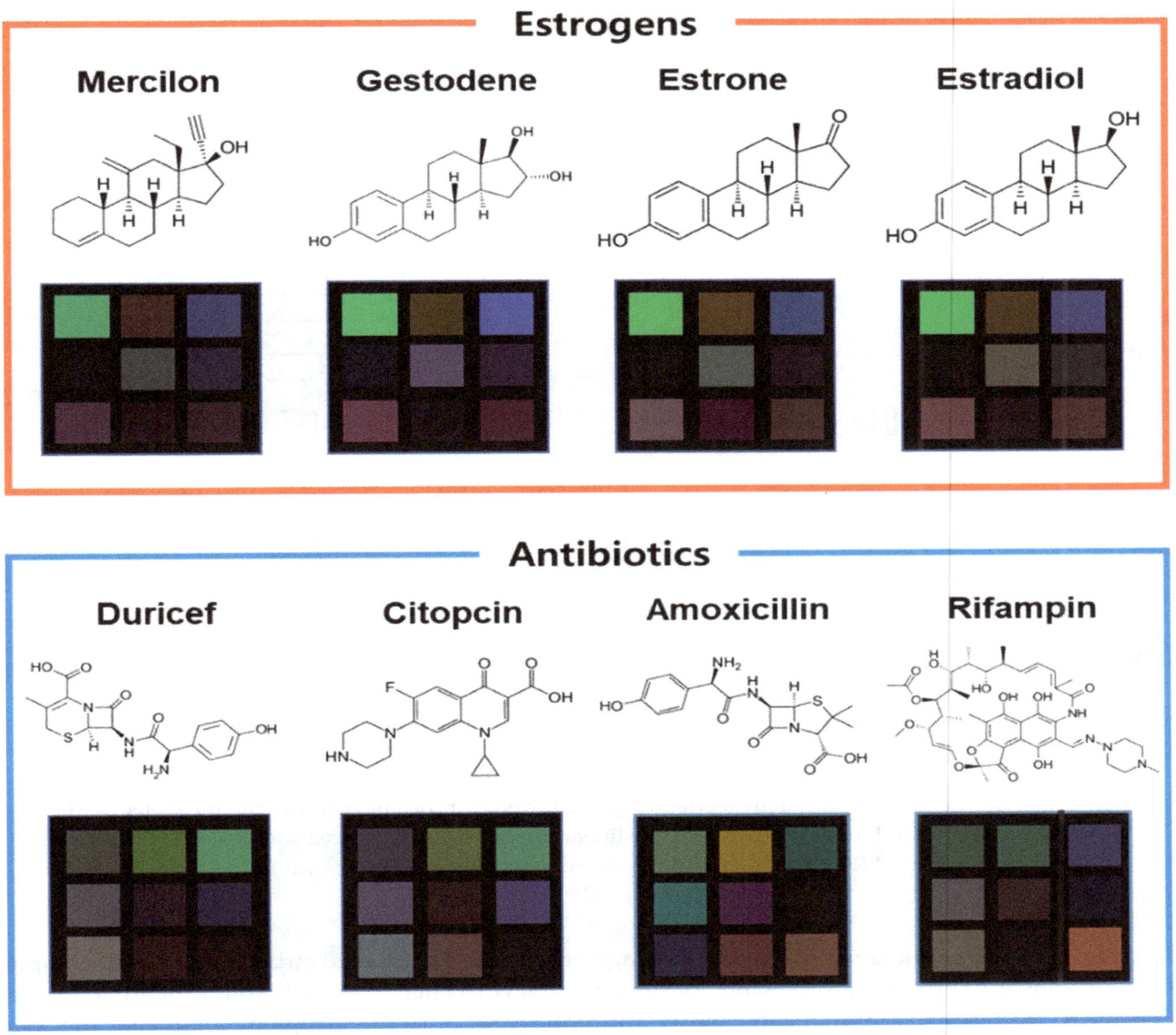

Figure 2. Image of the colorimetric sensor array chip after exposure to medical chemicals. The color pixels represent the mean value of the variation of the RGB values (in 8 bit) as a function of the M13 bacteriophage bundle's structural change.

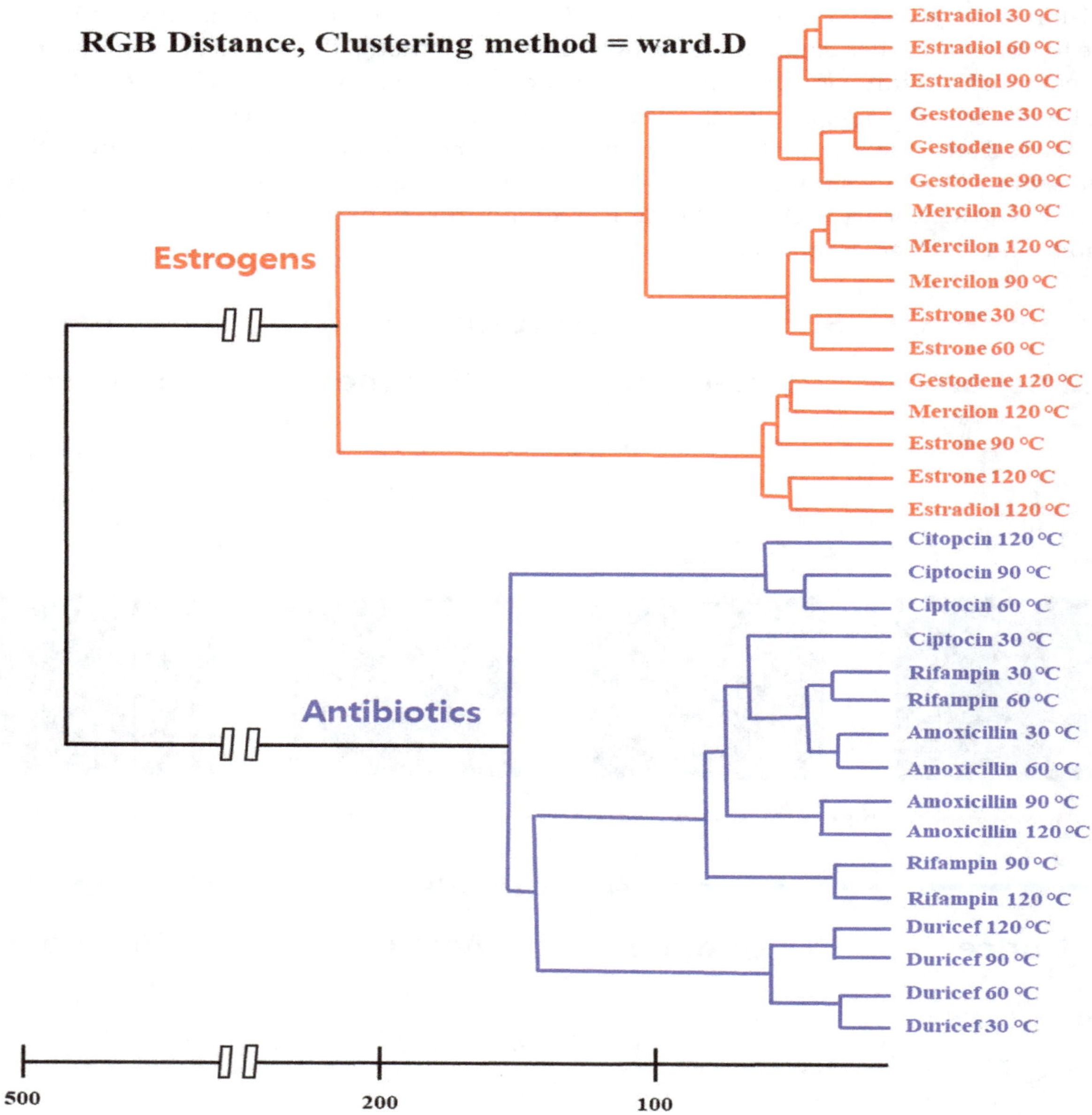

Figure 3. Hierarchical cluster analysis dendrogram for eight types of medical chemicals using the Ward.D linkage method, which is based on the linear model criterion of least squares. The Euclidean distance was set to the ΔRGB matrix which is made of the colorimetric shift values.

4. Conclusions

We fabricated a functional M13 bacteriophage-based photonic crystal structure and set a complex detection sensor device with three kinds of sensor chips with chemically inherent functional groups. We propose a sensor model that can classify antibiotics and hormone drugs by analyzing the color change amount of a 3 × 3 squared band, i.e., nine sections by setting three color bands per chip. In particular, commercial chemicals can be classified according to the product brand, so that the clustering results can slightly overlap. We see the possibility of integrating this method in a simple sensor kit for quality assurance and quality control of commercial drugs. This colorimetric sensor array model could be used to fabricate a very simple sensor platform consisting of a sensor chip 1 square centimeter wide, a chamber of about 30 cc capacity, and a small webcam for color change observation. The multi-array sensor system proposed in this work consists of nine different types of sensor chips, but in the future, by increasing the types of sensors, the selectivity of the multi-array sensor system could significantly ameliorate. Furthermore, our multi-array sensor platform could be used for user-friendly pre-analyses that could be simple and performed in real time, before instrument-based chromatographic analyses.

Author Contributions: Sensor fabrication, C.K. and Y.L.; Experimental set-up construction, Y.K. and V.D.; Preparation of genetically engineered M13 bacteriophage, C.K. and E.J.C.; Mass amplification of M13 bacteriophage, C.K. and N.-N.J.; Preparation of manuscript, C.K. and S.H.B.; Schematic illustrations, W.-G.K.; Hierarchical cluster analysis, H.L. and H.S.; Supervision and editing manuscript, D.-W.H. and J.-W.O. All authors have read and agreed to the published version of the manuscript.

References

1. Petrenko, V.A.; Vodyanoy, V.J. Phage display for detection of biological threat agents. *J. Microbiol. Methods* **2003**, *53*, 253. [CrossRef]
2. Goldman, E.R.; Pazirandeh, M.P.; Mauro, J.M.; King, K.D.; Frey, J.C.; Anderson, G.P. Phage-displayed peptides as biosensor reagents. *J. Mol. Recognit.* **2000**, *13*, 382. [CrossRef]
3. Smith, G.P. Filamentous fusion phage: Novel expression vectors that display cloned antigens on the virion surface. *Science* **1985**, *228*, 1315. [CrossRef] [PubMed]
4. Smith, G.P.; Petrenko, V.A. Phage display. *Chem. Rev.* **1997**, *97*, 391. [CrossRef] [PubMed]
5. Petrenko, V.A.; Smith, G.P. Phages from landscape libraries as substitute antibodies. *Protein Eng. Des. Sel.* **1996**, *9*, 797. [CrossRef] [PubMed]
6. Kim, C.; Lee, S.-Y.; Kim, W.-G.; Oh, J.-W. High sensitive and selective virus based structural colorimetric sensor. *Mol. Cryst. Liq. Cryst.* **2014**, *598*, 171. [CrossRef]
7. Moon, J.-S.; Lee, Y.; Shin, D.M.; Kim, C.; Kim, W.G.; Park, M.; Han, J.; Song, H.; Kim, K.; Oh, J.W. Identification of endocrine disrupting chemicals using a virus-based colorimetric sensor. *Chem. Asian. J.* **2016**, *11*, 3097. [CrossRef] [PubMed]
8. Moon, J.-S.; Kim, W.-G.; Shin, D.-M.; Lee, S.-Y.; Kim, C.; Lee, Y.; Han, J.; Kim, K.; Yoo, S.Y.; Oh, J.-W. Bioinspired M-13 bacteriophage-based photonic nose for differential cell recognition. *Chem. Sci.* **2017**, *8*, 921. [CrossRef] [PubMed]
9. Chung, W.-J.; Oh, J.-W.; Kwak, K.; Lee, B.Y.; Meyer, J.; Wang, E.; Hexemer, A.; Lee, S.-W. Biomimetic self-templating supramolecular structures. *Nature* **2011**, *478*, 364. [CrossRef] [PubMed]
10. Oh, J.-W.; Chung, W.-J.; Heo, K.; Jin, H.-E.; Lee, B.Y.; Wang, E.; Zueger, C.; Wong, W.; Meyer, J.; Kim, C.; et al. Biomimetic virus-based colourimetric sensors. *Nat. Commun.* **2014**, *5*, 3043. [CrossRef] [PubMed]
11. Mathew, A.G.; Cissell, R.; Liamthong, S. Antibiotic resistance in bacteria associated with food animals: A united states perspective of livestock production. *Foodborne Pathog. Dis.* **2007**, *4*, 115. [CrossRef] [PubMed]
12. Mills, L.J.; Chichester, C. Review of evidence: Are endocrine-disrupting chemicals in the aquatic environment impacting fish populations? *Sic. Total Environ.* **2005**, *343*, 1. [CrossRef] [PubMed]
13. Hamid, H.; Eskicioglu, C. Fate of estrogenic hormones in wastewater and sludge treatment: A review of properties and analytical detection techniques in sludge matrix. *Water Res.* **2012**, *46*, 5813. [CrossRef] [PubMed]
14. Kuch, H.M.; Ballschmiter, K. Determination of endocrine-disrupting phenolic compounds and estrogens in surface and drinking water by HRGC-(NCI)-MS in the picogram per liter range. *Environ. Sci. Technol.* **2001**, *35*, 3201. [CrossRef] [PubMed]
15. Rakow, N.A.; Sen, A.; Janzen, M.C.; Ponder, J.B.; Suslick, K.S. Molecular recognition and discrimination of amines with a colorimetric array. *Angew. Chem. Int. Ed.* **2005**, *44*, 4528. [CrossRef] [PubMed]
16. Choi, D.S.; Jin, H.-E.; Yoo, S.Y.; Lee, S.-W. Cyclic RGD peptide incorporation on phage major coat proteins for improved internalization by HeLa Cells. *Bioconjug. Chem.* **2014**, *25*, 216. [CrossRef] [PubMed]
17. Lee, B.Y.; Zhang, J.; Zueger, C.; Chung, W.-J.; Yoo, S.Y.; Wang, E.; Meyer, J.; Ramesh, R.; Lee, S.-W. Virus-based piezoelectric energy generation. *Nat. Nanotechnol.* **2012**, *7*, 351. [CrossRef] [PubMed]
18. Kim, W.-G.; Kim, K.; Ha, S.-H.; Song, H.; Yu, H.-W.; Kim, C.; Kim, J.-M.; Oh, J.-W. Virus-based full colour pixels using a microheater. *Sci. Rep.* **2015**, *5*, 13757. [CrossRef] [PubMed]
19. Anton, H. *Elementary Linear Algebra, Binder Ready Version*; John Wiley & Sons: Hoboken, NJ, USA, 2013.
20. Ward, J.H., Jr. Hierarchical grouping to optimize an objective function. *J. Am. Stat. Assoc.* **1963**, *58*, 236. [CrossRef]

Permissions

All chapters in this book were first published by MDPI; hereby published with permission under the Creative Commons Attribution License or equivalent. Every chapter published in this book has been scrutinized by our experts. Their significance has been extensively debated. The topics covered herein carry significant findings which will fuel the growth of the discipline. They may even be implemented as practical applications or may be referred to as a beginning point for another development.

The contributors of this book come from diverse backgrounds, making this book a truly international effort. This book will bring forth new frontiers with its revolutionizing research information and detailed analysis of the nascent developments around the world.

We would like to thank all the contributing authors for lending their expertise to make the book truly unique. They have played a crucial role in the development of this book. Without their invaluable contributions this book wouldn't have been possible. They have made vital efforts to compile up to date information on the varied aspects of this subject to make this book a valuable addition to the collection of many professionals and students.

This book was conceptualized with the vision of imparting up-to-date information and advanced data in this field. To ensure the same, a matchless editorial board was set up. Every individual on the board went through rigorous rounds of assessment to prove their worth. After which they invested a large part of their time researching and compiling the most relevant data for our readers.

The editorial board has been involved in producing this book since its inception. They have spent rigorous hours researching and exploring the diverse topics which have resulted in the successful publishing of this book. They have passed on their knowledge of decades through this book. To expedite this challenging task, the publisher supported the team at every step. A small team of assistant editors was also appointed to further simplify the editing procedure and attain best results for the readers.

Apart from the editorial board, the designing team has also invested a significant amount of their time in understanding the subject and creating the most relevant covers. They scrutinized every image to scout for the most suitable representation of the subject and create an appropriate cover for the book.

The publishing team has been an ardent support to the editorial, designing and production team. Their endless efforts to recruit the best for this project, has resulted in the accomplishment of this book. They are a veteran in the field of academics and their pool of knowledge is as vast as their experience in printing. Their expertise and guidance has proved useful at every step. Their uncompromising quality standards have made this book an exceptional effort. Their encouragement from time to time has been an inspiration for everyone.

The publisher and the editorial board hope that this book will prove to be a valuable piece of knowledge for researchers, students, practitioners and scholars across the globe.

List of Contributors

Satyavani Kaliamurthi, Gurudeeban Selvaraj and Keren Gu
Center of Interdisciplinary Science-Computational Life Sciences, College of Food Science and Engineering, Henan University of Technology, Zhengzhou 450001, China
College of Chemistry, Chemical Engineering and Environment, Henan University of Technology, Zhengzhou 450001, China

Sathishkumar Chinnasamy, Qiankun Wang and Asma Sindhoo Nangraj
The State Key Laboratory of Microbial Metabolism, College of Life Sciences and Biotechnology, Shanghai Jiao Tong University, Shanghai 200240, China

Dong-Qing Wei
Center of Interdisciplinary Science-Computational Life Sciences, College of Food Science and Engineering, Henan University of Technology, Zhengzhou 450001, China
The State Key Laboratory of Microbial Metabolism, College of Life Sciences and Biotechnology, Shanghai Jiao Tong University, Shanghai 200240, China

William CS Cho
Department of Clinical Oncology, Queen Elizabeth Hospital, Kowloon, Hong Kong

Jakub Dalibor Rybka and Alicja Plis
Center for Advanced Technology, Adam Mickiewicz University in Poznań, Umultowska 89C, 61-614 Poznań, Poland

Adam Aron Mieloch
Center for Advanced Technology, Adam Mickiewicz University in Poznań, Umultowska 89C, 61-614 Poznań, Poland
Faculty of Chemistry, Adam Mickiewicz University in Poznań, Umultowska 89B, 61-614 Poznań, Poland

Marcin Pyrski and Tomasz Pniewski
Institute of Plant Genetics, Polish Academy of Sciences, Strzeszyńska 34, 60-479 Poznań, Poland

Michael Giersig
Center for Advanced Technology, Adam Mickiewicz University in Poznań, Umultowska 89C, 61-614 Poznań, Poland
Institute of Experimental Physics, Freie Universität Berlin, Arnimallee 14, 14195 Berlin, Germany

Motohiro Akashi and Masaharu Takemura
Laboratory of Biology, Department of Liberal Arts, Faculty of Science, Tokyo University of Science, Shinjuku, Tokyo 162-8601, Japan

Carmen Yuste-Calvo, Lucía Zurita, César F. Cruz-Fernández, Flora Sánchez and Fernando Ponz
Centro de Biotecnología y Genómica de Plantas, Universidad Politécnica de Madrid-Instituto Nacional de Investigación y Tecnología Agraria y Alimentaria (CBGP, UPM-INIA), Campus Montegancedo, Autopista M-40, km 38. Pozuelo de Alarcón, 28223 Madrid, Spain

Mercedes López-Santalla and Marina I. Garín
Division of Hematopoietic Innovative Therapies, Centro de Investigaciones Energéticas, Medioambientales y Tecnológicas (CIEMAT) and Centro de Investigación Biomédica en Red de Enfermedades Raras (CIBER-ER), 28040 Madrid, Spain
Advanced Therapy Unit, Instituto de Investigación Sanitaria Fundación Jiménez Díaz (IIS-FJD/UAM), 28040 Madrid, Spain

Joel A. Finbloom, Ioana L. Aanei, Jenna M. Bernard, Sarah H. Klass, Susanna K. Elledge and Kenneth Han
Department of Chemistry, University of California, Berkeley, CA 94720, USA

Tomoko Ozawa and Mitchel S. Berger
Department of Neurological Surgery, University of California, San Francisco, CA 94158, USA

Theodore P. Nicolaides
Department of Pediatrics, NYU Langone Medical Center, New York, NY 10016, USA

Matthew B. Francis
Department of Chemistry, University of California, Berkeley, CA 94720, USA
Materials Sciences Division, Lawrence Berkeley National Laboratories, Berkeley, CA 94720, USA

Dominik Damm, Hannah Theobald, Klaus Überla and Vladimir Temchura
Institute of Clinical and Molecular Virology, Friedrich-Alexander University Erlangen-Nürnberg, 91054 Erlangen, Germany

Leonardo Rojas-Sánchez, Viktoriya Sokolova and Matthias Epple
Inorganic Chemistry and Center for Nanointegration Duisburg-Essen (CeNIDE), University of Duisburg-Essen, 45141 Essen, Germany

Richard T. Wyatt
Department of Immunology and Microbial Science, The Scripps Research Institute, La Jolla, CA 92037, USA

Nadav Brandes
The Rachel and Selim Benin School of Computer Science and Engineering, The Hebrew University of Jerusalem, Jerusalem 91904, Israel

Michal Linial
Department of Biological Chemistry, The Alexander Silberman Institute of Life Sciences, The Hebrew University of Jerusalem, Jerusalem 91904, Israel

Umberto Rosani and Paola Venier
Department of Biology, University of Padua, 35121 Padua, Italy

Maxwell Shapiro
Department of Applied Mathematics and Statistics, Stony Brook University, Stony Brook, NY 11794-5000, USA

Bassem Allam
School of Marine and Atmospheric Sciences, Stony Brook University, Stony Brook, NY 11794-5000, USA

Johanna Lechner, Felix Hartkopf, Pauline Hiort, Bernhard Y. Renard and Thilo Muth
Bioinformatics Unit (MF 1), Department for Methods Development and Research Infrastructure, Robert Koch Institute, 13353 Berlin, Germany

Andreas Nitsche and Marica Grossegesse
Centre for Biological Threats and Special Pathogens, Highly Pathogenic Viruses (ZBS1), Robert Koch Institute, 13353 Berlin, Germany

Joerg Doellinger
Centre for Biological Threats and Special Pathogens, Proteomics and Spectroscopy (ZBS 6), Robert Koch Institute, 13353 Berlin, Germany

Hamizah Suhaimi
Laboratory of Biotechnology, Department of Bioscience, Graduate School of Science and Technology, Shizuoka University, 836 Ohya, Suruga-ku, Shizuoka 422-8529, Japan

Rikito Hiramatsu
Laboratory of Biotechnology, Department of Applied Biological Chemistry, College of Agriculture, Graduate School of Integrated Science and Technology, Shizuoka University, 836 Ohya Suruga-ku, Shizuoka 422-8529, Japan

Jian Xu
Laboratory of Biotechnology, Research Institute of Green Science and Technology, Shizuoka University, 836 Ohya Suruga-ku, Shizuoka 422-8529, Japan

Tatsuya Kato and Enoch Y. Park
Laboratory of Biotechnology, Department of Bioscience, Graduate School of Science and Technology, Shizuoka University, 836 Ohya, Suruga-ku, Shizuoka 422-8529, Japan
Laboratory of Biotechnology, Department of Applied Biological Chemistry, College of Agriculture, Graduate School of Integrated Science and Technology, Shizuoka University, 836 Ohya Suruga-ku, Shizuoka 422-8529, Japan
Laboratory of Biotechnology, Research Institute of Green Science and Technology, Shizuoka University, 836 Ohya Suruga-ku, Shizuoka 422-8529, Japan

Laura E. Brettell
Hawkesbury Institute for the Environment, Western Sydney University; Locked bag 1797, Penrith 2751, NSW, Australia
School of Environment and life Sciences, University of Salford, Manchester, M5 4WT, UK

Stephen J. Martin
School of Environment and life Sciences, University of Salford, Manchester, M5 4WT, UK

Declan C. Schroeder
School of Biological Sciences, University of Reading, Reading RG6 6LA, UK
Veterinary Population Medicine, College of Veterinary Medicine, University of Minnesota, St Paul, MN 55108, USA

René Kallies, Rodolfo Brizola Toscan and Ulisses Nunes da Rocha
Helmholtz Centre for Environmental Research - UFZ, Department of Environmental Microbiology, 04318 Leipzig, Germany

Martin Hölzer
Friedrich Schiller University Jena, RNA Bioinformatics and High-Throughput Analysis, 07743 Jena, Germany
European Virus Bioinformatics Center, 07743 Jena, Germany

John Anders
Helmholtz Centre for Environmental Research - UFZ, Department of Environmental Microbiology, 04318 Leipzig, Germany
Bioinformatics Group, Department of Computer Science and Interdisciplinary Center for Bioinformatics, University Leipzig, 04081 Leipzig, Germany

Manja Marz
Friedrich Schiller University Jena, RNA Bioinformatics and High-Throughput Analysis, 07743 Jena, Germany
European Virus Bioinformatics Center, 07743 Jena, Germany
German Centre for Integrative Biodiversity Research (iDiv) Halle-Jena-Leipzig, 04103 Leipzig, Germany

Antonis Chatzinotas
Helmholtz Centre for Environmental Research - UFZ, Department of Environmental Microbiology, 04318 Leipzig, Germany
German Centre for Integrative Biodiversity Research (iDiv) Halle-Jena-Leipzig, 04103 Leipzig, Germany

In Woo Park, Kyung Won Kim, Yunhwa Hong, Hyun Ji Yoon, Yonghun Lee, Dham Gwak and Kwang Heo
Department of Nanotechnology and Advanced Materials Engineering, Sejong University, Seoul 05006, Korea

Heleri Kirsip
Department of Bioinformatics, University of Tartu, Tartu, 51010, Riia 23, Estonia

Aare Abroi
Institute of Technology, University of Tartu, Tartu, 50411, Nooruse 1, Estonia

Chuntae Kim, Won-Geun Kim, Yujin Lee and Yeji Kim
Department of Nanofusion Technology, Pusan National University, Busan 46241, Korea

Hansong Lee and Hokeun Sun
Interdisciplinary Program of Genomic Data Science, Pusan National University, Busan 46241, Korea
Department of Statistics, Pusan National University, Busan 46241, Korea

Vasanthan Devaraj and Eun Jung Choi
Research Center for Energy Convergence and Technology, Pusan National University, Busan 46241, Korea

Na-Na Jeong
BK21PLUS Program in Embodiment: Health-Society Interaction, Department of Public Health Sciences, Graduate School, Korea University, Seoul 02841, Korea

Sang Hong Baek
Laboratory of Cardiovascular Disease, Division of Cardiology, School of Medicine, The Catholic University of Korea, Seoul 06591, Korea

Dong-Wook Han
Department of Cogno-Mechatronics Engineering, College of Nanoscience & Nanotechnology, Pusan National University, Busan 46241, Korea

Jin-Woo Oh
Department of Nanofusion Technology, Pusan National University, Busan 46241, Korea
Research Center for Energy Convergence and Technology, Pusan National University, Busan 46241, Korea
Department of Nanoenergy Engineering, Pusan National University, Busan 46241, Korea

Index

www.ingramcontent.com/pod-product-compliance
Lightning Source LLC
LaVergne TN
LVHW060522210926
848485LV00005B/1127